T0212054

Lecture Notes in Physics

Founding Editors

Wolf Beiglböck

Jürgen Ehlers

Klaus Hepp

Hans-Arwed Weidenmüller

Volume 1008

Series Editors

Roberta Citro, Salerno, Italy

Peter Hänggi, Augsburg, Germany

Morten Hjorth-Jensen, Oslo, Norway

Maciej Lewenstein, Barcelona, Spain

Angel Rubio, Hamburg, Germany

Wolfgang Schleich, Ulm, Germany

Stefan Theisen, Potsdam, Germany

James D. Wells, Ann Arbor, MI, USA

Gary P. Zank, Huntsville, AL, USA

The series Lecture Notes in Physics (LNP), founded in 1969, reports new developments in physics research and teaching - quickly and informally, but with a high quality and the explicit aim to summarize and communicate current knowledge in an accessible way. Books published in this series are conceived as bridging material between advanced graduate textbooks and the forefront of research and to serve three purposes:

- to be a compact and modern up-to-date source of reference on a well-defined topic;
- to serve as an accessible introduction to the field to postgraduate students and non-specialist researchers from related areas;
- to be a source of advanced teaching material for specialized seminars, courses and schools.

Both monographs and multi-author volumes will be considered for publication. Edited volumes should however consist of a very limited number of contributions only. Proceedings will not be considered for LNP.

Volumes published in LNP are disseminated both in print and in electronic formats, the electronic archive being available at springerlink.com. The series content is indexed, abstracted and referenced by many abstracting and information services, bibliographic networks, subscription agencies, library networks, and consortia.

Proposals should be sent to a member of the Editorial Board, or directly to the responsible editor at Springer:

Dr Lisa Scalone
Springer Nature
Physics
Tiergartenstrasse 17
69121 Heidelberg, Germany
lisa.scalone@springernature.com

Ievgen Dubovyk • Janusz Gluza •
Gábor Somogyi

Mellin-Barnes Integrals

A Primer on Particle Physics Applications

 Springer

Ievgen Dubovyk
Institute of Physics
University of Silesia
Katowice, Poland

Janusz Gluza
Institute of Physics
University of Silesia
Katowice, Poland

Gábor Somogyi
Institute for Particle and Nuclear Physics
Wigner Research Centre for Physics
Budapest, Hungary

ISSN 0075-8450   ISSN 1616-6361   (electronic)
Lecture Notes in Physics
ISBN 978-3-031-14271-0   ISBN 978-3-031-14272-7   (eBook)
https://doi.org/10.1007/978-3-031-14272-7

This Springer imprint is published by the registered company Springer Nature Switzerland AG
The registered company address is: Gewerbestrasse 11, 6330 Cham, Switzerland

We dedicate this book to the memory of Dr. Tord Riemann.

Dr. Tord Riemann, 1951–2021, https://cerncourier.com/a/tord-riemann-1951-2021.

Photo thanks to Dr. Sabine Riemann.

We are thankful to Tord for his support and encouragement for writing up these notes.

Preface

On the left: Hjalmar Mellin (1854–1933) Finnish mathematician, Helsinki University of Technology. © Atelier Charles Rüs & Co., Hjalmar Mellin, Ident.Nr. 1992,32/003a.024, Kunstbibliothek, Staatlichen Museen zu Berlin under CC BY-NC-SA 3.0 DE, http://www.smb-digital.de/eMuseumPlus. On the right: Ernest William Barnes (1874–1953) British mathematician, liberal theologian and bishop. ©Foto: https://en.wikipedia.org/wiki/Ernest_Barnes

This book is about the mathematical basics of *Mellin-Barnes* integrals; however, it is motivated by the practical needs of precision calculations in particle physics.

In modelling physical phenomena, we aim at predictability and exactness. In particle physics, a compelling model exists, commonly called the Standard Model (SM), which describes elementary particles and their interactions. We know since the 1960s that gauge interactions, mathematically based on a product of $U(1)$, $SU(2)$, and $SU(3)$ groups, describe the electromagnetic, weak, and strong forces. We know the charges, spins, C and P parities, as well as the masses of elementary particles. We understand how fermions and bosons interact by chiral weak interactions and we understand the dynamics of the quarks and gluons that partake in the strong interaction. The SM describes innumerable physical phenomena observed in the laboratory and the cosmos, linking present-day experiments with the first 10^{-11}

second of the Universe. Looking at the list of Nobel laureates in physics, many of them are connected with particle physics. Their awards speak to the appreciation of the Standard Model theory builders and to the enormous efforts made to confirm the SM structure and predictions experimentally.

The SM is surprisingly precise in its predictions and complete in the sense that all of its theorized elementary particles have now been observed. The last missing piece of the SM, the Higgs boson was discovered in 2012. This discovery was made possible by the construction of the LHC accelerator and the ATLAS and CMS detectors, whose designs were guided by theoretical predictions based on the SM. Thus, we now believe that we can answer the theoretical question dating from the 1960s: "what gives mass to the electroweak bosons?".

However, this achievement does not stop the need for further exploration!

First, does the 125 GeV/c^2 particle that was discovered really play the role which theory says it does? Second, several unanswered mysteries call for new phenomena beyond the SM: (i) the apparent presence of dark matter in the Universe; (ii) the matter-antimatter asymmetry of the Universe; and (iii) the existence and pattern of neutrino masses to name three outstanding ones. Although the list of issues to be explored in the future is long,[1] there is at present no clear guidance on the energy scale or strength of the new phenomena. However, since the SM is complete, there is no freedom in its predictions and any observed deviation would be a discovery and a precious pointer into the unknown. An ambitious program of precision Higgs and electroweak measurements is thus in order. However, a well-prepared and well-motivated collider project requires precise theory input to guide the experiment and confront the results, thus optimizing the chances for significant scientific progress. Ever-increasing mathematical virtuosity is however required to reach the ever-improving level of experimental precision.

And this is a place where the theory and methods connected with Mellin-Barnes integrals described in the book enter.

To achieve the required precision of theoretical calculations, higher-order perturbative corrections in Quantum Field Theory must be calculated. This involves the computation of so-called multi-loop Feynman integrals as well as phase space integrals. However, currently there is no complete theory for Feynman integrals beyond one loop. This defines a challenge to theoreticians and calls for innovative methods and tools, both numerical and analytical.

This book focuses on one of these, the *Mellin-Barnes* (MB) method. This method has been applied to real applications at the frontier of research and has proven to be one of very few ones which could solve so far untouchable problems of precision physics. The book provides a self-contained overview of the necessary elements of complex analysis and special functions, an easy-to-follow introduction to the subject of MB representations with hands-on examples and a comprehensive view

---

[1] *Physics Briefing Book: Input for the European Strategy for Particle Physics Update 2020*, Richard Keith Ellis et al., CERN-ESU-004, arXiv:1910.11775.

on the analytical and numerical methods and tools related to MB integrals. It should be a good starting point for learning and using the MB method in practical research.

Katowice, Poland Ievgen Dubovyk
Cisownica, Poland Janusz Gluza
Budapest, Hungary Gábor Somogyi
October 2022

Acknowledgments

We would like to thank for collaborations on the papers connected with the MB subject and/or many interesting discussions and insights on the subject the following people:

Stefano Actis, Paolo Bolzoni, Michał Czakon, Andrei Davydychev, Claude Duhr, Wojciech Flieger, Ayres Freitas, Marek Gluza, Krzysztof Grzanka, Tomasz Jeliński, Krzysztof Kajda, Mikhail Kalmykov, David Kosower, Sven-Olaf Moch, Vladimir A. Smirnov, Bas Tausk, Johann Usovitsch, and Szymon Zięba.

This work has been supported in part by the Polish National Science Center (NCN) under grants 2017/25/B/ST2/01987 and 2020/37/B/ST2/02371.

Contents

Acronyms

1BL	First Barnes Lemma
2BL	Second Barnes Lemma
BSM	Beyond the Standard Model
CO	Collinear Divergence
CW	Cheng-Wu (Variables, Theorem)
DEs	Differential Equations
IBP	Integration-By-Parts
LT	Lefschetz Thimbles
IR	Infrared Divergence
LA	Loop-by-Loop Approach
MIs	Master Integrals
NLO	Next-to-Leading Order in Perturbation Theory
FI	Feynman Integrals
GA	Global Approach
GRMT	Generalized Ramanujan's Master Theorem
HPL	Harmonic Polylogarithms
MB	Mellin-Barnes
MPL	Multiple Polylogarithms
SD	Sector Decomposition
SM	Standard Model
UV	Ultraviolet Divergence
QCD	Quantum Chromodynamics
QED	Quantum Electrodynamics
QFT	Quantum Field Theory
$\mathbb{N}$	Set of Natural Numbers $(0, 1, 2, \ldots)$
$\mathbb{N}^+$	Set of Natural Numbers, Zero Excluded $(1, 2, \ldots)$
$\mathbb{Z}$	Set of Integers $(0, \pm1, \pm2, \ldots)$
$\mathbb{R}$	Set of Real Numbers
$\mathbb{C}$	Set of Complex Numbers
$\mathfrak{R}(z)$	Real Part of the Complex Number z
$\mathfrak{I}(z)$	Imaginary Part of the Complex Number z
γ	Euler's Constant
Γ	Gamma Function
ψ	Digamma Function (Polygamma Function of Order Zero)

$\psi^{(n)}$	nth Derivative of the Digamma Function ψ
$\mathrm{Li}_n(z)$	Classical Polylogarithm of Weight n
$S_{n,p}(z)$	Nielsen Generalized Polylogarithm
$(a)_n$	Pochhammer's Symbol
$\zeta(z)$	Euler-Riemann zeta function

Precision in Perturbative Particle Physics

1

Abstract

Precision in particle physics may lead to the discovery of anomalies which in turn may prompt the formulation of new paradigms in theory. For instance, particles thought to be elementary could turn out to be composite. In this chapter, we discuss classes of quantum corrections in particle physics which emerge in precision studies and discuss basic properties of Feynman integrals, notably their singularities. As a simple example, we sketch the standard calculation of dimensionally regulated one-loop Feynman integrals; then we introduce the basic idea of using Mellin-Barnes representations to solve Feynman integrals in Euclidean and Minkowskian kinematic regimes.

1.1 Loops and Real Quantum Corrections in Precision Physics

One of the most important ways of probing the laws of nature at the smallest distance scales is through the study of high-energy particle collisions. Indeed, the construction and verification of the SM has involved many such experiments of increasing energy, spanning decades. Today, a steady increase of the energy and the intensity of the colliding particles is indispensable in searching for unknown feeble interactions and exploring in a deeper way the structure of the smallest quantum objects, and ultimately the vacuum itself. For instance, particles currently thought of as elementary, i.e., having no internal structure, could turn out to be composite, being built out of yet undiscovered constituents. For example, a particle of mass $125\,\mathrm{GeV}$ (the Higgs boson) corresponds to the natural wavelength of $\sim 10^{-17}\,\mathrm{m}$, so unraveling its internal structure, if any, requires that distance scales of at least one or two magnitudes smaller are explored. This will be possible at future colliders as well as through the indirect analysis of precision quantum corrections and effective theories that will probe scales of new physics at tens of TeV level. This means

© The Author(s), under exclusive license to Springer Nature Switzerland AG 2022
I. Dubovyk et al., *Mellin-Barnes Integrals*, Lecture Notes in Physics 1008,
https://doi.org/10.1007/978-3-031-14272-7_1

that future colliders and studies will probe physics and particle substructures at the level of 10^{-17} m–10^{-19} m. Especially powerful information can be obtained by merging analyses at lepton and hadron colliders leading to the exploration of quantum structures at the unprecedented level of attometers (10^{-18} m) and below.

> In order to fully exploit the physics potential of future facilities (establish discovery limits, study the viable parameter space of new particle physics models, and so on), the experimental data will need to be confronted with precision calculations.

Precision calculations of quantum effects in high-energy particle collisions are performed using the methods of perturbative quantum field theory, pioneered by Dyson and Feynman among others. The application of perturbation theory is made possible by the relative smallness of all SM coupling constants at the energies relevant for these collisions, i.e., above a few tens of GeVs. (We do not consider here a problem of gauge theories like QCD which become non-perturbative at low energies, where they must be treated by different methods.) The perturbative expansion of physical quantities in the small couplings then leads to mathematical objects called Feynman integrals (FI) which must be computed explicitly to obtain numerical predictions. Feynman diagrams provide a very useful and convenient way of representing Feynman integrals and also provide a way to visualize particle collision processes. It is not our intent here to discuss how Feynman diagrams and the corresponding Feynman integrals arise in quantum field theory (QFT), and we leave these questions to the whole spectrum of excellent textbooks on the subject.

For any given particle collision process, the lowest nonvanishing perturbative order, called the Born approximation, involves diagrams with a fixed number of external lines; these represent the initial and final particles in the collision. In most cases, the diagrams of the Born approximation are tree diagrams, i.e., they do not contain internal loops. Tree diagrams are characterized by the property that cutting any internal line in the diagram makes the diagram disconnected: not every vertex can be reached from any other by traversing along internal lines. In passing we note that a given diagram has L loops if one can cut L different internal lines such that the diagram remains connected, but cutting $L + 1$ different internal lines makes the diagram disconnected. Higher orders in perturbation theory then lead to Feynman diagrams with more and more lines. These can either be internal to the diagram, forming closed loops, or they can exit the diagram, such that the number of external legs is increased. In the former case, the closed lines represent internal virtual particles and hence we speak of *virtual or loop corrections*. In the latter, the extra external lines represent the emission of extra real particles and thus such contributions are called *real corrections*. Obviously at higher perturbative orders, mixed corrections also appear, with extra real radiation from loop diagrams. Higher-order contributions in perturbation theory are also called *radiative corrections*, because they involve the emission or radiation of extra (virtual or real) particles.

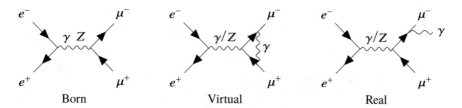

Fig. 1.1   Electron-positron annihilation into muon-antimuon pair. Left, Born approximation is given by a single diagram; center, sample diagram contributing to the one-loop correction; right, sample diagram contributing to the single real emission correction

Before moving on, let us make a small remark about real corrections. While the existence of loop corrections is more or less straightforward, the reason why real radiation corrections must be taken into account is perhaps less intuitive. A quick and non-rigorous explanation is as follows: Consider, e.g., the process of muon-antimuon pair production in electron-positron annihilation in QED, $e^+e^- \to \mu^+\mu^-$. The Born approximation is described by the single Feynman diagram labeled Born in Fig. 1.1. The one-loop correction to this process involves the emission and reabsorption of a virtual photon in the basic diagram. An example is shown in Fig. 1.1 (Virtual). On the other hand, the real correction involves the emission of an extra real photon, as in the rightmost picture of Fig. 1.1 (Real). But why should the process with an extra photon $e^+e^- \to \mu^+\mu^-\gamma$ be considered as a correction to the process we are studying? The answer lies in the fact that the emitted photon can have an arbitrarily small energy (soft limit) and the direction of its momentum can be arbitrarily close to that of one of the fermions (collinear limit). But in these cases, it is impossible in practice—and in the strict limits also in principle—to distinguish the two final states: that with just the muon-antimuon pair and that with the muon-antimuon pair and the extra photon. But the rules of quantum mechanics tell us that when computing a physical observable, say a cross section, we must sum over all final states that we do not distinguish. Thus, we must include also real corrections. This can be done in a systematic way [1]. The above argument also shows that only massless particles (in the SM photons and gluons) can enter as extra radiation in real corrections.

Finally we note that the integrals corresponding to virtual and real corrections exhibit various divergences in four space-time dimensions. We will see in the next section that these are due either to integrating over very large momenta in loops (UV divergences) or to singularities that arise when a massless particle becomes soft (IR divergences) or their momentum becomes collinear to that of another particle (collinear divergences). The latter situation may occur both in loop and real radiation diagrams. So a naive calculation leads to infinite results! The way of treating the infinities associated with UV divergences has been known since the time of Feynman, Dyson, and Tomonaga for QED, and thanks to 't Hooft and Veltman, it is also known in non-abelian gauge theories like the SM. It turns out that all such infinities are cancelled if one defines carefully the physical (i.e., measurable)

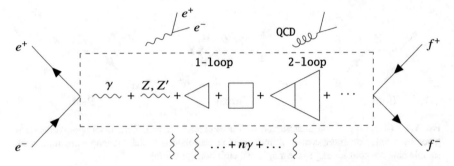

Fig. 1.2 A general scheme for $e^+e^- \rightarrow f\bar{f}$ process with typical real radiation effects (outside the dashed frame) and some chosen virtual effects (inside the dashed frame) at the lowest level (represented by γ, Z-boson propagators and Z' extra gauge boson which can be a part of some Beyond the Standard Model (BSM) extensions) and higher loop effects (represented by one-loop diagrams for vertex and box and two loops for the planar vertex)

couplings and masses of the theory. This procedure is called renormalization.[1] On the other hand, the infinities associated with soft and collinear particles are present even after renormalization and are in fact cancelled between virtual and real corrections. This is a somewhat more mathematical way of understanding why real radiation corrections must be included: they cancel the infrared and collinear singularities of the virtual corrections. Thus, the calculation of higher-order radiative corrections to a real process involves UV renormalization and also the evaluation of both virtual and real corrections. Figure 1.2 presents the typical picture: the $2 \rightarrow 2$ process of electron-positron annihilation which leads to fermion pair production, a process investigated in depth at LEP. In this figure we can find real emission and virtual loop contributions in the form of Feynman diagrams. In this book, we will consider the mathematical objects these diagrams give rise to and examine how they can be calculated using the Mellin-Barnes method.

1.2   Singularities of Amplitudes in QFT

As we have described above, the basic building blocks of higher-order perturbative corrections are loop and real emission diagrams. The extra emitted particles (virtual or real) carry momenta, and when computing a physical observable, these momenta must be integrated over. Indeed, consider an example of the L-loop diagram such as the one in Fig. 1.3. Since the diagram involves loops, momentum conservation does not fix the momenta of all internal lines. In fact, in an L-loop diagram, there are precisely L independent internal momenta that are not fixed. The Feynman rules

---

[1] Besides the couplings and masses, the normalization of fields must also be considered, but we do not enter into these subtleties here.

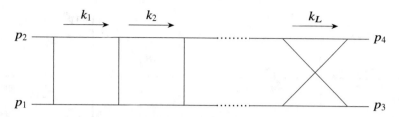

Fig. 1.3 An example of the L-loop diagram with four external momenta $p_1, \ldots, p_4$ and internal momenta $k_1, \ldots k_L$ which includes planar and non-planar sub-topologies

then instruct us to integrate over these loop momenta; thus the amplitude represented by the diagram in Fig. 1.3 will have the following general structure:

$$G_L \propto \int d^4k_1 \, d^4k_2 \ldots d^4k_L \frac{T(k)}{(q_1^2 - m_1^2 + i\delta)(q_2^2 - m_2^2 + i\delta)\ldots(q_N^2 - m_N^2 + i\delta)}. \tag{1.1}$$

Here q_i and m_i are the momenta and masses of the internal N lines, while $T(k)$ denotes a generic numerator which itself may depend on the loop momenta $k_1, \ldots k_L$. Importantly, the range of integration is the complete four-dimensional momentum space for each k_i, i.e., we must integrate over each component of all loop momenta from $-\infty$ to $+\infty$.

The Role of the Feynman Prescription for Propagators

Before moving on, let us clear up one point which may be cause for some confusion. In Eq. (1.1), we were careful to indicate the Feynman prescription for propagators by including the $+i\delta$ terms. The textbooks tell us that δ is a small positive quantity and that we must set $\delta \to 0^+$ at the end of the computation. However, if we take $\delta \to 0^+$, then the integrand clearly develops poles for real momenta and masses. Namely, let us consider the on-shell particle with *four*-momentum $p = (p^0, \mathbf{p})$ and a mass m. As

$$p^2 - m^2 = 0 \quad \Rightarrow \quad (p^0)^2 = \mathbf{p}^2 + m^2 \tag{1.2}$$

thus

$$p^0 = \pm E_p = \pm\sqrt{\mathbf{p}^2 + m^2}, \tag{1.3}$$

and the solutions for the energy of the particle lie on the real axis in the complex p^0 plane; see Fig. 1.4, left.

As we will show below on a simple example, one cannot simply integrate over the singularities; thus it seems like we are faced with a problem. However, the Feynman

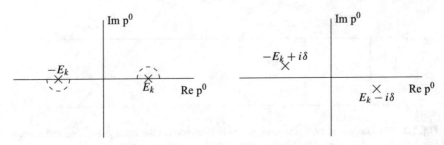

Fig. 1.4 On-shell poles at the complex p^0 plane before (left) and after (right) infinitesimal shifts

prescription tells us precisely how we must avoid these poles. In effect, it fixes the correct contour of integration to use in the complex p_0 plane. But this prescription can also be implemented in the following way: As the so-called on-shell relativistic relation $p^2 - m^2 = 0$ enters the propagator, we can avoid the singularity when integrating on the real axis by moving the kinetic energy solution E_p for p^0 to the complex plane by adding an infinitesimal parameter δ, schematically[2]

$$\frac{1}{p^2 - m^2} \longrightarrow \left[\text{singularity for } p^2 \to m^2 \right] \longrightarrow \frac{1}{p^2 - m^2 + i\delta}; \qquad (1.4)$$

see Fig. 1.4, right. The solutions for p^0 are then indeed complex, since $p^2 - m^2 + i\delta = (p^0)^2 - E_p^2 + i\delta = 0$, which implies $(p^0)^2 = (E_p^2 - i\delta)^2$, and so $p^0 = \pm\sqrt{E_p - i\delta} = \pm E_p \mp i\delta'$, where $E_p = +\sqrt{\mathbf{p}^2 + m^2}$ and $\delta' = \delta/(2E_p)$. Note that since $E_p \geq m > 0$, δ' is positive.

Thus, the singularities in the $\delta \to 0^+$ limit are only apparent and the corresponding poles are never integrated over. Indeed, taking this limit in final results for Feynman or phase space integrals never leads to singularities in δ. In this respect, they are quite different from the UV, IR and collinear (CO) singularities. We will see shortly that UV, IR, and CO singularities can be regularized in a way where removing the regularization corresponds to taking the regulator ϵ to zero. Then these singularities will show up as poles in ϵ.

Unfortunately the integral in Eq. (1.1) above is often ill-defined! From mathematical point of view, the heart of the problem which we encounter in calculating the integrals that appear in radiative corrections stems from the fact that the integrands have poles inside or on the borders of the integration region. This means that there are kinematic configurations where the denominators of the considered amplitudes and FI tend to zero. We will see shortly how one may make sense of these divergent

---

[2] "One of the most remarkable discoveries in elementary particle physics has been that of the existence of the complex plane."—Julian Schwinger.

integrals; however let us first show an amusing example of how naive integration can fail spectacularly when there are poles around.

Example of a Simple Singular Integral

As the first, purely mathematical example, let us consider the definite integral

$$I = \int_{-1}^{1} \frac{1}{x^2} dx . \tag{1.5}$$

Elementary integration gives $I = (-\frac{1}{x})|_{-1}^{1} = -2$, though the integrand in Eq. (1.5) is never negative. According to Riemann's definition of definite integrals, the result should be positive. The problem is of course that the region of integration includes $x = 0$ where the integrand is infinite and the integrand exhibits *a singularity*. Such an integral is called improper. As warned in [2], "You have to be ever alert for singularities when you are doing integrals; always, stay away from singularities. Singularities are the black holes of integrals; don't 'fall into' one (don't integrate across a singularity)."

We face a similar problem as in our example above for scattering amplitudes. Namely, let us consider the simple one-loop diagrams from the left of Fig. 1.5. This leads to the integral (after the earlier note on the Feynman prescription of the propagators, if not necessary, we will not make the $+i\delta$ terms explicit in the propagators; notation using capital roman letters will be explained in Sect. 1.3 where general integrals deviated from four dimensions will be discussed)

$$A_0 \propto \int \frac{d^4k}{k^2} \tag{1.6}$$

and

$$B_0 \propto \int \frac{d^4k}{k^2[(k-p)^2 - m^2]} . \tag{1.7}$$

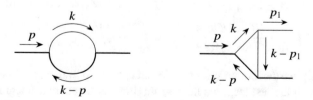

Fig. 1.5 One-loop diagrams exhibiting UV (left) and IR (right) singularities. The thick line denotes a particle of non-zero mass m, while the thin line represents a massless particle

Remember that the momenta of virtual particles in loops do not have to satisfy the mass-shell conditions (put differently, the loop integration over the four-momentum k is fully unconstrained), and so $[(k - p)^2 - m^2] = k^2 - 2k \cdot p$. (Notice that p is an external momentum, so the mass-shell condition is satisfied for p and thus $p^2 = m^2$.) Then for very large momenta $k \to \infty$, we find that our integrals behave in the following way:

$$A_0 \propto \int \frac{d^4k}{k^2} \xrightarrow{k \to \infty} \int \frac{d^4k}{k^2} \xrightarrow{\text{Wick}} \int_0^\infty \frac{|K|^3 d|K| d\Omega}{|K|^2} \propto \int_0^\infty |K| d|K|, \quad (1.8)$$

$$B_0 \propto \int \frac{d^4k}{k^2(k^2 - 2k \cdot p)} \xrightarrow{k \to \infty} \int \frac{d^4k}{(k^2)^2} \xrightarrow{\text{Wick}} \int_0^\infty \frac{|K|^3 d|K| d\Omega}{|K|^4} \propto \int_0^\infty \frac{d|K|}{|K|}. \quad (1.9)$$

In the second step, we have performed a Wick rotation, $k^0 \to iK^0$, $k^i \to K^i$, $(i = 1, 2, 3)$ in order to pass to Euclidean space in the loop momentum, so $|K|$ is the length of the Wick-rotated, Euclidean four-momentum K. Then, we passed to four-dimensional spherical coordinates and $d\Omega$ is the angular measure in four dimensions. In the last step, we have dropped the trivial angular integral that does not influence the behavior of the result at $k \to \infty$. Clearly A_0 and B_0 are UV divergent, quadratically and logarithmically singular at the upper limit of integration, respectively. Similar asymptotic dependencies of integrals on the loop momentum behavior will be observed for analogous d-dimensional integrals in Sect. 1.3.

Turning to the one-loop diagram on the right of Fig. 1.5, we find that it corresponds to the integral

$$C_0 = \int \frac{d^4k}{k^2(k - p_1)^2[(k - p)^2 - m^2]}. \quad (1.10)$$

An analysis similar to the one above shows that this integral is not UV divergent. However, considering now the limit of very small momenta $k \to 0$, we find that

$$C_0 = \int \frac{d^4k}{k^2(k^2 - 2k \cdot p_1)(k^2 - 2k \cdot p)} \xrightarrow{k \to 0} \int \frac{d^4k}{k^2(2k \cdot p_1)(2k \cdot p)}$$
$$\xrightarrow{\text{Wick}} \int \frac{|K|^3 d|K| d\Omega}{|K^2|(2K \cdot p_1)(2K \cdot p)} \propto \int_0^\infty \frac{d|K|}{|K|}. \quad (1.11)$$

As previously, we performed Wick rotation and passed to four dimensional spherical coordinates. In the last step, we have again dropped an angular integral that does not play a role in analyzing the small k behavior of the integral. Now we see that our integral is logarithmically singular at the lower limit of integration and

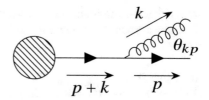

Fig. 1.6 Schematic amplitude for single real emission

so the integral is IR divergent. However, the integrand has yet another type of singularity: a collinear singularity when the internal momentum becomes collinear to the momentum of the neighboring massless leg. Indeed, let us parametrize the loop momentum k as $k = xp_1 + k_\perp$, where $k_\perp$ points in the transverse direction with respect to p_1 so that $k_\perp \cdot p_1 = 0$. The collinear limit is reached as $k_\perp \to 0$. Then we find (recall $p_1^2 = k_\perp \cdot p_1 = 0$)

$$C_0 = \int \frac{d^4k}{k^2(k^2 - 2k \cdot p_1)(k^2 - 2k \cdot p)} = \int \frac{d^4k_\perp}{k_\perp^2 (k_\perp^2)[k_\perp^2 - 2(xp_1 + k_\perp) \cdot p]}$$

$$\xrightarrow{k_\perp \to 0} \int \frac{d^4k_\perp}{k_\perp^2(2k_\perp^2)(-2xp_1 \cdot p)} \propto \int \frac{d^4k_\perp}{k_\perp^4} \xrightarrow{\text{Wick}} \int_0^\infty \frac{d|K_\perp|}{|K_\perp|}.$$

$$(1.12)$$

In the last step, we have performed the usual Wick rotation. As before, we find a logarithmic singularity for small $k_\perp$, i.e., in the collinear limit. We will see shortly how one can make sense out of the divergent integrals that we have encountered.

We note in passing that special treatment is required also for so-called threshold effects connected with conspired relations among masses of virtual states [3,4]. We will use thresholds for some specific constructions of MB representations and their numerical evaluation in Sects. 3.5 and 6.2.4.

Next, we note that real emission integrals involving massless particles can also be ill-defined due to the presence of poles in the integrand. To see this, let us consider the emission of a real massless particle in some particle scattering process; see Fig. 1.6. We assume that the emitted particle, denoted by a gluon in the figure, is massless, i.e., $k^2 = 0$. The solid line emerging from the rest of the diagram (denoted by the blob in the figure) represents some particle's propagator with momentum $p + k$ and mass m, thus the complete amplitude will contain the scalar part of this propagator (i.e., its denominator without the Lorentz structure in the propagator's numerator),

$$\mathcal{M} \propto \frac{1}{(p + k)^2 - m^2} = \frac{1}{2p \cdot k} = \frac{1}{E_k E_p} \frac{1}{1 - \beta_p \cos\theta_{kp}}, \qquad (1.13)$$

where $p = (E_p, \mathbf{p})$ and $k = (E_k, \mathbf{k})$, and we set $\beta_p = \sqrt{1 - \frac{m^2}{E_p^2}}$; see Problem 1.1. Notice that $E_p \geq m > 0$, but since k is massless, E_k can go to zero. From the above we see that the amplitude has an IR divergence when $E_k \to 0$ and a collinear (CO) divergence when $\theta \to 0$ and in addition $\beta_p = 0$, i.e., the internal particle is also massless:

(i) $E_k \to 0$ IR,
(ii) $\theta \to 0$ and $\beta_p = 0$ CO.

Thus, *infrared singularities are associated with vanishing particle energies, while collinear singularities arise when the daughter particles are emitted in the same direction.* We note in passing that if the internal particle is not massless, the $\theta \to 0$ collinear limit is not singular; however it produces what is called a "mass singularity." That is, the integration over the angle of emission is finite, but the result becomes singular if the internal mass is taken to zero.

We should warn the reader that the proper analysis of real emission singularities is more involved than the above simple calculation suggests. First of all, when we compute a physical quantity, what enters the real emission integrals is actually the squared amplitude, not the amplitude itself. Second, the integration measure also plays a role in the analysis of singularities. Indeed, the quantum mechanical rules for obtaining a cross section from a squared amplitude state that we must integrate over the momenta of final-state particles with the following phase space measure:

$$d\phi_n(p_1, \ldots, p_n; Q) = \prod_{k=1}^{n} \frac{d^4 p_k}{(2\pi)^3} \delta_+(p_k^2 - m_k^2)(2\pi)^4 \delta^4(p_1 + \ldots + p_n - Q) \,.$$

$$(1.14)$$

Here Q is the total incoming momentum, while the δ_+ function is defined as

$$\delta_+(p_k^2 - m_k^2) = \delta(p_k^2 - m_k^2)\Theta(p_k^0) \,. \tag{1.15}$$

A trivial way in which the measure enters the analysis of singularities is as follows: Since overall momentum is conserved, phase space is compact for any finite incoming momentum Q. This implies that we cannot have $p_k \to \infty$. This is very clear physically: the energy of any outgoing particle is clearly bounded from above. Hence, *real emission integrals do not have UV singularities.* A more subtle way in which the measure enters the analysis can be seen in the soft limit. Let us suppose that momentum k is massless. Then the corresponding one-particle measure becomes

$$\frac{d^4 k}{(2\pi)^3}\delta_+(k^2) = \frac{d^3 \mathbf{k}}{(2\pi)^3 2E_k} = \frac{|\mathbf{k}|^2 d|\mathbf{k}| \, d(\cos\theta) \, d\phi}{(2\pi)^3 2E_k} = \frac{E_k d E_k \, d(\cos\theta) \, d\phi}{2(2\pi)^3} \,.$$

$$(1.16)$$

In the first step, we performed the integration over the energy component of k using the δ_+ function; then we passed to three-dimensional spherical coordinates for the integration over the space-like components. Finally, we used $|k| = E_k$, which is a consequence of $k^2 = 0$ and the fact that the energy is non-negative. Now, considering the soft limit, i.e., $E_k \to 0$, we see that the integration measure carries one factor of E_k, so the integrand must diverge at least like E_k^{-2} in order to give a non-integrable singularity. This is precisely the case for the square of the real emission amplitude (recall the amplitude itself behaves as E_k^{-1} in the soft limit), and hence the phase space integration over the soft momentum produces an ill-defined integral. For example, if the gluon in Fig. 1.6 can be emitted from more than one leg, say those with momenta p_1 and p_2, then the matrix element will involve a sum of Feynman diagrams and hence a sum of terms like those in Eq. (1.13),

$$\mathcal{M} \propto \ldots + \frac{1}{2p_1 \cdot k} + \frac{1}{2p_2 \cdot k} + \ldots . \tag{1.17}$$

Then the squared matrix element $|\mathcal{M}|^2$ will contain terms such as

$$|\mathcal{M}|^2 \propto \ldots + \frac{1}{(2p_1 \cdot k)(2p_2 \cdot k)} + \ldots , \tag{1.18}$$

which lead to integrals of the form

$$\int d\phi_n(p_1, \ldots, p_n; Q) \frac{1}{(2p_1 \cdot k)(2p_2 \cdot k)} =$$

$$= \int \frac{E_k dE_k \, d(\cos\theta) \, d\phi}{2(2\pi)^3} \frac{1}{E_k^2 E_1 E_2} \frac{1}{(1 - \beta_1 \cos\theta_{k1})(1 - \beta_1 \cos\theta_{k2})} \times \ldots$$

$$= \frac{1}{2(2\pi)^3} \int_0^Q \frac{dE_k}{E_k} \frac{1}{E_1 E_2} \int d(\cos\theta) \, d\phi \frac{1}{(1 - \beta_1 \cos\theta_{k1})(1 - \beta_1 \cos\theta_{k2})} \times \ldots , \tag{1.19}$$

where the $\ldots$ denote the rest of the measure and integrand. The integration over E_k is indeed singular on the lower limit signaling the IR divergence. We see also that the evaluation of phase space integrals can involve integrations over angular variables, a topic we will return to in Sect. 3.14.

We have described the basic structures and integrals in the computation of higher-order perturbative corrections in particle physics: loop integrals and phase space integrals. We have seen that both are ill-defined in four space-time dimensions due to UV, IR, and CO divergences. While UV divergences can be removed by passing to renormalized quantities, IR and CO divergences cancel only after summing virtual and real corrections to some given process.

Note that there are technical requirements on the physical observable we want to compute in order for this cancellation to work, and this is the content of the Kinoshita-Lee-Nauenberg theorem [5,6]. Quantities for which the cancellation takes place are called IR and collinear-safe observables. Observables that are not IR and collinear-safe may still be defined and measured; however they cannot be computed order by order in perturbation theory. For example, the definition of jets using so-called seeded cone algorithms is not IR and collinear-safe beyond a certain perturbative order [7]. Nevertheless, they have been extensively used in past hadron collider experiments.

However, before any specific calculation can be performed, we must first give a precise meaning to the divergent integrals we encountered above. This procedure is called regularization and will be the topic of the next section.

1.3   Dimensional Regularization and Evaluation of Feynman Integrals

As we have just seen, in QFT we can disentangle three types of divergences depending on the masses of the parent and emitted particles in a given interaction vertex or the whole scattering amplitude: UV, IR, and CO. The presence of these singularities makes the naively defined amplitudes and cross sections ill-defined. Thus, our first task is to give a precise meaning to the various objects and integrals we have encountered so far.

There are several ways to modify our integrals such that the results become well-defined. This process is called regularization. It is important that in physical results we should be able to remove the regularization and obtain unambiguous predictions. One way of regularizing our integrals is through dimensional regularization. This method has proven to be very powerful in higher-order perturbative calculations so much so that it is ubiquitous in the modern literature. The idea is to formally modify the dimensionality of space-time from $d = 4$ to $d = 4 - 2\epsilon$, where ϵ is an infinitesimal space-time regulator[3] [8–10] though recently methods of direct integration in $d = 4$ space-time are developing [11, 12]. The MB method has been used mostly in conjunction with the former, dimensional regularization approach, and we will employ dimensional regularization throughout. In spite of that, it is not our intention here to give a comprehensive introduction to dimensional regularization, and we refer the reader to the many excellent QFT textbooks that discuss this topic in detail. However, the basic idea can be demonstrated on very simple integrals, which we discuss next.

---

[3] The factor of 2 appears because of practical reasons: many results take a simpler form with this choice; see, e.g., Eq. (1.40).

Regularization: An Elementary Example

Consider the very simple integral

$$I(\alpha) = \int_0^1 \frac{1}{x^\alpha} dx \,. \tag{1.20}$$

Obviously this integral is convergent for $\alpha < 1$ and can be evaluated in an elementary way, $I(\alpha) = \left. \frac{x^{1-\alpha}}{1-\alpha} \right|_0^1 = \frac{1}{1-\alpha}$. However for $\alpha = 1$, the integral is logarithmically singular on the lower limit of integration and indeed the $\alpha \to 1$ limit of the result is ill-defined. In order to give meaning to the integral even for $\alpha = 1$, we can add a regulator to the measure in the following way: We set $dx \to x^{-\epsilon} dx$ and define the regularized integral

$$I_{\text{reg}}(\alpha; \epsilon) = \int_0^1 \frac{1}{x^\alpha} x^{-\epsilon} dx \,. \tag{1.21}$$

The original integral is recovered for $\epsilon \to 0$. The integration is of course still elementary and we find $I_{\text{reg}}(\alpha; \epsilon) = \left. \frac{x^{1-\alpha-\epsilon}}{1-\alpha-\epsilon} \right|_0^1 = \frac{1}{1-\alpha-\epsilon}$. Now, notice the following: On the one hand, when $\alpha < 1$, i.e., if the original integral is convergent, we may simply take the $\epsilon \to 0$ limit in the regularized result and recover the original one: $I_{\text{reg}}(\alpha; \epsilon = 0) = I(\alpha)$. On the other hand, for $\alpha \to 1$, we obtain $I_{\text{reg}}(\alpha = 1; \epsilon) = \frac{1}{-\epsilon}$. Hence, the regularized integral is well-defined for $\alpha = 1$, and moreover, whenever the original integral is convergent, we recover the correct result after removing the regularization.

Perhaps it is worth emphasizing that regularization did not magically make our divergent integral finite, since we recover the singularity in the form of a pole in ϵ if we try to remove the regulator in $I_{\text{reg}}(\alpha = 1; \epsilon)$. However, we now have a well-defined expression that we can associate to the divergent integral $I(\alpha = 1)$. The point is that in a real computation in QFT perturbation theory, although divergent integrals show up at intermediate stages of the calculation, the final physical results are of course finite. Thus, regularization is a tool used to make all intermediate results well-defined in a consistent way. The cancellation of ϵ poles in physical results is in fact a strong check on the correctness of the calculations.

Let us now turn to properly defining the dimensionally regularized Feynman integrals whose evaluation via the MB method will be the focus of this book. The L-loop Feynman integral in $d = 4 - 2\epsilon$ dimensions with N internal lines of masses m_i carrying momenta q_i and E external legs with momenta p_e can be written in the following way:

$$G_L[T(k)] = \frac{e^{\epsilon \gamma L}(2\pi \mu)^{(4-d)L}}{(i\pi^{d/2})^L} \int \frac{d^d k_1 \ldots d^d k_L \, T(k)}{(q_1^2 - m_1^2)^{\nu_1} \ldots (q_N^2 - m_N^2)^{\nu_N}}. \qquad (1.22)$$

Notice in particular the d-dimensional measures $d^d k_i$: the integrals are now defined with all L loop momenta living in d-dimensional momentum space.[4] The integration region is all of this $L \times d$-dimensional space. The numerator $T(k)$ is in general a tensor in the integration variables:

$$T(k) = 1, k_l^\mu, k_l^\mu k_n^\nu, \ldots . \qquad (1.23)$$

Integrals with $T(k) = 1$ are referred to as basic scalar integrals. Internal four-momenta k_i are connected with virtual particles propagating inside closed loops, and each closed loop i corresponds to one k_i, as shown schematically in Fig. 1.3. Each virtual particle of mass m_i carries the four-momentum q_i, which are in general functions of both internal k_i and external four-momenta p_i; see, e.g., the right-hand diagram in Fig. 1.5. The powers ν_i in denominator are integer numbers which are not necessarily equal to one. For instance, in the so-called Integration-By-Parts (IBPs) procedure [13], differentiation of the FI with respect to internal momenta leads to relations among integrals with different powers of scalar propagators, leading to a reduction of the FI to a small number of so-called master integrals (MIs). Hence, it is generally useful to consider arbitrary propagator powers as we have written. The overall prefactor is chosen in a way to simplify the final expressions as we will see shortly. For example, the reason for factoring $e^{\epsilon \gamma L}$ in Eq. (1.22) becomes clear with a discussion of the expansion of gamma functions in ϵ; see Eq. (1.39). Second, the parameter μ has dimensions of mass, so including the factor of $(2\pi \mu)^{(4-d)L}$ restores the overall mass dimension of the integral to its original four dimensional value from the modified d-dimensional one. Finally, the factor $(i\pi^{d/2})^L$ in the integration measure is conventional; it could well be taken as $(2\pi)^{dL}$, as discussed in [12]. The integrals in $G_L[T(k)]$ of Eq. (1.22) over the internal momenta k_i in d-dimensional space can be parametrized in many ways, and we will examine this in detail in Sect. 3.1. There we will show an example how it can be evaluated with Feynman parametrization. In Chap. 3 we will give examples with corresponding evaluations by MB approach.

However, let us first present a simple example of how such loop integrals can be computed directly. The most basic integrals in loop calculations arise from one-loop

---

[4] External momenta can be defined to be either four- or d-dimensional, and this choice leads to different dimensional regularization schemes such as conventional dimensional regularization (CDR), 't Hooft-Veltman (HV) scheme, dimensional reduction (DR), etc. See, e.g., [12] for an overview.

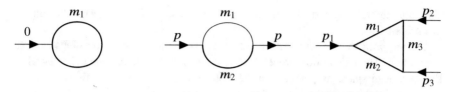

Fig. 1.7 Some one-loop Feynman diagrams with external momenta p_i and internal masses m_i which can be written using the basic integral Eq. (1.24)

diagrams. Figure 1.7 shows some basic examples: a so-called tadpole,[5] bubble, and triangle.

Omitting here the factor $e^{\gamma \epsilon}$ given in Eq. (1.22) for a moment, let us define

$$I_a^b(M^2) \equiv \frac{(2\pi\mu)^{4-d}}{i\pi^2} \int d^d Q \frac{(Q^2)^b}{(Q^2 - M^2)^a}. \tag{1.24}$$

For $a = 1$ and $b = 0$, we get the simplest integral, the tadpole

$$A_0(m^2) = \mathrm{T1l1m} \equiv I_1^0\left(m^2\right). \tag{1.25}$$

The notation by symbols $A, B, C, \dots$ for tadpoles, bubbles, triangles, $\dots$ is due to Passarino and Veltman [17]. We also call these functions one-point (1-PF), two-point (2-PF), three-point (3-PF) functions, and so on, indicating the number of legs which represent the scattered objects with external four-momenta $p_1, p_2, p_3, \dots$. Returning to I_a^b, integrals with indices a and b greater than one appear during IBP reduction [13] and also in derivatives of more complicated multipoint diagrams with respect to external kinematics, such as the bubble and triangle shown in Fig. 1.7. They also appear as building blocks of higher-point integrals, e.g., for B_0 and C_0, we have

$$B_0(p^2; m_1^2, m_2^2) = \int_0^1 dx \, I_2^0\left(M^2\right), \tag{1.26}$$

$$C_0(p_2^2, p_3^2, p_1^2; m_1^2, m_2^2, 0) = 2 \int_0^1 dx \int_0^x dy \, I_3^0\left(M^2\right). \tag{1.27}$$

---

[5] The name tadpole is due to Sidney Coleman. They constitute vacuum effects in one space-time point and do have physical meaning if the corresponding external field has a non-zero vacuum expectation value; such is the case, e.g., for the Higgs field. Tadpole diagrams have been used for the first time in [14] to explain symmetry breaking in the strong interaction. They are also considered in the context of the free energy density or pressure of the QCD plasma, finite temperature effective potentials, and phase transitions; see, for instance, [15, 16].

That the integrals B_0 and C_0 can be written as function of Eq. (1.24) is the subject of Problem 1.4.

Having seen that the integral in Eq. (1.24) is the basic one-loop object, let us compute it. The calculation proceeds in two steps. First, we change the integral to the Euclidean space by performing the Wick rotation $Q \to iq$. This is merely a change of integration variables and the integral becomes

$$I_a^b(M^2) = (-1)^{b-a} \frac{(2\pi\mu)^{4-d}}{i\pi^2} \int d^d q \frac{(q^2)^b}{(q^2 + M^2)^a}. \tag{1.28}$$

The propagator in Eq. (1.28) is not zero, unless $M = 0$ and $q^2 = 0$. Now we have to focus on the d-dimensional measure $d^d q$. The integrand in Eq. (1.28) is spherically symmetric (independent of angular coordinates), and therefore we can separate out the angular part of the integral by going to spherical coordinates

$$d^d q = q^{d-1} dq d\Omega,$$

where, in d dimensions, we can write

$$d\Omega = d\theta_1 \sin^{d-2}\theta_1 d\theta_2 \sin^{d-3}\theta_2 \ldots d\theta_{d-1}, \quad d \geq 2. \tag{1.29}$$

We will consider this measure in detail in Sect. 3.14 where real radiation integrals are discussed. The integral over the angles in Eq. (1.29) can be evaluated analytically for general d, and we find (see Problem 3.18)

$$\int d\Omega = \frac{2\pi^{\frac{d}{2}}}{\Gamma(\frac{d}{2})}. \tag{1.30}$$

This is a situation when we see for the first time how the gamma function appears.[6] We will discuss its definition and properties in detail in Chap. 2. We will also need an extension of this function to the complex plane, which will also be done in the next chapter.

Using the result for the angular integral, we can continue our evaluation of $I_a^b(M^2)$,

$$I_a^b(M^2) = (-1)^{b-a} \frac{(2\pi\mu)^{4-d}}{i\pi^2} \frac{2\pi^{\frac{d}{2}}}{\Gamma\frac{d}{2}} \int d\left(\frac{q}{M}\right) \left(\frac{q}{M}\right)^{d-1} \frac{M^d}{M^{2a}} \frac{1}{\left(1 + \frac{q^2}{M^2}\right)^a}$$

$$= (-1)^{b-a} \frac{(2\pi\mu)^{4-d}}{i\pi^2} \frac{\pi^{\frac{d}{2}}}{\Gamma(\frac{d}{2})} 2 \int_0^\infty dt\, t^{d-1}(1+t^2)^{-a}. \tag{1.31}$$

---

[6] The gamma function also appears in many basic formulae and many different areas of physics like statistical physics or in derivation of the Casimir effect. For more physical cases, see [18].

The last one-dimensional integral can be performed in terms of the so-called beta function. By definition

$$B(m, n) = \frac{\Gamma(m)\Gamma(n)}{\Gamma(m+n)} = \int_0^\infty dx\, x^{m-1}(1+x)^{-m-n}, \quad \Re(x), \Im(y) > 0,$$

(1.32)

thus after setting $t^2 \to x$ we find that the last integral reduces simply to a beta function with

$$m = \frac{d}{2}, \quad \text{and} \quad n = a - m = a - \frac{d}{2},$$

(1.33)

and we find the result

$$I_a^b(M^2) = (-1)^{b-a} \frac{(2\pi\mu)^{4-d}}{i\pi^2} \frac{\pi^{\frac{d}{2}}}{\Gamma(\frac{d}{2})} \frac{1}{(M^2)^{a-\frac{d}{2}}} \frac{\Gamma(\frac{d}{2})\Gamma(a-\frac{d}{2})}{\Gamma(a)}.$$

(1.34)

Specializing to the case of $a = 1$ and $b = 0$ for the tadpole, from the general formula in Eq. (1.24), we get[7]

$$A_0(m^2) = m^2 \left(\frac{4\pi\mu^2}{m^2}\right)^\epsilon \frac{\Gamma(\epsilon)}{1-\epsilon}$$

$$= m^2 \left[\frac{1}{\epsilon} - \gamma + \ln(4\pi) + 1 - \ln\frac{m^2}{\mu^2}\right] + O(\epsilon).$$

(1.35)

Note that in Eq. (1.35) the Euler gamma constant γ (see Chap. 2) and $\ln(4\pi)$ constants are present. Adding the factor $e^{\gamma\epsilon}$, as in the original definition of Feynman integrals in Eq. (1.22), we find

$$\bar{A}_0(m^2) = \frac{(2\pi\mu)^{2\epsilon}}{i\pi^2} e^{\gamma\epsilon} \int \frac{d^d q}{q^2 - m^2 + i\epsilon} = m^2 \left(\frac{4\pi\mu^2}{m^2}\right)^\epsilon e^{\gamma\epsilon} \frac{\Gamma(\epsilon)}{1-\epsilon}$$

$$= \frac{1}{\epsilon} + 1 + \left(1 + \frac{1}{2}\zeta_2\right)\epsilon + \left(1 + \frac{1}{2}\zeta_2 - \frac{1}{3}\zeta_3\right)\epsilon^2 + \dots$$

(1.36)

where we have set in addition

$$4\pi\mu^2 = 1,$$

$$m^2 = 1.$$

(1.37)

---

[7] This coincides with definitions of the one-loop functions which are commonly used in public packages, e.g., LoopTools [19].

We see that the γ can be absorbed by the appropriate prefactor. Since

$$\Gamma(\epsilon) = \frac{1}{\epsilon} - \gamma + \frac{1}{2}\left[\gamma^2 + \zeta(2)\right]\epsilon + \frac{1}{6}\left[-\gamma^3 - 3\gamma^2\zeta(2) - 2\zeta(3)\right]\epsilon^2 + \cdots \qquad (1.38)$$

we have that[8]

$$e^{\epsilon\gamma}\,\Gamma(\epsilon) = \frac{1}{\epsilon} + \frac{1}{2}\zeta(2)\epsilon - \frac{1}{3}\zeta(3)\epsilon^2 + \cdots \qquad (1.39)$$

As already indicated in Eq. (1.25), in this book we will use also another notation for FI based on the number of massive and massless lines as done, for example, in [20] where in addition T stands for tadpoles, SE for self-energies, V for vertices, and B for boxes, so

$$\bar{A}_0(m^2 = 1) \equiv \mathtt{T111m} = \frac{1}{\epsilon} + 1 + \left(1 + \frac{\zeta_2}{2}\right)\epsilon + \left(1 + \frac{\zeta_2}{2} - \frac{\zeta_3}{3}\right)\epsilon^2 + \cdots \qquad (1.40)$$

As discussed in [21], one-loop Feynman integrals of higher multiplicity (i.e., more than five legs) and higher ranks (numerators) can be expressed in terms of a basis which includes one-point functions, tadpoles; two-point functions, self-energies; three-point functions, vertices; and four-point functions, box diagrams.

In addition, QED integrals at the one-loop level lead to Euler dilogarithms discussed in Chap. 2 or simpler functions (at least up to finite terms in ϵ); for electroweak integrals we meet more general hypergeometric functions; see, e.g., [22] and Sect. 2.7. The general one-loop integrals were tackled systematically since the 1990s; see, e.g., [23]. In [24, 25], the class of generalized hypergeometric functions for massive one-loop Feynman integrals with unit indices ($\nu_1, \ldots \nu_N = 1$ in Eq. (1.22)) was determined and studied with a novel approach based on dimensional difference equations:

(i) $_2F_1$ Gauss hypergeometric functions are needed for self-energies.

(ii) F_1 Appell functions are needed for vertices.

(iii) F_S Lauricella-Saran functions are needed for boxes.

Finally, general massive one-loop one- to four-point functions with unit indices at arbitrary kinematics were determined in [26], where also the numerics of the generalized hypergeometric functions was worked out.

---

[8] For the origin of Euler's gamma constant γ, see the chapter "Gamma's Birthplace" in [11]. It was for long not clear if γ is irrational and G. H. Hardy offered to vacate his Savilian Chair at Oxford to anyone who could prove gamma to be irrational. The emergence of the (Euler-Riemann) zeta function is discussed in this book also.

Below we gather how these simplest scalar integrals behave in four dimensions in the limits of the internal momentum $k \to \infty$ (UV) and $k \to 0$ (IR).

$$A_0 = \frac{1}{(i\pi^{d/2})} \int \frac{d^d k}{D_1} \quad \to \text{UV-divergent}: \quad \sim \frac{d^4 k}{k^2}$$
$$\to \text{IR-finite}$$

$$B_0 = \frac{1}{(i\pi^{d/2})} \int \frac{d^d k}{D_1 D_2} \quad \to \text{UV-divergent} \quad \sim \frac{d^4 k}{k^4}$$
$$\to \text{IR-finite}$$

$$C_0 = \frac{1}{(i\pi^{d/2})} \int \frac{d^d k}{D_1 D_2 D_3} \quad \to \text{UV-finite} \quad \sim \frac{d^4 k}{k^6}$$
$$\to \text{IR-divergent}$$

The propagators are

$$D_1 = k^2 - m_1^2, \quad D_2 = (k + p_1)^2 - m_2^2; \quad D_3 = (k + p_1 + p_2)^2 - m_3^2.$$

After defining FI and showing a traditional way how to solve FI in dimensional regularization procedure where singularities of the integrals are encoded in dimensional regulator ϵ, we come to the idea of Mellin-Barnes representations and integrals.

1.4   Basic Idea of Mellin-Barnes Representations

According to Slater [27], the basic idea of representing a function by a contour integral with gamma functions in the integrand is due to the nineteenth-century work by S. Pincherle. It has been developed further by R. Mellin and E.W. Barnes. In 1895 Mellin [28] introduced theorem where

$$f(x) = \frac{1}{2\pi i} \int_{c-i\infty}^{c+i\infty} x^{-s} g(s) ds, \tag{1.41}$$

then

$$g(s) = \int_0^\infty x^{s-1} f(x) dx. \tag{1.42}$$

The Mellin transform function $g(s)$ is a locally integrable function where x is a positive real number and s is complex in general.

Five years after the work by Mellin, the paper "The theory of the gamma function" appeared [29], followed by the 1907s series of papers by Barnes [30–32] in which the so-called Barnes contour integrals have been explored. They are of the type [27]

$$I_C = \int_C \frac{dz}{2\pi i} \frac{\Gamma(a+z)\Gamma(b+z)\Gamma(-z)}{\Gamma(c+z)} (-s)^z. \tag{1.43}$$

It can be shown that this integral (when convergent with $|s| < 1$ and $|arg(-s)| < \pi|$) is equivalent to the hypergeometric function $_2F_1$ which will be discussed in Sect. 2.7.

There are different definitions of what is called the Mellin-Barnes integral. For instance, in recent textbook [18], the integral in Eq. (1.41) is directly called Mellin-Barnes as "any integral in the complex plane whose integrand contemplates at least one gamma function." For applications in particle physics calculations, we are interested in a slightly modified form of the above equations. What is nowadays commonly called the Mellin-Barnes representation[9] in its simplest form is a representation of some power λ of a sum of two terms A and B, in terms of an integral on the complex plane

$$\boxed{\frac{1}{(A+B)^\lambda} = \frac{1}{\Gamma(\lambda)} \frac{1}{2\pi i} \int_{-i\infty}^{+i\infty} dz \Gamma(\lambda+z)\Gamma(-z) \frac{B^z}{A^{\lambda+z}}.} \tag{1.44}$$

The exact conditions for validity of this formula and a proof will be given in Sect. 3.3. This integral follows from Eq. (1.41) when $f(x) = \frac{x^\lambda}{(1+x)^\lambda}$ and $g(s) = \frac{\Gamma[-s]\Gamma[s+\lambda]}{\Gamma[\lambda]}$

$$\frac{x^\lambda}{(1+x)^\lambda} = \frac{1}{2\pi i} \int_{c-i\infty}^{c+i\infty} \frac{\Gamma[-s]\Gamma[s+\lambda]}{\Gamma[\lambda]} x^{-s} ds. \tag{1.45}$$

---

[9] In many old textbooks, they are called just Barnes integrals. Nevertheless, you may find Mellin-Barnes integrals in books by L.J. Slater [33] or A.W. Babister [34] or more recently in a context of Feynman integrals in works by N.I. Usyukina, A. Davydychev, B. Arbuzov, E. Boos, and V. Smirnov; see, e.g., [8, 35, 36].

The proof for this equation is exactly the same as will be given for Eq. (1.44) in Sect. 3.3. Replacing x by A/B, Eq. (1.44) follows.

The relation in Eq. (1.44) has an immediate application to physics, for instance, a massive propagator can be written as

$$\frac{1}{(p^2 - m^2)^a} = \frac{1}{\Gamma(a)} \frac{1}{2\pi i} \int_{-i\infty}^{+i\infty} dz \, \Gamma(a+z)\Gamma(-z) \frac{(-m^2)^z}{(p^2)^{a+z}}. \tag{1.46}$$

The upshot of this change is that a mass parameter m merges with a kinematic variable p^2 into the ratio $\left(-\frac{m^2}{p^2}\right)^z$ and the integral effectively becomes massless.

In the context of phase space integrals, the basic Mellin-Barnes formula can be employed to bring the integrand to a form where the integration over the original phase space variables can be performed trivially. Of course, the price to pay is that we must introduce MB integrations; however this representation is many times much more convenient for further work than the original one. As an example, consider the integral

$$\int d\phi_3(p_1, p_2, p_3, Q) \frac{1}{(p_1 \cdot p_2)(p_1 \cdot p_2 + p_1 \cdot p_3)}, \tag{1.47}$$

where all momenta are massless, i.e., $p_1^2 = p_2^2 = p_3^2 = 0$. It can be shown (see, e.g., [37]) that in $d = 4 - 2\epsilon$ dimensions, this integral is proportional to

$$\int_0^1 dx \int_0^{1-x} dy \, x^{-\epsilon} y^{-\epsilon} (1 - x - y)^{-\epsilon} \frac{1}{x(x+y)}. \tag{1.48}$$

After performing the trivial change of variable $y \to (1 - x)y$, we find

$$\int_0^1 dx \int_0^1 dy \, x^{-1-\epsilon} (1 - x)^{1-2\epsilon} y^{-\epsilon} (1 - y)^{-\epsilon} \frac{1}{x + (1 - x)y}. \tag{1.49}$$

Were it not for the last factor, this integral would be trivial to evaluate in terms of the beta function of Eq. (1.32). However, we can use the basic Mellin-Barnes formula to convert the sum in the denominator into factors of x and $(1 - x)y$ with generic powers, which allows to perform the integration over x and y immediately. Then we obtain the MB representation

$$\int_{-i\infty}^{+i\infty} dz \frac{\Gamma(1-\epsilon)\Gamma(-z_1)\Gamma(z_1+1)\Gamma(-\epsilon - z_1)\Gamma(z_1 - \epsilon)}{\Gamma(1 - 3\epsilon)}, \tag{1.50}$$

which can be further manipulated (e.g., expanded in ϵ or evaluated analytically and numerically), as we will show in detail in the later chapters.

Obviously, the basic MB formula in Eq. (1.44) can be applied multiple times if the problem demands, and generally we encounter multi-dimensional MB integrals.

In fact, in this book we will refer to any multiple complex contour integral of the form

$$
\int_{-i\infty}^{+i\infty} \cdots \int_{-i\infty}^{+i\infty} \prod_{j=1}^{n} \frac{dz_j}{2\pi i} f(z_1, \ldots, z_n, x_1, \ldots, x_p, a_1, \ldots, a_q) \frac{\prod_k \Gamma(A_k + V_k)}{\prod_l \Gamma(B_l + W_l)},
$$

(1.51)

as an MB integral. In this expression the x_j are fixed parameters (e.g., kinematic invariants and masses), while the a_j are expressions of the form $a_i = n_i + b_i \epsilon$, with $n_i \in \mathbb{N}$ and $b_i \in \mathbb{R}$. Furthermore, A_k and B_l are linear combinations of the a_i, while V_k and W_l are linear combinations of the integration variables z_i. Finally, f is an analytic function; in practice it is a product of powers of the x_i with exponents that are linear combinations of a_i and z_i.

The use of the Mellin transform and Mellin-Barnes contour integrals in QFT dates back to the 1960s. Indeed, the Mellin transform was used for the analysis of scattering amplitudes and the asymptotic expansion of FIs already at that time; see, for instance, [38] and [39, 40]. Mellin-Barnes contour integrals were investigated for finite three-point functions in [35], followed by related works [36, 41, 42]. However, in terms of mass production of new results in the field, a real breakthrough came toward the end of the last millennium when the infrared divergent massless planar two-loop box was solved analytically using MB methods [43], followed in the same year by the non-planar case [44]. These results were based on the Feynman parametrization of FI, and the MB representations were constructed by the repeated application of the basic formula of Eq. (1.44) to the elements of the F and U polynomials (called Symanzik polynomials) associated to the given FI.[10] This procedure was automatized initially in [45] and will be discussed in full length in Chap. 3. We note that the gamma functions that appear in Eq. (1.44), and hence the MB representations of FI and phase space integrals, play a pivotal role, changing the original singular structure of propagators in four-momentum space (see Eq. (1.4)) into another one, connected with singularities of the gamma functions. We will explore this in Chap. 4.

1.5   Analytical and Numerical Approaches to Mellin-Barnes Integrals

Regarding the evaluation of Mellin-Barnes integrals, there is no one universal method or program covering all cases at the frontier of perturbation theory.

Concerning analytical approaches, given that the MB representation involves integrations over complex variables, the use of Cauchy's residue theorem to evaluate

---

[10] Feynman parametrization and the representation of FI via the Symanzik polynomials will be discussed in depth in Chap. 3.

these integrals comes immediately to mind. We will follow this approach in Chap. 5, where we will show in detail how one can obtain a series representation of MB integrals by summing over residues. However, we should note that so far, due to the complex nature of the series which arise and the difficulties in establishing their convergence, mostly expansion of MB integrals in some kinematic variables (see Sect. 5.4.1) has been explored. Nevertheless, some new developments in understanding the structure of these series (see [46] and Sect. 5.4) allow to hope that a full chain of actions shown in Eq. (1.52) with a more efficient application of the approach based on convergence of series to physics studies will result.

$$\boxed{\textbf{Feynman Integrals} \hookrightarrow (MB)\textbf{ContourIntegrals} \hookrightarrow (Residues)\textbf{Series}}$$

(1.52)

Other approaches for obtaining analytic solutions involve transforming the MB integrals into real integrals of so-called Euler type and evaluating those. We will discuss this approach in Sect. 5.3, where we will show that it leads naturally to iterated integrals. Under suitable circumstances, such integrals can be evaluated analytically in terms of certain classes of functions that can be thought of as generalizations of the logarithm, so-called multiple polylogarithms.

We note that in these analytical approaches, the level of complexity, at a given loop order, is defined by the number of virtual massive particles that appear in the Feynman integrals. For one-loop integrals, the most complicated mathematical objects that appear in their series expansions in ϵ *up to finite terms* are Euler dilogarithm functions that are special cases of multiple polylogarithms. However, in last years it has become obvious that analytic solutions to multiloop integrals, which are always the best solutions to have, involve also functions that can no longer be expressed with multiple polylogarithms, such as elliptic functions and iterated integrals of modular forms. In fact, the class of functions that arise may go even beyond the elliptic case [47]; see also Sect. 2.4.

As an interesting aside, we note that elliptic curves, modularity, and modular forms were some of the key ingredients to solve a long-standing and very famous problem in number theory, Fermat's Last Theorem [48]. It is spectacular that the precision of calculations in high energy physics reached the level of sophistication in contemporary cutting-edge mathematical studies.

However, in our opinion, due to the many masses and large multiplicities (many external legs) that arise at higher orders in perturbation theory for precision studies in the SM, numerical integration methods may be the most promising avenues for addressing incoming challenges. In the context of MB integrals, one key problem is the construction of MB representations with the lowest possible dimension. This will be explored in the main part of Chap. 3. Numerical evaluation of MB integrals will be discussed thoroughly in Chap. 6.

1.6   Mellin and Barnes Meet Euclid and Minkowski

As discussed in Sect. 1.2, the propagators in scattering amplitudes exhibit singularities. We can deal with them by relocating the problem into the complex
plane, as schematically denoted in Eq. (1.4). However, there are cases where it
is convenient to avoid singularities altogether by considering artificial kinematic
conditions. For example, we may choose to consider some four momenta squared
to be negative, e.g., $p^2 \rightarrow -p^2$ in Eq. (1.4). In such cases, obviously, solutions for
E_k discussed in Sect. 1.2 do not need special treatment in the complex plane. This
kinematic configuration, which does not generate such singularities of amplitudes, is
realized in Euclidean space with Euclidean kinematic conditions. The real *physical*
kinematics is of course defined in Minkowski space and is called the Minkowskian
kinematics. Obviously, the problem of singularities is related to the Minkowskian
nature of kinematics in special relativity. By the way, we have already used
Euclidean space for solving the integral in Eq. (1.24) in a standard way in Sect. 1.2:
the Wick rotation effectively transforms Minkowskian momenta into Euclidean
ones.

In the context of multiloop calculations and the MB method, considering
Euclidean kinematic has two main advantages. First, numerical cross-checks of
the obtained solutions for scattering amplitudes or specific multiloop integrals can
be obtained relatively easily. The reason for this is that in Euclidean kinematics the
integrands behave "well" in a certain sense (they are smooth functions with no rapid
oscillations); hence numerical computations are straightforward. Thus, packages
for manipulating MB integrals that use general-purpose numerical integrators,
such as MB.m [49], can be used directly. Second, obtaining analytic solutions
can be sometimes easier using Euclidean kinematics. One example is the fixing
of boundary conditions in the differential equation method of solving Feynman
integrals. We can then *analytically continue* the results (functions) from Euclidean
to Minkowski space, checking the results numerically in the Euclidean kinematic
regime. Analytic continuation of complex functions will be discussed in Chap. 2.

When Is Euclidean (*Unphysical*) Kinematics Useful?

Evaluation of integrals in Euclidean kinematics is especially useful if we want to:

1. Check individual integrals numerically against analytic solutions,
2. Check results obtained with different approaches where analytic solutions are not
   always available.

Fig. 1.8 Typical assignment of kinematic variables for vertex diagrams. A black blob can include multiloop corrections like the three-point vertex in Fig. 1.7

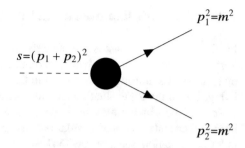

Having said that, we will show in Sect. 6.2.4 how to use the knowledge of the threshold behavior of Symanzik polynomials to calculate MB integrals numerically, directly in Minkowskian kinematics. We will avoid the bad behavior of integrands by a special construction of the MB integrals which restores their Euclidean behavior.

To finish this section, let us show an example of the Euclidean and Minkowskian kinematic space useful for analytic and numerical solutions of Feynman integrals.

Euclidean and Minkowskian Kinematics

As kinematics depends on the number of external particles present in a given process, it is useful to define it using kinematic invariants. In the simple case of 2 → 2 scattering processes, these are known as Mandelstam invariants. Transformations between external four-momenta and kinematic invariants are defined in many packages used in high energy physics.

Here we give an example based on the `KinematicsGen.m` package for a vertex kinematics in which invariants are defined in a standard way (see Fig. 1.8) as squares of external momenta p_1 and p_2 (connected with two outgoing legs) and their sum squared s (connected with the incoming leg).

```
invariants = {p1^2 -> m^2, p2^2 -> m^2, p1*p2 -> -m^2 + s/2};
invEucl = {m->1, s->1};
invMink = {m->1, s->-1};
```

$$(1.53)$$

In Eq. (1.53) we show verbatim some input kinematic parameters related to Fig. 1.7 with a notation which is taken from a `Mathematica` script file[11] `run_script_1loop_QED_vertex_mink`, available in Section "MBnumerics," "Related and auxiliary Software" at [51].

---

[11] In this book we will often use examples evaluated with the `Mathematica` computer algebra system (CAS) [50]. `Mathematica` is frequently used in particle physics and for the construction and analysis MB integrals, as will be described in the following chapters and the Appendix.

1.7   A Simple Example as an Invitation to the MB Topic

In Fig. 1.9 a scheme of possible actions is given based on the one-loop example. Various analytical, semi-analytical, and numerical analyses for FI based on the MB method will be thoroughly discussed in Chaps. 5–6. "Other methods" highlighted in Fig. 1.9 for the FI evaluation include notably the powerful differential equations (DEs) method which is intensively developing. For recent studies and software used in analytical and numerical calculations, see [52–54]. For other exploratory methods in FI computation, see a review [55].

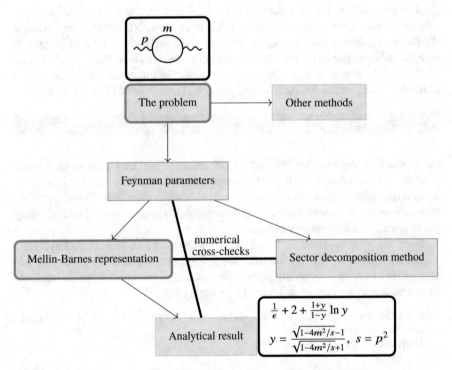

Fig. 1.9 A scheme of actions which will be discussed in the book, the one-loop example. We highlight MB and sector decomposition (SD) methods which are explored presently at the two- and three-loop levels [10]. The analytical result will be discussed and derived using the MB method in Sects. 5.1.2 and 5.1.7; see Eq. (5.177)

Fig. 1.10 One-loop
correction to the three-point
vertex. Thick lines denote
massive particles with mass
m

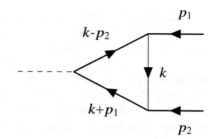

Let us consider the simple example of a one-loop correction to a three-point
vertex as shown in Fig. 1.10. The corresponding Feynman integral is given in
Eq. (1.54), where we have anticipated that the Laurent series expansion of the
result[12] in ϵ, $d = 4 - 2\epsilon$ starts at $O(\epsilon^{-1})$,

$$
\begin{aligned}
V(s) &= \frac{e^{\epsilon\gamma}}{i\pi^{d/2}} \int \frac{d^d k}{[(k+p_1)^2 - m^2][k^2][(k-p_2)^2 - m^2]} \\
&= \frac{V_{-1}(s)}{\epsilon} + V_0(s) + \cdots
\end{aligned}
\tag{1.54}
$$

In order to illustrate the issues we meet when considering both analytic and
numerical solutions of such integrals, it is enough to look at the leading divergent
integral $V_{-1}(s)$, which, translated to the MB representation, takes the following form,
with $m = 1$, $s = (p_1 + p_2)^2$ (the construction of this MB representation and the
corresponding `Mathematica` file will be discussed in Chap. 5):

$$
V_{-1}(s) = -\frac{1}{2s} \int\limits_{-\frac{1}{2}-i\infty}^{-\frac{1}{2}+i\infty} \frac{dz}{2\pi i} \underbrace{(-s)^{-z}}_{Part\ I} \overbrace{\frac{\Gamma^3(-z)\Gamma(1+z)}{\Gamma(-2z)}}^{Part\ II}.
\tag{1.55}
$$

Parts I and *II* exhibit some general features of MB integrals that are important to
understand when seeking either numerical or analytic solutions. In particular, if
the kinematic variable s is positive, the exponential function in *Part I* is highly
oscillating along the path of integration (Problem 1.6), while the gamma functions
appearing in *Part II* exhibit singularities (infinite values) for real integer values of
the arguments. These are obvious obstacles that prevent a straightforward numerical
evaluation, and in general we find a fragile stability for integrations over products
and ratios of gamma functions. We will learn about gamma functions starting from
Chap. 2. There are many subtleties connected also with finding analytic solutions of

---

[12] The exact result in ϵ includes the hypergeometric function $_2F_1$ and can be obtained with
`Mathematica`; see the file `MB_miscellaneous_Springer.nb` in the auxiliary material
in [56] and [57].

MB integrals. For instance, for diagrams with massless propagators only, the gamma functions in MB integrals have arguments that involve the integration variable z in the form $\Gamma(\ldots \pm z)$. However, in massive cases some of the gamma functions will contain in their argument the integration variable z multiplied by 2, as in the denominator of Eq. (1.55). The application of Cauchy's residue theorem to such integrals then leads to the appearance of so-called inverse binomial sums for massive cases. Nonetheless, for many basic integrals, like the one considered here, the analytical results are known and can provide a good theoretical laboratory for investigating more complicated integrals. Indeed, the analytical result for $V_{-1}(s)$ can be written as

$$V_{-1}(s) = \frac{1}{2} \sum_{n=0}^{\infty} \frac{s^n}{\binom{2n}{n}(2n+1)} = \frac{2 \arcsin(\sqrt{s}/2)}{\sqrt{4-s}\sqrt{s}} = -\frac{y \ln y}{1-y^2} \tag{1.56}$$

This result arises by applying the residue theorem to the MB integral and summing up the residues of Eq. (1.55). This procedure, as well as the sums that arise, will be discussed in detail in Sect. 5.1. The conformal variable y will be also discussed there.

Decomposition of V312m

Actually, FI can be decomposed to MIs (T111m and SE212m) using IBPs, a procedure mentioned in Sect. 1.3. In our case, we can use the dedicated public IBPs software AIR [58], Fire [59], Kira [60], LiteRed [61], or Reduze [62] to get

$$\text{V312m} = \frac{\epsilon - 1}{\epsilon(s-4)} \text{T111m} + \frac{1 - 2\epsilon}{\epsilon(s-4)} \text{SE212m}. \tag{1.57}$$

Adding expansions for T111m in Eq. (1.40) and for SE212m given in Fig. 1.9 (see also the result for ϵ^0 term in Sect. 5.1.7), we get

$$\text{V312m} = \frac{y H(0,y)}{\epsilon(y^2-1)} - \frac{y\left(12H(-1,0,y) - 6H(0,0,y) + \pi^2\right)}{6(y^2-1)}, \tag{1.58}$$

where functions H are harmonic polylogarithms which will be discussed in next chapter. By inspection (compare Eq. (1.56) with ϵ^{-1} term in Eq. (1.58)), we can see that $H(0,y) \equiv \ln y$. The decomposition in Eq. (1.57) and numerical cross-checks are given in the file MB_miscelaneous_Springer.nb in the auxiliary material in [56].

Problems

Problem 1.1 Derive Eq. (1.13).

Hint: This is a straightforward calculation by exploring the four-vector notation below Eq. (1.13) and the kinematics as in Fig. 1.6.

Problem 1.2 Prove a general formula for Feynman parametrization of products of denominators

$$\frac{1}{A_1^{n_1} A_2^{n_2} \ldots} = \frac{\Gamma(n_1 + \ldots + n_m)}{\Gamma(n_1) \ldots \Gamma(n_m)} \tag{1.59}$$

$$\int_0^1 dx_1 \ldots \int_0^1 dx_m \frac{x_1^{n_1-1} \ldots x_m^{n_m-1} \delta(1 - x_1 \ldots - x_m)}{(x_1 A_1 + \ldots + x_m A_m)^{n_1 + \ldots n_m}}$$

Hint: This relation can be proved by induction; see, e.g., [63].

Problem 1.3 Using Eq. (1.59) show in particular that

$$\frac{1}{AB} = \int_0^1 dy \frac{1}{[Ay + B(1 - y)]^2} \tag{1.60}$$

$$\frac{1}{ABC} = \Gamma[3] \int_0^1 dx \int_0^1 dy \int_0^1 dz \frac{\delta(1 - x - y - z)}{(Ax + By + Cz)^3}$$

$$= \Gamma[3] \int_0^1 dx \int_0^{1-x} dy \frac{1}{[Ax + By + C(1 - x - y)]^3} \tag{1.61}$$

$$= \Gamma(3) \int_0^1 dx \int_0^1 dy \frac{x}{[A(1 - x) + Bxy + Cx(1 - y)]^3}. \tag{1.62}$$

The relation in Eq. (1.60) is known in the literature as "the Feynman integration trick" [63].

Problem 1.4 Apply relations derived in the previous problem to the definition of the B_0 and C_0 functions (see Fig. 1.7)

$$B_0(p^2; m_1^2, m_2^2) = \frac{(2\pi\mu)^{4-d}}{i\pi^2} \int \frac{d^d q}{(q^2 - m_1^2 + i\epsilon)[(q+p)^2 - m_2^2 + i\epsilon]}.$$

(1.63)

$$C_0(p_2^2, p_3^2, p_1^2; m_1^2, m_2^2, 0) = \frac{(2\pi\mu)^{4-d}}{i\pi^2} \int \frac{d^d k}{d_1 d_2 d_3},$$

(1.64)

to prove that that these functions can be written in form of Eqs. (1.26) and (1.27).

Problem 1.5 Check kinematic invariants written in the first line of Eq. (1.53), and find kinematic invariants needed to generate 4-PF, e.g., two-loop topology given in Fig. 4.2 in Chap. 4.
Hint: You can use `kinematicGen.m` [51].

Problem 1.6 Using known to you numerical methods and packages, discuss the problem of accuracy for the *Part 1* of the integral in Eq. (1.55).
Hint: You may follow discussions in [49] and [64].

References

1. D.R. Yennie, S.C. Frautschi, H. Suura, The infrared divergence phenomena and high-energy processes. Ann. Phys. **13**, 379–452 (1961). https://doi.org/10.1016/0003-4916(61)90151-8
2. P. Nahin, *Inside Interesting Integrals*, 2nd edn. (Springer, Cham, 2020). https://doi.org/10.1007/978-1-4939-1277-3
3. L. Landau, On analytic properties of vertex parts in quantum field theory. Nucl. Phys. **13**, 181–192 (1959). https://doi.org/10.1016/0029-5582(59)90154-3
4. N. Nakanishi, Parametric integral formulas and analytic properties in perturbation theory. Prog. Theor. Phys. Supplement **18**, 1 (1961). http://ptps.oxfordjournals.org/content/18/1.full.pdf
5. T. Kinoshita, Mass singularities of Feynman amplitudes. J. Math. Phys. **3**, 650–677 (1962)
6. T.D. Lee, M. Nauenberg, Degenerate systems and mass singularities. Phys. Rev. **133**, B1549–B1562 (1964)
7. G.P. Salam, G. Soyez, A practical seedless infrared-safe cone jet algorithm. JHEP **05**, 086 (2007). arXiv:0704.0292, https://doi.org/10.1088/1126-6708/2007/05/086
8. V. Smirnov, *Evaluating Feynman Integrals*. Springer Tracts in Modern Physics (Springer, Berlin, 2004)
9. S. Weinzierl, The Art of computing loop integrals, Fields Inst. Commun. **50**, 345–395 (2007). arXiv:hep-ph/0604068
10. A. Blondel, et al., Standard model theory for the FCC-ee Tera-Z stage, in *Mini Workshop on Precision EW and QCD Calculations for the FCC Studies : Methods and Techniques,*

vol. 3/2019 of CERN Yellow Reports: Monographs (CERN, Geneva, 2018). arXiv:1809.01830, https://doi.org/10.23731/CYRM-2019-003

11. C. Gnendiger, et al., To d, or not to d: recent developments and comparisons of regularization schemes. Eur. Phys. J. C **77**(7), 471 (2017). arXiv:1705.01827, https://doi.org/10.1140/epjc/s10052-017-5023-2

12. S. Weinzierl, Feynman Integrals, arXiv:2201.03593

13. K. Chetyrkin, F. Tkachov, Integration by parts: The algorithm to calculate β functions in four loops. Nucl. Phys. **B192**, 159–204 (1981)

14. S.R. Coleman, S. Glashow, Departures from the eightfold way: Theory of strong interaction symmetry breakdown. Phys. Rev. **134**, B671–B681 (1964). https://doi.org/10.1103/PhysRev.134.B671

15. C.G. Boyd, D.E. Brahm, S.D.H. Hsu, Resummation methods at finite temperature: The Tadpole way. Phys. Rev. D **48**, 4963–4973 (1993). arXiv:hep-ph/9304254, https://doi.org/10.1103/PhysRevD.48.4963

16. T. Luthe, Y. Schröder, Five-loop massive tadpoles. PoS **LL2016**, 074 (2016). arXiv:1609.06786, https://doi.org/10.22323/1.260.0074

17. G. Passarino, M. Veltman, One loop corrections for e^+e^- annihilation into $\mu^+\mu^-$ in the Weinberg model. Nucl. Phys. **B160**, 151 (1979). https://doi.org/10.1016/0550-3213(79)90234-7

18. E. de Oliveira, *Solved Exercises in Fractional Calculus*. Studies in Systems, Decision and Control (Springer International Publishing, 2019). https://doi.org/10.1007/978-3-030-20524-9

19. T. Hahn, M. Perez-Victoria, Automatized one loop calculations in four-dimensions and D-dimensions. Comput. Phys. Commun. **118**, 153–165 (1999). arXiv:hep-ph/9807565, https://doi.org/10.1016/S0010-4655(98)00173-8

20. M. Czakon, J. Gluza, T. Riemann, Master integrals for massive two-loop Bhabha scattering in QED. Phys. Rev. **D71**, 073009 (2005). arXiv:hep-ph/0412164

21. G. 't Hooft, M. Veltman, Scalar one loop integrals. Nucl. Phys. **B153**, 365–401 (1979). https://doi.org/10.1016/0550-3213(79)90605-9

22. J. Fleischer, J. Gluza, A. Lorca, T. Riemann, First order radiative corrections to Bhabha scattering in d dimensions. Eur. J. Phys. **48**, 35–52 (2006). arXiv:hep-ph/0606210, https://doi.org/10.1140/epjc/s10052-006-0008-6

23. O. V. Tarasov, Connection between Feynman integrals having different values of the space-time dimension. Phys. Rev. D **54**, 6479–6490 (1996). arXiv:hep-th/9606018, https://doi.org/10.1103/PhysRevD.54.6479

24. O.V. Tarasov, Application and explicit solution of recurrence relations with respect to space-time dimension. Nucl. Phys. Proc. Suppl. **89**, 237–245 (2000). arXiv:hep-ph/0102271

25. J. Fleischer, F. Jegerlehner, O. Tarasov, A New hypergeometric representation of one loop scalar integrals in d dimensions. Nucl. Phys. **B672**, 303–328 (2003). arXiv:hep-ph/0307113, https://doi.org/10.1016/j.nuclphysb.2003.09.004

26. K.H. Phan, T. Riemann, Scalar one-loop Feynman integrals as meromorphic functions in space-time dimension d. Phys. Lett. B **791**, 257–264 (2019). arXiv:1812.10975, https://doi.org/10.1016/j.physletb.2019.02.044

27. L.J. Slater, *Generalized Hypergeometric Functions* (Cambridge University Press, 1966)

28. R.H. Mellin, Om definita integraler. Acta Soc. Sci. Fenn. **20**(7), 1 (1895)

29. E.W. Barnes, The theory of the gamma function. Messenger Math. **29**(2), 64 (1900)

30. E.W. Barnes, The asymptotic expansion of integral functions defined by generalised hypergeometric series. Proc. Lond. Math. Soc. **s2-5**(1), 59–116 (1907). https://doi.org/10.1112/plms/s2-5.1.59

31. E.W. Barnes, A new development of the theory of the hypergeometric functions. Proc. Lond. Math. Soc. **s2-6**(1), 141–177 (1908). https://doi.org/10.1112/plms/s2-6.1.141

32. E.W. Barnes, On functions defined by simple types of hypergeometric series. Trans. Camb. Phil. **20**, 253 (1907)

33. L.J. Slater, *Confluent Hypergeometric Functions* (Cambridge University Press, Cambridge-New York, 1960); table errata: Math. Comp. **30**(135), 677–678 (1976)
34. A.W. Babister, *Transcendental Functions Satisfying Nonhomogeneous Linear Differential Equations* (The Macmillan Co., New York, 1967)
35. N. Usyukina, On a representation for three point function. Teor. Mat. Fiz. **22**, 300–306 (1975) (in Russian)
36. E.E. Boos, A.I. Davydychev, A method of evaluating massive Feynman integrals. Theor. Math. Phys. **89**, 1052–1063 (1991) [Teor. Mat. Fiz. **89**, 56 (1991)]. https://doi.org/10.1007/BF01016805
37. A. Gehrmann-De Ridder, T. Gehrmann, G. Heinrich, Four particle phase space integrals in massless QCD. Nucl. Phys. B **682**, 265–288 (2004). arXiv:hep-ph/0311276, https://doi.org/10.1016/j.nuclphysb.2004.01.023
38. J.D. Bjorken, T.T. Wu, Perturbation theory of scattering amplitudes at high energies. Phys. Rev. **130**, 2566–2572 (1963). https://doi.org/10.1103/PhysRev.130.2566
39. M.C. Bergere, Y.-M.P. Lam, Asymptotic expansion of Feynman amplitudes. Part 1: The convergent case. Commun. Math. Phys. **39**, 1 (1974). https://doi.org/10.1007/BF01609168
40. H. Cheng, T. Wu, *Expanding Protons: Scattering at High Energies* (MIT Press, Cambridge, MA, 1987)
41. A. Davydychev, Recursive algorithm of evaluating vertex type Feynman integrals. J. Phys. **A25**, 5587–5596 (1992)
42. N.I. Usyukina, A.I. Davydychev, Exact results for three and four point ladder diagrams with an arbitrary number of rungs. Phys. Lett. **B305**, 136–143 (1993). https://doi.org/10.1016/0370-2693(93)91118-7
43. V. Smirnov, Analytical result for dimensionally regularized massless on-shell double box. Phys. Lett. **B460**, 397–404 (1999). arXiv:hep-ph/9905323
44. B. Tausk, Non-planar massless two-loop Feynman diagrams with four on- shell legs. Phys. Lett. **B469**, 225–234 (1999). arXiv:hep-ph/9909506
45. J. Gluza, K. Kajda, T. Riemann, AMBRE - a Mathematica package for the construction of Mellin-Barnes representations for Feynman integrals. Comput. Phys. Commun. **177**, 879–893 (2007). arXiv:0704.2423, https://doi.org/10.1016/j.cpc.2007.07.001
46. B. Ananthanarayan, S. Banik, S. Friot, S. Ghosh, Multiple series representations of N-fold Mellin-Barnes integrals. Phys. Rev. Lett. **127**(15), 151601 (2021). arXiv:2012.15108, https://doi.org/10.1103/PhysRevLett.127.151601
47. L. Adams, S. Weinzierl, Feynman integrals and iterated integrals of modular forms. Commun. Num. Theor. Phys. **12**, 193–251 (2018). arXiv:1704.08895, https://doi.org/10.4310/CNTP.2018.v12.n2.a1
48. A.J. Wiles, Modular elliptic curves and Fermat's last theorem. Ann. Math. **141**, 443–551 (1995). https://doi.org/10.2307/2118559
49. M. Czakon, Automatized analytic continuation of Mellin-Barnes integrals. Comput. Phys. Commun. **175**, 559–571 (2006). arXiv:hep-ph/0511200, https://doi.org/10.1016/j.cpc.2006.07.002
50. S. Wolfram, *The Mathematica Book* (Wolfram Media/Cambridge University Press, 2003)
51. AMBRE webpage: http://jgluza.us.edu.pl/ambre, Backup: https://web.archive.org/web/20220119185211/http://prac.us.edu.pl/~gluza/ambre/
52. I. Dubovyk, A. Freitas, J. Gluza, K. Grzanka, M. Hidding, J. Usovitsch, Evaluation of multi-loop multi-scale Feynman integrals for precision physics. arXiv:2201.02576
53. X. Liu, Y.-Q. Ma, AMFlow: A mathematica package for Feynman integrals computation via auxiliary mass flow. arXiv:2201.11669
54. F.F. Cordero, A. von Manteuffel, T. Neumann, Computational challenges for multi-loop collider phenomenology, in *2022 Snowmass Summer Study* (2022). arXiv:2204.04200
55. G. Heinrich, Collider physics at the precision frontier. Phys. Rept. **922**, 1–69 (2021). arXiv:2009.00516, https://doi.org/10.1016/j.physrep.2021.03.006
56. https://github.com/idubovyk/mbspringer, http://jgluza.us.edu.pl/mbspringer.

57. J. Gluza, T. Riemann, A new treatment of mixed virtual and real IR-singularities. PoS **RADCOR2007**, 007 (2007). arXiv:0801.4228, https://doi.org/10.22323/1.048.0007
58. C. Anastasiou, A. Lazopoulos, Automatic integral reduction for higher order perturbative calculations. JHEP **07**, 046 (2004). arXiv:hep-ph/0404258
59. A.V. Smirnov, Algorithm FIRE – Feynman integral REduction. JHEP **10**, 107 (2008). arXiv: 0807.3243, https://doi.org/10.1088/1126-6708/2008/10/107
60. P. Maierhofer, J. Usovitsch, P. Uwer, Kira–A Feynman integral reduction program. Comput. Phys. Commun. **230**, 99–112 (2018). arXiv:1705.05610, https://doi.org/10.1016/j.cpc.2018. 04.012
61. R.N. Lee, LiteRed 1.4: a powerful tool for reduction of multiloop integrals. J. Phys. Conf. Ser. **523**, 012059 (2014). arXiv:1310.1145, https://doi.org/10.1088/1742-6596/523/1/012059
62. C. Studerus, Reduze-Feynman Integral Reduction in C++, Comput. Phys. Commun. **181**, 1293–1300 (2010). arXiv:0912.2546, https://doi.org/10.1016/j.cpc.2010.03.012
63. E. Zeidler, *Quantum Field Theory II: Quantum Electrodynamics: A Bridge between Mathematicians and Physicists* (Springer Science & Business Media, 2009). https://doi.org/10.1007/ 978-3-540-85377-0
64. I. Dubovyk, J. Gluza, T. Riemann, J. Usovitsch, Numerical integration of massive two-loop Mellin-Barnes integrals in Minkowskian regions. PoS **LL2016**, 034 (2016). arXiv:1607.07538

Complex Analysis

2

Abstract

Mellin-Barnes (MB) integrals are defined in the complex plane as contour integrals over integrands involving the exponential and gamma functions in their most elementary form. Thus we consider basic features of the complex exponential, logarithm, gamma, polygamma, and hypergeometric functions as well as certain generalizations of these functions that appear in particle physics applications. We also discuss general complex integrals and the notion of residues, along with Cauchy's theorem, and introduce the notion of MB representations. We recall the basic notions and constructions of complex analysis to the level which allows to study the structure and solutions of MB integrals.

2.1 Complex Numbers and Complex Functions

Besides their relevance in most fields of mathematics, complex functions have also become indispensable tools in the natural sciences and engineering. *We could say that complex numbers and functions are auxiliary, technical tools there.* However, quantum mechanics is a clear exception since the wave function is complex. This is already at the heart of the Schrödinger equation, $i\hbar \frac{\partial \Psi}{\partial t} = \hat{H}\Psi$. If Ψ were real, the Hermitian operator $\hat{H}$ acting on a real Ψ would give a real number, which would contradict the left side of the Schödinger equation where the time derivative acting on a real Ψ would also be real, multiplied by i. In this book, complex analysis is basic for MB integrals studies.

At a very basic level, a complex number is just a number of the form $a + bi$, where a and b are real numbers, while i is simply a quantity satisfying $i^2 = -1$.

© The Author(s), under exclusive license to Springer Nature Switzerland AG 2022
I. Dubovyk et al., *Mellin-Barnes Integrals*, Lecture Notes in Physics 1008,
https://doi.org/10.1007/978-3-031-14272-7_2

Obviously i cannot be a real number, as $r^2 \geq 0$ for all $r \in \mathbb{R}$, and it is called the *imaginary unit*. The set of all complex numbers is denoted by $\mathbb{C}$. As a set, we have simply $\mathbb{C} = \{a + bi \,|\, a, b \in \mathbb{R}\}$. Complex numbers were initially considered as "subtle as they are useless" (Cardano, 1545), an "amphibian between existence and nonexistence" (Leibnitz, 1702), "impossible," and "imaginary." Nevertheless, using just the defining property $i^2 = -1$, it is straightforward enough to perform basic arithmetic with such numbers, e.g.,

$$(a + bi) \pm (c + di) = (a \pm c) + (b \pm d)i, \tag{2.1}$$

$$(a + bi)^2 = (a^2 - b^2) + 2abi \tag{2.2}$$

$$(a + bi) \cdot (c + di) = (ac - bd) + (ad + bc)i, \tag{2.3}$$

$$\frac{a + bi}{c + di} = \frac{(ac + bd)}{c^2 + d^2} + \frac{(bc - ad)}{c^2 + d^2} i, \tag{2.4}$$

and so on. In fact, the notion of complex numbers arose through the study of cubic equations, whose solution through radicals sometimes contains negative numbers under square roots even when all roots of the cubic are real. These real solutions could be computed using nothing more than the rudimentary understanding of complex numbers above. For a more historical account on the origin of complex numbers, we refer the reader to the excellent textbook [1], where interesting details can be found, e.g., an account of the idea just mentioned, due to Bombelli, to solve cubic equations by constructing complex arithmetic.

A breakthrough came with Wessel, Argand, and Gauss who interpreted complex numbers $z = a + bi$ geometrically as points or vectors in the xy-plane having Cartesian coordinates (a, b). This plane is denoted by C and is called a complex plane. The coordinates a and b are called the *real part* and *imaginary part* of the complex number z and are denoted as $\Re(z) \equiv a$ and $\Im(z) \equiv b$. However, in this geometric viewpoint, it is equally natural to describe a complex number by its *absolute value* and *argument*, which are nothing but the coordinates of the corresponding vector in two-dimensional polar coordinates; see Fig. 2.1. Clearly the absolute value and argument are related to the real and imaginary parts by $R = |z| = \sqrt{a^2 + b^2}$ and $\tan \varphi = b/a$, and conversely $\Re(z) = R \cos \varphi$, while $\Im(z) = R \cos \varphi$. Note that the argument is only defined up to multiples of 2π, a fact that is obvious from the geometric interpretation. We will see a more direct way of representing a complex number with its absolute value and argument shortly.

Now that we have the set of complex numbers at our disposal, we can define complex functions as maps from some subset $X \subseteq \mathbb{C}$ to some other subset $Y \subseteq \mathbb{C}$, where each element of X is mapped to exactly one element of Y. As a very basic example, consider squaring a complex number z: to each $z \in \mathbb{C}$ of the form $a + bi$, we assign its square, $z^2 \equiv z \cdot z = (a^2 - b^2) + 2abi$. Obviously this is a function on all of $\mathbb{C}$, as there is a unique assignment of the square for each complex number. Besides such simple arithmetic constructions, another well-known procedure for defining functions is through power series, the most obvious examples being the

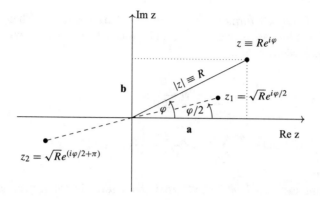

Fig. 2.1 Geometrical representation of the complex number with parameters R and φ, $z = Re^{i\varphi} = |z|(\cos\varphi + i\sin\varphi)$, $|z| \equiv R = \sqrt{\mathbf{a}^2 + \mathbf{b}^2}$, $\varphi = \arctan(\mathbf{a}/\mathbf{b})$. Two solutions z_1, z_2 of Eq. (2.16) with $n = 2$ are shown

exponential and trigonometric functions. Similar to the real case, let us define

$$\exp(z) \overset{\text{def}}{=} \sum_{k=0}^{\infty} \frac{z^k}{k!}, \quad \text{for all } z \in \mathbb{C}. \tag{2.5}$$

It can be shown that this power series is absolutely convergent for all $z \in \mathbb{C}$ and defines the *complex exponential* of z. The notation $e^z = \exp(z)$ is also commonly used and we will use both. It can be proven that the usual properties of the exponential function continue to hold for complex arguments, so

$$e^{z+w} = e^z e^w, \quad z, w \in \mathbb{C},$$
$$\left(e^z\right)^n = e^{nz}, \quad n \in \mathbb{N},$$
$$e^0 = 1,$$
$$e^z \neq 0, \quad z \in \mathbb{C}, z \neq 0$$
$$\frac{de^z}{dz} = e^z, \quad z \in \mathbb{C}. \tag{2.6}$$

Furthermore, for a real number t, the series representation of e^{it} can be written in the following way (note that the rearrangement of the series is allowed because of absolute convergence):

$$e^{it} = \left(1 - \frac{t^2}{2!} + \frac{t^4}{4!} - \cdots\right) + i\left(\frac{t}{1!} - \frac{t^3}{3!} + \frac{t^5}{5!} - \cdots\right), \tag{2.7}$$

which implies Euler's formula $e^{it} = \cos t + i \sin t$ for real t. Then, motivated by this observation, we define the sine and cosine of a complex variable z by the corresponding power series,

$$\cos(z) \stackrel{\text{def}}{=} \sum_{k=0}^{\infty}(-1)^k \frac{z^{2k}}{(2k)!},$$

$$\sin(z) \stackrel{\text{def}}{=} \sum_{k=0}^{\infty}(-1)^k \frac{z^{2k+1}}{(2k+1)!}, \quad \text{for all } z \in \mathbb{C}. \tag{2.8}$$

These definitions then lead directly to Euler's formula for a general complex number, which connects the exponential and trigonometric functions of a complex variable:

$$e^{iz} = \cos z + i \sin z. \tag{2.9}$$

This equation implies in particular

$$e^{i\pi} = -1 \quad \text{and} \quad e^{2i\pi} = 1 \tag{2.10}$$

and more generally $e^{2ni\pi} = 1$ for all $n \in \mathbb{Z}$. This implies that the exponential function is in fact periodic in the complex plane with a period of $2\pi i$,

$$e^{z+2\pi i} = e^z e^{2\pi i} = e^z \cdot 1 = e^z. \tag{2.11}$$

Incidentally, Euler's relation allows us to express a complex number z directly with its absolute value and argument. Recalling that $\mathfrak{R}(z) = R\cos\varphi$ while $\mathfrak{I}(z) = R\cos\varphi$, we immediately find

$$z = Re^{i\varphi}, \tag{2.12}$$

a relation that will prove to be very useful in computing fractional powers, so in particular n^{th} roots of complex numbers.

Another familiar procedure for defining interesting mappings is to consider the inverse of particular functions. Starting with a simple example, let us consider the inverse of the function $f(z) = z^2$. That is, for a complex z, we now seek another complex number, denoted $\sqrt{z}$, such that $(\sqrt{z})^2 = z$. In order to find such a complex number, consider the representation of z in terms of its absolute value and argument: $z = Re^{i\varphi}$. Then, it is immediately obvious that the number $\sqrt{z} = \sqrt{R}e^{i\frac{\varphi}{2}}$ has the desired property. Indeed

$$\left(\sqrt{z}\right)^2 = \left(\sqrt{R}e^{i\frac{\varphi}{2}}\right)^2 = \left(\sqrt{R}\right)^2 \left(e^{i\frac{\varphi}{2}}\right)^2 = Re^{2i\frac{\varphi}{2}} = Re^{i\varphi}. \tag{2.13}$$

We note that by definition the absolute value of a complex number is real and non-negative, $R \geq 0$, so its real square root $\sqrt{R}$ always exists and is real. However, notice that the square root is not unique. Clearly for any number $\sqrt{z}$ all other numbers of the form $\sqrt{z}e^{in\pi}$ also have the property that their square is z (recall $e^{2n\pi i} = 1$). Since

$$e^{in\pi} = (-1)^n, \quad n \in \mathbb{Z}, \tag{2.14}$$

(this is easily seen from the geometric interpretation of complex numbers or Euler's formula), we have precisely two distinct solutions: $+\sqrt{z}$ and $-\sqrt{z}$. This is not unexpected and follows the pattern for non-negative real numbers. We note in particular that

$$\sqrt{-1} = +i \quad \text{or equally well} \quad \sqrt{-1} = -i. \tag{2.15}$$

Before moving on, let us note that the generalization of the above considerations to general n^{th} roots is rather straightforward. Indeed, given some complex number z, we are now looking for a complex number $z^{1/n}$ such that $\left(z^{1/n}\right)^n = z$. It is easy to see that the number $z^{1/n} = \sqrt[n]{R}e^{i\frac{\varphi}{n}}$ has this property. However, as before, this is not the only solution. In fact if $z^{1/n}$ is an nth root of z, then so is any other number of the form $z^{1/n}e^{2m\pi i/n}$, for all $m \in \mathbb{Z}$. From the geometric interpretation, it is quite clear that the numbers $e^{2m\pi i/n}$ are distinct only for $m = 0, 1, 2, \ldots, n-1$. (Indeed, the associated vectors point to the vertices of a regular n-sided polygon inscribed in the unit circle, such that one vertex, corresponding to $m = 0$, is on the positive real axis.) Thus we find that in general

$$z^{1/n} = \sqrt[n]{R}e^{\left(i\frac{\varphi}{n} + \frac{2\pi m}{n}\right)}, \quad m = 0, 1, 2, \ldots, n-1. \tag{2.16}$$

There are then in fact n separate complex n^{th} roots of a complex number. In Fig. 2.1 the complex number is presented as a vector in the complex plane. This makes it easy, among other things, to visualize fractional powers, as exemplified on the case of the square root and its two solutions z_1, z_2 given in Eq. (2.16) for $n = 2$. See [1,2] for more useful concepts of visualizations connected with complex numbers and functions.

As we have seen, the inverse mappings of functions can be multivalued. Returning to the square function, its inverse is evidently a two-valued relationship, since the equation $w^2 = z$ does not have a unique solution. Instead it has two solutions, $\pm\sqrt{z}$. Thus, in order to be able to think of the mapping

$$w = f(z) = \pm\sqrt{z} = z^{1/2} \tag{2.17}$$

as a function, we must make some restrictions. For example, for a non-negative real numbers x, we can simply declare that

$$y = g(x) = +\sqrt{x} = x^{1/2} \tag{2.18}$$

that is, we *define* the square root to be the non-negative number y for which $y^2 = x$. This restriction then defines a single-valued mapping, i.e., a function, on the non-negative real numbers. Nevertheless, we stress that this is a choice (however natural), and we could have defined the square root of a non-negative real number to be the non-positive number y for which $y^2 = x$. A way of thinking about this restriction that will become useful very shortly is to imagine that we have erected a "barrier" at the single point $x = 0$. This idea must be refined when working with complex numbers. To see how, let us think about how the value of $f(z)$ changes as the point z moves along the unit circle in the complex plane. For definiteness, let us start at the point $z_0 = -1$ and move in the clockwise direction. We can write

$$z = e^{i\varphi} \quad \text{so} \quad w = z^{1/2} = e^{i\frac{\varphi}{2}}, \quad \pi \geq \varphi \geq -\pi . \tag{2.19}$$

We have chosen the argument φ to run between π and $-\pi$ and clearly $\varphi = \pm\pi$ both correspond to the same point $z_0 = -1$. However, as z moves all the way around the circle from $\varphi = \pi$ to $\varphi = -\pi$ and we return to our starting position, w traces out only half of a circle. Thus a continuous motion in the complex plane changes the value of the square root from $e^{i\frac{\pi}{2}} = +i$ to $e^{-i\frac{\pi}{2}} = -i$.[1]

This problem arises because the number $z = 0$ is special: it has just one square root, while every other complex number $z \neq 0$ has two distinct square roots. This of course already happens for the reals, but in that case we can circumvent the problem by setting a barrier at the single point $x = 0$. In the complex case though, we must make sure that *no continuous path* can completely encircle the special point $z = 0$. This requires that we use a bigger "barrier" and this is commonly done by introducing a *branch cut*.

In our case, we can, for example, have the "cut" extend from the branch point[2] at $z = 0$ along the negative real axis to the point at infinity, so that the argument of the variable z in this cut plane is restricted to the range (notice the sharp inequality) $-\pi < \varphi \leq \pi$. With this restriction, the square root mapping is single-valued, and we call this the *principal value* or the main branch of the argument. Notice though that the price we pay for making the function single-valued is that it is not

---

[1] Note that you can compute the value of $e^{i\frac{\pi}{2}}$ and $e^{-i\frac{\pi}{2}}$ unambiguously from Euler's relation.

[2] A branch point of a multivalued map is a point such that *the function is discontinuous* when going around an arbitrarily small circuit around this point.

continuous "across the cut." What this means is that we obtain two different values for the function if we approach a point on the cut from the two sides. For example, in our case, the cut is along the negative real axis, so we can approach some negative number, say $-R$ ($R > 0$), from the upper or lower complex half plane. So let us take $z = -R \pm i\epsilon$ with $\epsilon > 0$ and consider the limit $\epsilon \to 0$. We find

$$\sqrt{z} = \sqrt{-R \pm \epsilon} = \sqrt{Re^{\pm i(\pi - \epsilon')}} = \sqrt{R}e^{\pm i(\pi/2 - \epsilon'/2)} \tag{2.20}$$

$$= \sqrt{R}\left[\sin(\epsilon'/2) \pm i\cos(\epsilon'/2)\right] \xrightarrow{\epsilon' \to 0} \pm i\sqrt{R},$$

where we have used that a number of the form $-R \pm i\epsilon$ with a small but positive ϵ can be expressed as $Re^{\pm i(\pi - \epsilon')}$ with some ϵ' which is also small and positive. Notice that in this expression, the arguments are between $-\pi$ and π in accordance with our choice above. Thus, the limit is either $+i\sqrt{R}$ or $-i\sqrt{R}$ depending on how we approach the cut and so the function is discontinuous. Incidentally, because of this discontinuity, we also have some freedom to define the value of the function along the cut itself [3]. For example, in our prescription, we have chosen to allow $-\pi < \varphi \le \pi$, so negative real numbers must be represented as $x = |x|e^{i\pi}$, and we find $\sqrt{x} = +i\sqrt{|x|}$. But we could also have chosen the same cut along the negative real axis but required that $-\pi \le \varphi < \pi$. This would have meant that the square root of a negative real number x evaluates to $\sqrt{x} = -i\sqrt{|x|}$. All of these conventions are part of the principle value prescription, and it should be clear that choices different from those here can be made both for the position of the cut and the precise definition of the function on the cut. In fact, there is no general convention about the definition of the principal value, and another common choice is to take the argument in the interval $\varphi \in [0, 2\pi)$, which corresponds to cutting along the positive real axis. As discussed in [4], this ambiguity is a perpetual source of misunderstandings and errors. Indeed we have seen that even the value of the innocent-looking $\sqrt{-1}$ can be different based on the specific convention. Thus, it is very important to fix the notation, and we should be aware of possible differences in software, e.g., in Mathematica $\sqrt{-1} = +i$. We give an example of how to control numerical evaluation of results in Minkowskian kinematics by putting small imaginary parts to the input variable; see a pitfall note in the next section and the file MB_miscellaneous_Springer.nb in the auxiliary material in [5].

Last, let us briefly turn to the notion of Riemann sheets. Staying with our example of the square root, we have seen that having introduced a cut along the negative real axis, we can still choose the value of $\sqrt{-1}$ as either $+i$ or $-i$. Thus, in order to give a complete description of the square root function, we can consider two copies of the cut z plane and on one plane define the square root of -1 to be $e^{i\pi/2} = +i$, while on the other we define the square root of -1 to be $e^{-i\pi/2} = -i$. We call these two copies of the cut plane *sheets*. Then, we see that the now single-valued function $w = z^{1/2}$ maps the first sheet to the right half of the w plane while the second sheet is mapped to the left half of the w plane. The two disconnected sheets can then be "glued together" along the cuts to form a single

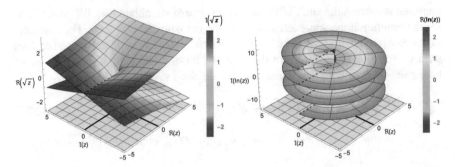

Fig. 2.2 Riemann surfaces for the functions $w = \sqrt{z}$ (left) and $w = \ln(z)$ (right). The gray planes at the bottom represent the complex z plane, while the colored sheets are the Riemann surfaces on which the functions are continuous and one-to-one. Dashed lines show the cuts along which the different branches are glued

Riemann surface on which $w = f(z) = z^{1/2}$ can be defined as a continuous (in fact holomorphic, i.e., complex differentiable) function whose image is the entire w plane (except $w = 0$). In Fig. 2.2 we present the Riemann-sheet plots for $w = \sqrt{z}$ and $w = \ln(z)$ generated with `Mathematica`; see also the source in the file `MB_miscellaneous_Springer.nb` in the auxiliary material in [5].

We will not discuss the details of this procedure any further here, but the interested reader will find more examples connected with Riemann sheets and branch cuts with nice visualizations in the textbook [6].

2.2   The Complex Logarithm

The basic MB formula in Eq. (1.44) involves just two types of functions: the power function A^z and the gamma function $\Gamma(z)$. The power function for generic complex base and exponent is defined through the complex logarithm function, $\ln z$,

$$z^w \equiv e^{w \ln z}, \quad z, w \in \mathbb{C}. \tag{2.21}$$

But logarithms of generally complex arguments also appear explicitly, e.g., in Taylor expansions of quantities raised to $d = 4 - 2\epsilon$-dependent powers

$$a^\epsilon \equiv e^{\epsilon \ln(a)} = 1 + \ln(a)\,\epsilon + \frac{1}{2}\ln^2(a)\,\epsilon^2 + \dots . \tag{2.22}$$

Thus we will first study the complex logarithm and its generalizations in this and the next section. Then we will consider the gamma function in detail.

To begin, notice that the logarithm is the inverse of the exponential function e^z that we have studied above. That is, $w = \ln z$ is by definition a complex number such that $z = e^w$. If the complex number z is given in polar form, $z = Re^{i\varphi}$, then one such number is $w = \ln R + i\varphi$. However, this is not the only such number and evidently any number of the form $\ln R + i(\varphi + 2n\pi)$, $n \in \mathbb{Z}$ is also a logarithm of z. Thus, we are in the familiar situation that the inverse function is not single-valued. Similarly to the case of the square root function discussed previously, in order to define a single-valued function, we must make a branch cut and impose a principal value prescription. It is easy to check that again $z = 0$ is a branch point: the logarithm changes by $2\pi i$ as we move on a circle of arbitrary size around this point. The branch cut will then extend in this case as well from the point $z = 0$ to the point at infinity, and in our principle value prescription, we will choose the cut along the negative real axis. Thus, we demand that the imaginary part of the logarithm be in the interval between $-\pi$ and π. Of course, the logarithm is then discontinuous across the cut. Indeed, for $R > 0$ and $\epsilon > 0$, we find with a calculation very similar to the one for the square root that in the limit $\epsilon \to 0$

$$\ln(-R \pm i\epsilon) \xrightarrow{\epsilon \to 0} \ln R \pm i\pi, \tag{2.23}$$

$$\ln(+R \pm i\epsilon) \xrightarrow{\epsilon \to 0} \ln R. \tag{2.24}$$

To finish the definition, we must choose the value of the logarithm along the negative real axis, and we employ the widely used prescription that the argument of the logarithm takes values in the interval $(-\pi, +\pi]$. In practice, this amounts to setting the logarithm of a negative real number $-R$, $R > 0$ to $\ln(-R) = \ln(R) + i\pi$. We note that with this choice, the function is continuous as the cut is approached coming around the finite endpoint of the cut in the counter-clockwise direction.

Summing up the above considerations, in the general case, the logarithm can be defined as

$$\ln z = \ln(a + ib) \equiv \ln\left(\rho e^{i\varphi}\right) = \ln \rho + i\varphi, \tag{2.25}$$

$$\rho = \sqrt{a^2 + b^2}, \tag{2.26}$$

$$\varphi = \arctan \frac{b}{a} + \pi\theta(-a)\mathrm{sign}(b), \quad \Re(z), \Im(z) \neq 0. \tag{2.27}$$

Here we assume that $a \neq 0$ and $b \neq 0$. Then, the ratio $\frac{b}{a}$ which appears in the argument of the arctan exists and is real. Moreover, we choose the branch of the arctan function that takes values in the interval $(-\pi/2, +\pi/2)$. Furthermore θ is the step function with $\theta(a) = 1$ if $a > 0$ and $\theta(a) = 0$ if $a < 0$. Last sign is the sign of its argument, i.e., $\mathrm{sign}(b) = 1$ if $b > 0$ and $\mathrm{sign}(b) = -1$ if $b < 0$. Turning to the special cases, if $a \neq 0$ and $b = 0$, we have simply

$$\ln z = \ln(a) = \ln|a| + i\pi\theta(-a), \quad \Im(z) = 0, \tag{2.28}$$

while for $a = 0$ and $b \neq 0$, we find

$$\ln z = \ln(ib) = \ln |b| + i\frac{\pi}{2}\text{sign}(b)\,, \qquad \mathfrak{R}(z) = 0\,. \tag{2.29}$$

Finally, for $a = b = 0$, i.e., at $z = 0$, the logarithm function is not defined.

The Logarithm Function in `Mathematica`: A Pitfall

The implementation of the logarithm function in `Mathematica` follows the principal value prescription given above. In particular, the function is discontinuous across the negative real axis as shown by, e.g., the numerical evaluations[3]

```
In[1]:= Log[-2 + 0.001 I]
Out[1]:= 0.6931473055599296 + 3.14109265363146 I
In[2]:= Log[-2 - 0.001 I]
Out[2]:= 0.6931473055599296 - 3.14109265363146 I
```

On the cut itself, `Mathematica` evaluates the logarithm symbolically as

```
In[3]:= Log[-2]
Out[3]:= I Pi + Log[2]
```

The fact that the function is discontinuous across the negative real axis can lead to interesting pitfalls when computing with non-exact (i.e., machine precision) numbers. For example, consider the variable

$$x = \frac{\sqrt{1 - \frac{4m^2}{M^2}} - 1}{\sqrt{1 - \frac{4m^2}{M^2}} + 1}\,. \tag{2.30}$$

For $m > M/2$, the square root is imaginary and, in fact, x is just a complex phase, i.e., $x = e^{i\theta}$ for some real θ. For example, choosing $m = 1$ and $M = 1.25$, we find

```
In[4]:= xval = (Sqrt[1-4 (m^2/M^2)]-1)/(Sqrt[1-4(m^2/M^2)]+ 1)
   ↪   /. {m -> 1, M -> 1.25}
Out[4]:= 0.21875 + 0.9757809372497497 I
```

Now, let us evaluate the logarithm of $(1 - x)^2/x$ at this particular point in two ways—first, by substituting this expression directly into the logarithm function and, second, by substituting the partial fractioned form of the expression (`xval` is defined above):

---

[3] In this book we use "minted" TeX style to highlight the `Mathematica` In and Out cells to LaTeX [7].

```
In[5]:= {Log[(1 - x)^2/x], Log[1/x - 2 + x]} /. x -> xval
Out[5]:= {0.44628710262841936 - 3.141592653589793*I,
    0.44628710262841953 + 3.141592653589793*I}
```

Although the two expressions are exactly equivalent mathematically, we obtain different numerical results! The solution to this conundrum becomes obvious if we evaluate the arguments themselves

```
In[6]:= {(1 - x)^2/x, 1/x - 2 + x} /. {x -> xval}
Out[6]:= {-1.5624999999999998 - 1.1102230246251565*^-16*I,
    ↪   -1.5625 + 0.*I}
```

Apparently $(1-x)^2/x$ is real and negative, since the imaginary parts are zero up to machine precision, but due to the finite resolution of the number representation, the formally equivalent expressions evaluate to numbers whose tiny imaginary parts have the opposite signs! Thus, in one case, due to numerical rounding errors, the logarithm is actually evaluated below the cut, giving the incorrect imaginary part. Such subtleties must be kept in mind when using numerical software.

Turning to some basic properties of the complex logarithm, first it is immediately obvious that the relation $\ln(xy) = \ln(x) + \ln(y)$, well-known from the real case, cannot hold in general in the complex case. Indeed, already for two negative reals $x, y < 0$, the left-hand side is real, while the right-hand side will have an imaginary part equal to 2π. A similar situation occurs for the relation $\ln(x/y) = \ln(x) - \ln(y)$, where for non-zero real numbers x and y with different signs (i.e., such that $x/y < 0$), the left-hand side has an imaginary part of $+\pi$, while the imaginary part of the right-hand side is either $+\pi$ if $x < 0$ and $y > 0$ or $-\pi$ if $x > 0$ and $y < 0$. Indeed, the correct relations read

$$\ln(wz) = \ln w + \ln z + 2n\pi i, \qquad n \in \mathbb{Z}, \tag{2.31}$$

$$\ln\left(\frac{w}{z}\right) = \ln w - \ln z + 2n\pi i, \qquad n \in \mathbb{Z}. \tag{2.32}$$

That is, the logarithm of a product (quotient) of complex numbers is the sum (difference) of the logarithms, *plus a multiple of* $2\pi i$, such that the argument of the result is in the interval $(-\pi, \pi]$. In practice there are two solutions to this situation. In numerical evaluation, assuming we have in general a product or quotient of two complex numbers, we can evaluate the product or quotient of the two complex numbers first and then use the relations in Eqs. (2.25)–(2.29) to compute the logarithm. For analytical results, we can split the logarithm of a product (quotient) of two complex numbers to the sum (difference) of two logarithms as in the positive real case, adding a proper relation for an additional phase. Surprisingly, it is not entirely straightforward to write such a relation, though, and this question is taken up in Problem 2.1.

Complex Quadrature Under the Logarithm

A commonly occurring situation is that we encounter the logarithm of a quadratic expression, $\ln(ax^2 + bx + c)$. Since the equation

$$ax^2 + bx + c = 0 \qquad (2.33)$$

has two roots, say x_1 and x_2, we can write

$$ax^2 + bx + c = a(x - x_1)(x - x_2). \qquad (2.34)$$

However, as just discussed, when the factors of this product are not all positive real, we must be careful with using the logarithmic identities on this product. In particular, if the roots are real, we must be careful as we vary x as the sign of one of the factors changes as we cross the roots (assume here that x is real).

One way to proceed in this situation is to assign a small imaginary part to the quadratic expression. Then we consider the equation

$$ax^2 + bx + c - i\epsilon = 0, \qquad (2.35)$$

which has the roots (to first order in the small quantity ϵ)

$$\bar{x}_1 = x_1 + i \frac{\epsilon}{a(x_1 - x_2)}, \qquad \bar{x}_2 = x_2 - i \frac{\epsilon}{a(x_1 - x_2)}. \qquad (2.36)$$

Let us assume for simplicity that the original roots are ordered such that $x_1 > x_2$, so that $x_1 - x_2 > 0$. Then we can simply redefine the small quantity ϵ by this positive quantity and write

$$\bar{x}_1 = x_1 + i \frac{\epsilon}{a}, \qquad \bar{x}_2 = x_2 - i \frac{\epsilon}{a}. \qquad (2.37)$$

To derive our final expression, we need one more consideration. Namely, let A and B be real numbers. Then the following holds:

$$\ln(AB - i\epsilon) = \ln(A - i\epsilon') + \ln(B - i\frac{\epsilon}{A}), \qquad (2.38)$$

where the signs of ϵ' and ϵ are equal. Using the above, we obtain the following useful relation:

$$\ln(ax^2 + bx + c - i\epsilon) = \ln(a - i\epsilon') + \ln(x - x_1 - i\frac{\epsilon}{a}) + \ln(x - x_2 + i\frac{\epsilon}{a}), \qquad (2.39)$$

where again $x_{1,2}$ are roots of $x^2 + bx + c = 0$ and $x_1 > x_2$.

2.3 Generalizations of the Logarithm Function: Classical Polylogarithms

In the previous section, we have introduced the complex logarithm as the inverse of the exponential function. However, there are several other ways to think about the logarithm which generalize naturally to a set of functions that plays a very important role in modern Feynman integral calculations. Here we briefly explore these generalizations.

To start, consider computing the integral of a rational function of a single variable, say x,

$$\int dx \, \frac{P(x)}{Q(x)}, \tag{2.40}$$

where $P(x)$ and $Q(x)$ are polynomials in x. We know from algebra that after performing polynomial division and partial fraction decomposition (over the complex numbers), the integrand can be reduced to the sum of a polynomial plus functions of the form $\frac{1}{(x-a)^n}$, where n is a positive integer and a is a (in general complex) constant. Hence, the problem can be reduced to the computation of the following integrals:

$$\int dx \, x^n \quad \text{and} \quad \int dx \, \frac{1}{(x-a)^n}, \qquad n \in \mathbb{N}. \tag{2.41}$$

These integrals can almost all be expressed in terms of just rational functions. However the second integral for $n = 1$ cannot. In fact, this integral can be used to define a new function, the logarithm of x,

$$\int \frac{dx}{x-a} = \ln(x-a) + C. \tag{2.42}$$

2.3.1 The Arcus Tangent and the Logarithm

In basic courses of calculus, it is shown that the integral of the rational function $R(x) = \frac{1}{1+x^2}$ is just the arcus tangent function, which we already encountered in Eq. (2.27),

$$\int \frac{dx}{1+x^2} = \arctan(x) + C. \tag{2.43}$$

Usually we choose to work on the branch of the arcus tangent function where its values fulfill the condition

$$-\frac{\pi}{2} < \arctan(x) < +\frac{\pi}{2},$$

and we formally set $\arctan(\pm\infty) = \pm\frac{\pi}{2}$.

However, following our above discussion on integrating rational functions, we expect that the integral in Eq. (2.43) should be expressible in terms of at most logarithms. This is of course true and it is easy to derive the relation between the arcus tangent and logarithm functions. Indeed, by performing partial fraction decomposition over the complex numbers, we have

$$
\int \frac{dx}{1+x^2} = \int dx \left[\frac{1}{2i(x-i)} - \frac{1}{2i(x+i)} \right]
$$

$$
= \frac{1}{2i} \ln(x-i) - \frac{1}{2i} \ln(x+i) \tag{2.44}
$$

$$
= -\frac{i}{2} \ln \left(\frac{1+ix}{1-ix} \right),
$$

where we have used $\ln(a) - \ln(b) = \ln(a/b)$ to simplify the result. Incidentally, we note that $\arctan(z) = -i \operatorname{arctanh}(iz)$, where the areatangens hyperbolicus function gives the inverse of the hyperbolic tangent of the complex number z, i.e., $\operatorname{arctanh}(z) = \tanh^{-1}(z)$. Indeed, Mathematica evaluates

```
In[7]:=  (-I)*ArcTanh[I*z]
Out[7]:= ArcTan[z]
```

As a heuristic check of the correctness of Eq. (2.44), we may compare the Taylor expansions of $\arctan(x)$ and $-\frac{i}{2} \ln \left(\frac{1+ix}{1-ix} \right)$ for small x. We find

```
In[8]:= Normal[Series[ArcTan[x], {x, 0, 5}]]
Out[8]:=  x - x^3/3 + x^5/5
```

and

```
In[9]:= Normal[Series[(-I/2)*Log[(1 + I*x)/(1 - I*x)], {x, 0,
 ↪ 5}]]
Out[9]:=  x - x^3/3 + x^5/5
```

The expansion can of course be carried to higher orders and we obtain the same result at any expansion order.

Of course, it turns out that the logarithm function appearing in Eq. (2.42) is the same function that we have been studying previously. However, this way of thinking about the logarithm immediately offers a path toward generalization: let us now look at integrating combinations of rational functions and the logarithm of x. By once more performing partial fraction decomposition, it is clear that we encounter integrals of the form

$$
\int dx\, x^n \ln(x) \quad \text{and} \quad \int dx \frac{\ln(x)}{(x-a)^n}, \qquad n \in \mathbb{N}. \tag{2.45}
$$

Again we find that almost all of these integrals can be expressed in terms of rational functions and the logarithm, e.g.,

$$\int dx\, x^n\, \ln(x) = \frac{x^{n+1}}{(n+1)^2}\left[(n+1)\ln(x) - 1\right] + C, \qquad n \in \mathbb{Z},\ n \neq -1,$$

$$(2.46)$$

$$\int dx\, \frac{\ln(x)}{x} = \frac{1}{2}\ln^2(x) + C,$$

$$(2.47)$$

$$\int dx\, \frac{\ln(x)}{(x-a)^2} = \frac{\ln(x-a)}{a} - \frac{x\ln(x)}{a(x-a)} + C,$$

$$(2.48)$$

$$\int dx\, \frac{\ln(x)}{(x-a)^3} = \frac{1}{2a}\left[\frac{1}{a-x} + \frac{(x-2a)x\ln(x)}{a(x-a)^2} - \frac{\ln(x-a)}{a}\right] + C.$$

However, the integral

$$\int dx\, \frac{\ln x}{x-a}, \qquad a \neq 0,$$

$$(2.49)$$

cannot be expressed in terms of just rational functions and logarithms and leads to a genuinely new function. This new function is called *Spence's function*, or more commonly (Euler) *dilogarithm*, and is customarily defined through the integral

$$\mathrm{Li}_2(x) \equiv -\int_0^x \frac{dt}{t}\ln(1-t).$$

$$(2.50)$$

The function $\mathrm{Li}_2(x)$ cannot be expressed by logarithms and rational functions.

For real values of the argument, the Euler dilogarithm behaves as shown in Fig. 2.3. The function has a cut along the positive real axis starting at $x = 1$ and for $x > 1$ develops an imaginary part, as can be seen in the figure. Approaching $x = -\infty$, $\mathrm{Li}_2(x)$ behaves like $\ln^2(|x|)$, which explains the name dilogarithm.

2.3.2 The General Dilogarithmic Integral

Having introduced the dilogarithm function, we can now solve more general integrals such as

$$\int \frac{dx}{mx+b}\ln(nx+a).$$

$$(2.51)$$

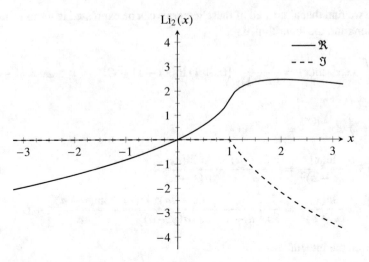

Fig. 2.3 The real and imaginary part of the dilogarithm $\text{Li}_2(x)$ at real values

In order to bring this integral to the form where we can apply Eq. (2.50), we need to perform some transformations of the integration variables. First, let us set $mx + b = t$ and change the variable of integration from x to t. We find

$$\int \frac{dx}{mx+b} \ln(nx+a) = \frac{1}{m} \int \frac{dt}{t} \ln\left(\frac{am - bn + nt}{m}\right). \tag{2.52}$$

Next, we use $t = -\frac{am-bn}{n} u$. This transformation is chosen such that the argument of the logarithm factorizes

$$\frac{1}{m} \int \frac{dt}{t} \ln\left(\frac{am - bn + nt}{m}\right) = \frac{1}{m} \int \frac{du}{u} \ln\left(\frac{(am - bn)(1 - u)}{m}\right). \tag{2.53}$$

Now we come to an interesting point. In order to proceed, we must relate the logarithm in the integrand to $\ln(1 - u)$. This is formally straightforward, since the logarithm of a product is the sum of the logarithms.[4] However, the general point to take away is that when we define new functions, we must also find ways of bringing expressions involving those new functions to standard forms, so that the definitions can actually be applied. We will come back to this point later and also in Sect. 5.3.

---

[4] Here we put aside the possible subtleties associated with imaginary parts; see Eq. (2.31) and the discussion there.

Returning to our integral, let us proceed formally and decompose the logarithm

$$\frac{1}{m} \int \frac{du}{u} \ln\left(\frac{(am - bn)(1 - u)}{m}\right) = \frac{1}{m} \int \frac{du}{u} \left[\ln\left(\frac{am - bn}{m}\right) + \ln(1 - u)\right].$$

$$(2.54)$$

We see that we can now apply Eqs. (2.42) and (2.50) to perform the integrations and we find the result

$$\frac{1}{m} \int \frac{du}{u} \left[\ln\left(\frac{am - bn}{m}\right) + \ln(1 - u)\right] = \frac{1}{m} \left[\ln\left(\frac{am - bn}{m}\right) \ln u - \mathrm{Li}_2(u)\right] + C.$$

$$(2.55)$$

To be absolutely clear, here we have used the fact that Eq. (2.50) implies that the primitive function of $-\frac{\ln(1-u)}{u}$ is the dilogarithm, $\mathrm{Li}_2(u)$. Last, we can replace u with our original variable x. Using $u = \frac{n(mx+b)}{-am+bn}$, we find the final result

$$\int \frac{dx}{mx + b} \ln(nx + a) = \frac{1}{m} \left[\ln\left(\frac{am - bn}{m}\right) \ln\left(\frac{n(mx + b)}{-am + bn}\right) - \mathrm{Li}_2\left(\frac{n(mx + b)}{-am + bn}\right)\right] + C.$$

$$(2.56)$$

Finally, we note that in the `Mathematica` language, the dilogarithm function $\mathrm{Li}_2(x)$ is represented by `PolyLog[2,x]`, and we can evaluate our integral immediately:

```
In[10] := Integrate[Log[n*x + a]/(m*x + b), x]
Out[10] := (Log[(n*(b + m*x))/((-a)*m + b*n)]*Log[a + n*x] +
 ↪ PolyLog[2, (m*(a + n*x))/(a*m - b*n)])/m
```

Apparently the result is different from what we obtained in Eq. (2.56)! However there is no contradiction. It can be shown using some properties of the dilogarithm (viz., Eq. (2.66)) that the difference is a constant, independent of x.[5]

2.3.3 Classical Polylogarithms

Repeating the arguments that have led us from the logarithm to the dilogarithm, we arrive at the *classical polylogarithms* $\mathrm{Li}_n(x)$. Indeed, it is easy to check that by a proper application of integration by parts, all integrals of the form

$$\int dx\, x^n \mathrm{Li}_2(x), \quad \text{and} \quad \int \frac{dx}{(x - a)^n} \mathrm{Li}_2(x), \quad n \neq -1, \quad (2.57)$$

---

[5] At least for parameters where both expressions are real. Recall we were not careful with the imaginary parts in our derivation. See Problem 2.2 for a discussion of the general case.

can be evaluated in terms of rational functions, logarithms and dilogarithms. However, the integral

$$\int \frac{dx}{x} \mathrm{Li}_2(x) \equiv \mathrm{Li}_3(x) + C, \tag{2.58}$$

leads to a function, the trilogarithm $\mathrm{Li}_3(x)$, that cannot be expressed with the functions already introduced. In this way, we are led to introduce the *classical polylogarithm of weight n* $\mathrm{Li}_n(x)$ by the recursive definition[6]

$$\boxed{\mathrm{Li}_n(x) \equiv \int_0^x \frac{dt}{t} \mathrm{Li}_{n-1}(t).} \tag{2.59}$$

The above equation holds also for $n = 2$ if we define

$$\mathrm{Li}_1(x) \equiv -\ln(1-x). \tag{2.60}$$

We have defined classical polylogarithms above via the integral in Eq. (2.59). However, it turns out that they may also be represented as simple infinite sums. This is quite relevant in connection with the Mellin-Barnes method, since as we have already mentioned, one possible way of solving MB integrals is by converting them to sums using Cauchy's residue theorem. Hence let us briefly turn to this alternative way of defining classical polylogarithms. We begin by noting that

$$-\ln(1-x) = \sum_{k=1}^{\infty} \frac{x^k}{k}. \tag{2.61}$$

The sum on the right-hand side is convergent for $|x| < 1$ and in fact can be used as a starting point to define the logarithm function in yet one more approach. Now, it can be shown that the simple generalization

$$\mathrm{Li}_n(x) = \sum_{k=1}^{\infty} \frac{x^k}{k^n} \tag{2.62}$$

leads precisely to the classical polylogarithm of weight n, which we have introduced in Eq. (2.59). Both representations are important and can be used to good effect when solving MB integrals analytically, as will be discussed in detail in Chap. 5.

---

[6] In the next section, we will see that Eq. (2.59) is just a special case of the iterated integral that defines a class of functions known as multiple polylogarithms. However, $\mathrm{Li}_2(x)$ and its generalizations $\mathrm{Li}_n(x)$ were historically the first functions of this class to be studied; thus in order to distinguish them from the more general cases to be studied below, we refer to them as *classical polylogarithms*.

Classical polylogarithms obey a number of important functional relations. For example, it is always possible to relate $\mathrm{Li}_n(1/z)$ to $\mathrm{Li}_n(z)$,

$$\mathrm{Li}_1\left(\frac{1}{z}\right) = -\ln\left(1 - \frac{1}{z}\right) = -\ln(1 - z) + \ln(-z) = \mathrm{Li}_1(z) + \ln(-z), \quad (2.63)$$

$$\mathrm{Li}_2\left(\frac{1}{z}\right) = -\mathrm{Li}_2(z) - \frac{1}{2}\ln^2(-z) - \zeta(2), \quad (2.64)$$

$$\mathrm{Li}_3\left(\frac{1}{z}\right) = \mathrm{Li}_3(z) - \frac{1}{6}\ln^3(-z) - \zeta(2)\ln(-z), \quad (2.65)$$

and so on. In general $\mathrm{Li}_n(z) + (-1)^n\mathrm{Li}_n(1/z)$ can be written as an n-th degree polynomial in $\ln(-z)$. Some other useful relations for classical polylogarithms are

$$\mathrm{Li}_2(1 - z) = -\mathrm{Li}_2(z) + \zeta(2) - \ln(z)\ln(1 - z), \quad (2.66)$$

$$\mathrm{Li}_2\left(1 - \frac{1}{z}\right) = -\mathrm{Li}_2(1 - z) - \frac{1}{2}\ln^2(z), \quad (2.67)$$

and

$$\mathrm{Li}_3\left(1 - \frac{1}{z}\right) = -\mathrm{Li}_3(z) - \mathrm{Li}_3(1 - z) + \zeta(3) + \frac{1}{6}\ln^3(z) + \zeta(2)\ln(z) \quad (2.68)$$

$$- \frac{1}{2}\ln^2(z)\ln(1 - z). \quad (2.69)$$

The square relation

$$\mathrm{Li}_n(z) + \mathrm{Li}_n(-z) = 2^{1-n}\mathrm{Li}_n(z^2) \quad (2.70)$$

is also sometimes employed in practical calculations. These relations are often useful for bringing polylogarithmic integrals to the standard form of Eq. (2.59) as well as for transforming the arguments to specific regions for numerical evaluation. For example, the convergence of the simple series expansion of the dilogarithm in Eq. (2.62) is already quite bad for $|z| > 1/2$. Hence an efficient evaluation of the dilogarithm (and more generally of $\mathrm{Li}_n(z)$) transforms the argument z to a region where the modulus and real part are bound: $|z| \leq 1$ and $\Re(z) \leq 1/2$ using, e.g., Eqs. (2.64) and (2.66).

Turning to specific values of polylogarithms, it is clear from the definition of $\mathrm{Li}_n(z)$ either as an integral (Eq. 2.59) or a sum (Eq. 2.62) that $\mathrm{Li}_n(0) = 0$. Moreover, $\mathrm{Li}_n(1) = \zeta(n)$ follows immediately from the representation of $\mathrm{Li}_n(z)$ as a sum. Some other special values are also easy to determine, e.g., for the dilogarithm we have

$$\text{Li}_2(0) = 0, \tag{2.71}$$

$$\text{Li}_2(1) = \frac{\pi^2}{6} = \zeta(2) \approx 1.64493, \tag{2.72}$$

$$\text{Li}_2(-1) = -\frac{1}{2}\text{Li}_2(1) = -\frac{\pi^2}{12} \approx -0.822467, \tag{2.73}$$

$$\text{Li}_2\left(\frac{1}{2}\right) = \frac{\pi^2}{12} - \frac{1}{2}\ln^2(2) \approx 0.582241, \tag{2.74}$$

$$\text{Li}_2(\pm i) = iG - \frac{\pi^2}{48} \approx -0.205617 + 0.915965i. \tag{2.75}$$

The constant G in Eq. (2.75) above is Catalan's constant, $G = \sum_{n=0}^{\infty} \frac{(-1)^n}{(2n+1)^2}$. It is conjectured that Catalan's number is irrational and transcendental [8].

In addition to the functional relations discussed above, classical polylogarithms admit a number of other representations as integrals and series. For example, the integral

$$\frac{(-1)^{n-1}}{(n-2)!} \int_0^1 \frac{dt}{t} \ln^{n-2}(t) \ln(1-zt) \tag{2.76}$$

simply evaluates to the classical polylogarithm $\text{Li}_n(z)$. Its straightforward generalization

$$S_{n,p}(z) = \frac{(-1)^{n+p-1}}{(n-1)!p!} \int_0^1 \frac{dt}{t} \ln^{n-1}(t) \ln^p(1-zt) \tag{2.77}$$

defines the so-called *Nielsen generalized polylogarithm* $S_{n,p}(z)$. Clearly we have $S_{n-1,1}(z) = \text{Li}_n(z)$. We note that in Mathematica the Nielsen generalized polylogarithm function $S_{n,p}(z)$ is represented by PolyLog[n,p,z], while PolyLog[n,z] gives the classical polylogarithm of weight n.

Turning to series expansions, an efficient method may be found in [3] and [9, 10]. For $\text{Li}_2(z)$ we have [3]

$$\text{Li}_2(z) = \sum_{j=0}^{\infty} \frac{B_j}{(j+1)!} [-\ln(1-z)]^{j+1} \tag{2.78}$$

$$= -\ln(1-z) - \frac{1}{4}\ln^2(1-z) + 4\pi \sum_{j=1}^{\infty} \zeta(2j)\frac{(-1)^j}{2j+1}\left[\frac{\ln(1-z)}{2\pi}\right]^{2j+1}.$$

Here the B_j are Bernoulli numbers, $B_0 = 1$, $B_1 = -1/2$, etc. In Mathematica B_n are represented as `BernoulliB[n] = BernoulliB[n, 0]`, and historical facts about these numbers can be found in Chapter "Gamma as a Decimal" in [11]. This expansion with Bernoulli numbers ensures rapid convergence. One of many possible ways of defining the Bernoulli numbers B_n is through a generating function,

$$\frac{t}{e^t - 1} = \sum_{m=0}^{\infty} B_m \frac{t^m}{m!}$$

Other useful series expansions for $\text{Li}_n(z)$ are also given in [3], and we reproduce one here for the special case of $n = 3$,

$$\text{Li}_3(z) = \sum_{j=0}^{\infty} \frac{C_3(j)}{(j+1)!} [-\ln(1-z)]^{j+1}, \tag{2.79}$$

with

$$C_3(j) = \sum_{k=0}^{j} \binom{j}{k} \frac{B_{j-k} B_k}{1+k}, \tag{2.80}$$

so that $C_3(0) = 1$, etc. Using these expansions for $\text{Li}_2(z)$ and $\text{Li}_3(z)$, we observe typically that retaining n terms in the summation gives $n \pm 1$ digit accuracy. We mention that just as for the logarithm, the evaluation of polylogarithms on their cuts (the positive real axis beginning at $z = 1$) requires the adoption of a particular convention and hence must be treated with special care. We refer to [3] for a discussion of these issues.

One More Dilogarithmic Integral

A special integral frequently met in one-loop Feynman integral calculations is (see [12])

$$\int_0^1 \frac{dx}{x - x_0} [\ln(x - x_A) - \ln(x_0 - x_A)] = \text{Li}_2\left(\frac{x_0}{x_0 - x_A}\right) - \text{Li}_2\left(\frac{x_0 - 1}{x_0 - x_A}\right), \tag{2.81}$$

which is valid for arbitrary complex x_A and real x_0. The formula demonstrates the need of a subtraction at $x = x_0$ in order to make the integral well-defined for $0 \le x_0 \le 1$.

The functions and some expansions which have been discussed here are important in systematic order by order ϵ studies of FI [13]. Nowadays suitable packages exist in $Mathematica$ or $Fortran$ [3, 14–17]; see Appendix A for details.

2.4   Multiple Polylogarithms and Beyond

The logarithm and all of its generalizations discussed so far are just special cases of a set of functions known as multiple polylogarithms. Similarly to the classical polylogarithms, multiple polylogarithms[7] (MPLs) can be defined recursively for $n \geq 1$, via the iterated integral [18, 19]

$$G(a_1, \ldots, a_n; z) = \int_0^z \frac{dt}{t - a_1} G(a_2, \ldots, a_n; t), \tag{2.82}$$

with $G(z) = G(; z) = 1.$[8] Here the $a_i \in \mathbb{C}$ are constants and z is a complex variable. For the special case when all of the a_is are zero, we define

$$G(\mathbf{0}_n; z) = \frac{1}{n!} \ln^n z, \tag{2.83}$$

where $\mathbf{0}_n$ denotes a set of n zeros. The number n is called the *weight* of the polylogarithm. It is clear from the definition that MPLs contain the ordinary logarithm as well as the classical polylogarithms as special cases. In particular we have

$$G(\mathbf{a}_n; z) = \frac{1}{n!} \ln^n \left(1 - \frac{z}{a}\right), \tag{2.84}$$

$$G(\mathbf{0}_{n-1}, 1; z) = -\mathrm{Li}_n(z), \tag{2.85}$$

where $\mathbf{a}_n = (a, \ldots, a)$ denotes a sequence of as of length n. We note the definition above corresponds to what is commonly used in the physics literature (see, e.g., the usage in the $PolyLogTools$ package; ref. [22]); however the notation for MPLs used in the mathematical literature (e.g., [23]) differs slightly from this:

$$I(a_0; a_1, \ldots, a_n; a_{n+1}) = \int_{a_0}^{a_{n+1}} \frac{dt}{t - a_n} I(a_0; a_1, \ldots, a_{n-1}; t) \tag{2.86}$$

---

[7] MPLs are also known as generalized polylogarithms (GPLs); see Appendix A.3 for details and packages.

[8] The above definition was already present in the works of Poincaré, Kummer, and Lappo-Danilevskij [20] as "hyperlogarithms" as well as implicitly in Chen's work on iterated integrals [21].

and $I(a_0; ; a_1) = 1$. The two definitions are related by (note the reversal of the arguments)

$$G(a_n, \ldots, a_1; a_{n+1}) = I(0, a_1, \ldots, a_n; a_{n+1}).$$ (2.87)

The iterated integrals defined by Eq. (2.86) are slightly more general than those defined in Eq. (2.82), since they allow for a generic base point of integration a_0. Nevertheless, it is easy to see that every integral with a generic base point can be converted into a combination of integrals with base point zero. Even so, some properties of MPLs are more straightforward to express in the I notation, and its use in the mathematical literature justifies its mention here.

In addition to the integral representation of Eq. (2.59), classical polylogarithms can also be defined by their series representation as in Eq. (2.62). Above we have presented the generalization of the integral definition to MPLs. However, there is also a way to generalize the series definition,

$$\text{Li}_{m_1, \ldots, m_k}(z_1, \ldots, z_k) = \sum_{0 < i_1 < i_2 \cdots < i_k} \frac{z_1^{i_1}}{i_1^{m_1}} \frac{z_2^{i_2}}{i_2^{m_2}} \cdots \frac{z_k^{i_k}}{i_k^{m_k}}$$

$$= \sum_{i_k=1}^{\infty} \frac{z_k^{i_k}}{i_k^{m_k}} \sum_{i_{k-1}=1}^{i_k-1} \cdots \sum_{i_1=1}^{i_2-1} \frac{z_1^{i_1}}{i_1^{m_1}}.$$ (2.88)

This definition makes sense whenever the sums converge (e.g., for $|z_i| < 1$). The number k is called the *depth* of the MPL. The G and Li functions define essentially the same class of functions and are related by

$$\text{Li}_{m_1, \ldots, m_k}(z_1, \ldots, z_k) = (-1)^k G\left(\mathbf{0}_{m_k-1}, \frac{1}{z_k}, \ldots, \mathbf{0}_{m_1-1}, \frac{1}{z_1 \ldots z_k}; 1\right)$$ (2.89)

or alternatively ($a_i \neq 0$)

$$G(\mathbf{0}_{m_1-1}, a_1, \ldots, \mathbf{0}_{m_k-1}, a_k; z) = (-1)^k \text{Li}_{m_k, \ldots, m_1}\left(\frac{a_{k-1}}{a_k}, \ldots, \frac{a_1}{a_2}, \frac{z}{a_1}\right).$$ (2.90)

We will make use of both the integral and sum representation of MPLs when we discuss various approaches of obtaining analytic solutions to MB integrals in Chap. 5.

Regarding the basic properties of MPLs, we note first of all $G(a_1, \ldots, a_n; z)$ is divergent whenever $z = a_1$, which is easy to check using the integral representation in Eq. (2.82). Similarly, $G(a_1, \ldots, a_n; z)$ is analytic at $z = 0$ (i.e., is given by a convergent power series around $z = 0$) whenever $a_n \neq 0$, which is consistent with the series representation in Eq. (2.88). Due to the singularities at $z = a_i$ in the integral representation, MPLs in general have a very complicated branch cut structure.

Second, if the rightmost index a_n of $G(a_1, \ldots, a_n; z)$ is non-zero, then the function is invariant under the simultaneous rescaling of all of its arguments for any non-zero complex number k,

$$G(ka_1, \ldots, ka_n; kz) = G(a_1, \ldots, a_n; z), \qquad a_n \neq 0. \tag{2.91}$$

MPLs also satisfy the so-called Hölder convolution. Whenever $a_1 \neq 1$ and $a_n \neq 0$, we have for all non-zero complex numbers p that

$$G(a_1, \ldots, a_n; 1) = \sum_{k=0}^{n} (-1)^k G\left(1 - a_k, \ldots, 1 - a_1; 1 - \frac{1}{p}\right) G\left(a_{k+1}, \ldots, a_n; \frac{1}{p}\right). \tag{2.92}$$

These examples make it clear that MPLs satisfy many functional relations. These of course include the functional relations for classical polylogarithms. For example, the Hölder convolution for $a_1 = 0$ and $a_2 = 1$ reduces simply to Eq. (2.66) with $z = \frac{1}{p}$.

Third, the product of two MPLs of weight n_1 and n_2 with the same argument z can be written as a linear combination of MPLs of weight $n_1 + n_2$ with argument z,

$$G(a_1, \ldots, a_{n_1}; z)G(a_{n_1+1}, \ldots, a_{n_1+n_2}; z) = \sum_{\sigma \in \Sigma(n_1, n_2)} G(a_{\sigma(1)}, \ldots, a_{\sigma(n_1+n_2)}; z), \tag{2.93}$$

where $\Sigma(n_1, n_2)$ denotes the set of *shuffles* on $n_1 + n_2$ elements. A shuffle is simply a permutation of the set $(a_1, \ldots, a_{n_1+n_2})$ that leaves the orderings of $(a_1, \ldots, a_{n_1})$ and $(a_{n_1+1}, \ldots, a_{n_1+n_2})$ unchanged. Formally we have

$$\Sigma(n_1, n_2) = \{\sigma \in S_{n_1+n_2} | \sigma^{-1}(1) < \ldots < \sigma^{-1}(n_1) \\ \text{and} \quad \sigma^{-1}(n_1 + 1) < \ldots < \sigma^{-1}(n_1 + n_2)\}, \tag{2.94}$$

where $S_{n_1+n_2}$ denotes the group of permutations on $n_1 + n_2$ elements.

The Shuffle Product of MPLs

Let us illustrate the concept of the shuffle product by a few examples:

$$G(a; z)G(b; z) = G(a, b; z) + G(b, a; z),$$
$$G(a; z)G(b, c; z) = G(a, b, c; z) + G(b, a, c; z) + G(b, c, a; z),$$
$$G(a, b; z)G(c, d; z) = G(a, b, c, d; z) + G(a, c, b, d; z) + G(a, c, d, b; z)$$
$$+ G(c, a, b, d; z) + G(c, a, d, b; z) + G(c, d, a, b; z).$$

This property turns the set of MPLs into a *shuffle algebra*, i.e., a vector space equipped with shuffle multiplication. The vector space structure simply means that we can take finite linear combinations of MPLs with complex coefficients, while the shuffle product is defined in Eq. (2.93). The proof of the shuffle algebra relations is relatively straightforward and follows from the definition of MPLs as iterated integrals. In Problem 2.3, you are asked to verify the shuffle algebra relation for the product of two MPLs, both of weight one.

Besides evaluating products of MPLs with the same argument, the shuffle algebra is also useful for extracting the singularities of MPLs at $z = 0$. Indeed, as noted above, $G(a_1, \ldots, a_n; z)$ is analytic (i.e., can be expanded into a power series) at $z = 0$ if $a_n \neq 0$. In case $a_n = 0$, we can employ the shuffle algebra to rewrite $G(a_1, \ldots, a_n; z)$ in terms of functions whose rightmost index is non-zero and MPLs of the form $G(\mathbf{0}_k; z) = \frac{1}{k!} \ln^k(z)$. This procedure thus makes the (logarithmic) singularities around z manifest.

Extracting Singularities of MPLs

To illustrate the extraction of singularities around $z = 0$ using the shuffle algebra, consider, e.g., the function $G(a, 0, 0; z)$. Clearly the rightmost index is zero and hence the function has logarithmic singularities at $z = 0$. Using the shuffle algebra, we compute

$$G(a, 0, 0; z) = G(0, 0; z)G(a; z) - G(0, a, 0; z) - G(0, 0, a; z)$$

$$= G(0, 0; z)G(a; z) - G(0, 0, a; z) - [G(0, a; z)G(0; z) - 2G(0, 0, a; z)]$$

$$= G(0, 0; z)G(a; z) - G(0; z)G(0, a; z) + G(0, 0, a; z). \tag{2.95}$$

Fourth, there is another algebra structure defined on MPLs. As noted above, the shuffle algebra structure follows from the definition of MPLs as iterated integrals. However, their definitions as sums induce another type of multiplication. We will have more to say about this in Sect. 5.1; however, to illustrate the basic idea, consider the product of two MPLs of depth one:

$$\text{Li}_1(z_1)\text{Li}_1(z_2) = \sum_{i_1=1}^{\infty} \frac{z_1^{i_1}}{i_1} \sum_{i_2=1}^{\infty} \frac{z_2^{i_2}}{i_2} = \sum_{0 < i_1, i_2} \frac{z_1^{i_1} z_2^{i_2}}{i_1 i_2}$$

$$= \sum_{0 < i_1 < i_2} \frac{z_1^{i_1} z_2^{i_2}}{i_1 i_2} + \sum_{0 < i_2 < i_1} \frac{z_1^{i_1} z_2^{i_2}}{i_1 i_2} + \sum_{0 < i_1 = i_2} \frac{z_1^{i_1} z_2^{i_2}}{i_1 i_2} \tag{2.96}$$

$$= \text{Li}_{1,1}(z_1, z_2) + \text{Li}_{1,1}(z_2, z_1) + \text{Li}_2(z_1 z_2)$$

In the second line, we have simply organized the double sum over i_1 and i_2 into three sums: in the first, $i_1 < i_2$, in the second $i_1 > i_2$, while the last sum represents the "diagonal" elements of the double sum, where $i_1 = i_2$. These sums are then immediately recognized as Li functions; see Eq. (2.88).[9] Products of MPLs of higher depth can be handled in a similar fashion. The algebra generated this way is called the *stuffle algebra* (also called the *quasi-shuffle algebra*) on MPLs. We emphasize that the stuffle algebra structure is completely independent of the shuffle algebra. The former is related to the definition of MPLs as iterated integrals, while the latter is connected with the sum representation. We mention that both shuffle and stuffle products preserve the weight. This means that the shuffle or stuffle product of two MLPs of weight w_1 and w_2 is a linear combination of MPLs of weight $w_1 + w_2$, as seen in the examples above. We say that the shuffle and stuffle algebras are *graded* by the weight. However, notice that the stuffle product does not preserve the depth of the sums, as evident from Eq. (2.96). Instead the depths of terms in the product are bounded by the sum of depths of the factors. In this case we say that the stuffle algebra is *filtered* by the depth.

Fifth, while the shuffle and stuffle algebras certainly generate many functional relations among MPLs, there are many more complicated relations that do not arise in this way. In particular, relations that change the arguments of the functions, such as the Hölder convolution, cannot be covered. However, such functional equations are very often crucial in evaluating Feynman integrals in terms of MPLs. We have seen a very simple example of this phenomenon already in Eq. (2.53), where we needed to use the functional relation $\ln(ab) = \ln a + \ln b$ to proceed with the calculation. Indeed, in Feynman integral calculations, the following problem often arises: We would like to evaluate a multi-dimensional integral, but after integration over some variable, say t_1, we find MPLs of the form $G(a_1(t_2), \ldots, a_n(t_2); z(t_2))$, where the a_i's as well as z are functions of t_2, the next integration variable. In order to proceed, we must first bring this function to "canonical form," i.e., one where t_2 only appears in the form $G(a'_1, \ldots, a'_n, t_2)$ so that we perform the integration over t_2 using the definition of MPLs (Eq. 2.82). As this rewriting will generally involve the change of arguments of the functions, we cannot hope to proceed using only the shuffle and stuffle algebra structures, but rather we need a general and flexible framework for deriving such relations. This framework is given by the *Hopf algebra* structure of MPLs. Even a cursory description of this framework goes beyond the scope of this chapter; however we mention that it can be used to find, and in some sense circumvent, functional equations among MPLs. We refer to [24, 25] for discussions of this topic.

Finally, as the name and notation imply, classical polylogarithms, Nielsen generalized polylogarithms, and the harmonic polylogarithms of Remiddi and Vermaseren [26] are all special cases of multiple polylogarithms. First of all,

---

[9] To be absolutely clear, we note that the last sum on the second line of Eq. (2.96) is simply

$$\sum_{0 < i_1 = i_2} \frac{z_1^{i_1} z_2^{i_2}}{i_1 i_2} = \sum_{0 < i_1} \frac{(z_1 z_2)^{i_1}}{i_1^2} = \text{Li}_2(z_1 z_2).$$

This property turns the set of MPLs into a *shuffle algebra*, i.e., a vector space equipped with shuffle multiplication. The vector space structure simply means that we can take finite linear combinations of MPLs with complex coefficients, while the shuffle product is defined in Eq. (2.93). The proof of the shuffle algebra relations is relatively straightforward and follows from the definition of MPLs as iterated integrals. In Problem 2.3, you are asked to verify the shuffle algebra relation for the product of two MPLs, both of weight one.

Besides evaluating products of MPLs with the same argument, the shuffle algebra is also useful for extracting the singularities of MPLs at $z = 0$. Indeed, as noted above, $G(a_1, \ldots, a_n; z)$ is analytic (i.e., can be expanded into a power series) at $z = 0$ if $a_n \neq 0$. In case $a_n = 0$, we can employ the shuffle algebra to rewrite $G(a_1, \ldots, a_n; z)$ in terms of functions whose rightmost index is non-zero and MPLs of the form $G(\mathbf{0}_k; z) = \frac{1}{k!} \ln^k(z)$. This procedure thus makes the (logarithmic) singularities around z manifest.

Extracting Singularities of MPLs

To illustrate the extraction of singularities around $z = 0$ using the shuffle algebra, consider, e.g., the function $G(a, 0, 0; z)$. Clearly the rightmost index is zero and hence the function has logarithmic singularities at $z = 0$. Using the shuffle algebra, we compute

$$
\begin{aligned}
G(a, 0, 0; z) &= G(0, 0; z)G(a; z) - G(0, a, 0; z) - G(0, 0, a; z) \\
&= G(0, 0; z)G(a; z) - G(0, 0, a; z) - [G(0, a; z)G(0; z) - 2G(0, 0, a; z)] \\
&= G(0, 0; z)G(a; z) - G(0; z)G(0, a; z) + G(0, 0, a; z).
\end{aligned}
\tag{2.95}
$$

Fourth, there is another algebra structure defined on MPLs. As noted above, the shuffle algebra structure follows from the definition of MPLs as iterated integrals. However, their definitions as sums induce another type of multiplication. We will have more to say about this in Sect. 5.1; however, to illustrate the basic idea, consider the product of two MPLs of depth one:

$$
\begin{aligned}
\mathrm{Li}_1(z_1)\mathrm{Li}_1(z_2) &= \sum_{i_1=1}^{\infty} \frac{z_1^{i_1}}{i_1} \sum_{i_2=1}^{\infty} \frac{z_2^{i_2}}{i_2} = \sum_{0 < i_1, i_2} \frac{z_1^{i_1} z_2^{i_2}}{i_1 i_2} \\
&= \sum_{0 < i_1 < i_2} \frac{z_1^{i_1} z_2^{i_2}}{i_1 i_2} + \sum_{0 < i_2 < i_1} \frac{z_1^{i_1} z_2^{i_2}}{i_1 i_2} + \sum_{0 < i_1 = i_2} \frac{z_1^{i_1} z_2^{i_2}}{i_1 i_2} \\
&= \mathrm{Li}_{1,1}(z_1, z_2) + \mathrm{Li}_{1,1}(z_2, z_1) + \mathrm{Li}_2(z_1 z_2)
\end{aligned}
\tag{2.96}
$$

classical polylogarithms are clearly just depth-one MPLs and from Eq. (2.89), we find

$$\mathrm{Li}_n(z) = -G(\mathbf{0}_{n-1}, 1; z). \tag{2.97}$$

Nielsen generalized polylogarithms can be expressed as

$$S_{n,p}(z) = (-1)^p G(\mathbf{0}_n, \mathbf{1}_p; z) = \mathrm{Li}_{1,\ldots,1,n+1}(\underbrace{1, \ldots, 1}_{p-1}, z). \tag{2.98}$$

Finally, harmonic polylogarithms [27] (HPLs) correspond the case when all a_i are 0 or ± 1. HPLs are equal to MPLs up to a sign and we have ($a_i \in \{0, \pm 1\}$)

$$H_{a_1,\ldots,a_k}(z) = (-1)^p G(a_1, \ldots, a_k; z). \tag{2.99}$$

Here p is the number of elements of the set of indices $\{a_1, \ldots, a_n\}$ which are equal to $+1$. Thus, we have, e.g., $H_{1,1,0}(z) = (-1)^2 G(1, 1, 0; z) = G(1, 1, 0; z)$ since two of the indices are $+1$, but $H_{1,-1,0}(z) = (-1)^1 G(1, -1, 0; z) = -G(1, -1, 0; z)$ as now only one index is equal to $+1$. HPLs can also be written in terms of Li using Eq. (2.90); however, the relation is best expressed by using a different notation for HPLs. The notation we have introduced above, where each index is either 0 or ± 1, is referred to as the "a"-notation. In contrast, the "m"-notation is given as follows: First, let us assume that the rightmost index in the "a"-notation is non-zero. Then, starting from the vector of indices $(a_1, \ldots, a_k)$, we construct a new vector $(m_1, \ldots, m_l)$, where we drop all zeros and increase the absolute value of each non-zero element by the number of zeros immediately preceding it. The sign of each index remains unchanged. So in this way, e.g., $(0, 1, 1)$ becomes $(2, 1)$, while, e.g., $(0, 0, 1, 0, -1)$ becomes $(3, -2)$ and so on. We can then extend this notation to allow for any number of rightmost zeros, so, e.g., $(0, -1, 0, 0, 1, 0, 0)$ goes to $(-2, 3, 0, 0)$. The "m"-notation is introduced here because it makes the relation between HPLs and Li functions particularly simple:

$$H_{m_1,\ldots,m_k}(z) = \mathrm{Li}_{m_k,\ldots,m_1}(1, \ldots, 1, z), \tag{2.100}$$

Notice that the order of indices is reversed when expressing HPLs in terms of Li. We note that the `Mathematica` package `HPL` [14] implements both notations for HPLs as well as appropriate functions to transform between the two.

The list of available analytical and numerical packages for evaluation of classical, Nielsen generalized, multiple polylogarithms and series summations is given in Appendices A.2 and A.3.

To finish this section, we note that although MPLs are a large class of functions, it is known that not all Feynman integrals can be evaluated in terms of them. In fact these days, Feynman integrals evaluating to HPLs or MPLs are in some sense considered to be the easiest Feynman integrals. However, the seemingly not very

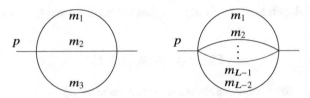

Fig. 2.4 The two-loop banana diagram (also called the sunrise or the sunset diagram, depending on the author's mood) with generic internal masses (on left) and its generalization to the $L - 1$ loop integral (on right)

complicated graph of Fig. 2.4 already cannot be evaluated in terms of MPLs when all masses of the internal lines are different.

In fact, this integral already at two loops leads to *elliptic integrals* and *elliptic polylogarithms*. The study of these new classes of functions going beyond MPLs is currently an active area of research. In general a space of functions for Feynman integrals is extremely rich (we discussed GPLs); there are elliptic functions, modular forms, and integrals over Calabi-Yau varieties [28]. Interestingly, an upper bound on this complexity exists; it has been discussed in [29] that Feynman integrals are special cases of an A-hypergeometric function and the scattering equations are given by the hyperdeterminant of a multi-dimensional array, within the theory developed by Gel'fand, Zelevinskii, and Kapranov [30]. For the state of the art, see [25, 31].

2.5   Gamma Function

The gamma function is present in the basic MB formula of Eq. (1.44), but it also appears in the definitions of a wide range of special functions, such as the (generalized) hypergeometric function and its particular cases which include Jacobi, Gegenbauer, Chebyshev, or Legendre polynomials. These polynomials occur frequently both in classical and quantum physics [32]; hence the gamma function is ubiquitous in science and engineering.

Introducing the gamma function, we start from factorial $n!$ which stands for the product of consecutive integers starting from 1, $n! = 1 \cdot 2 \cdots n$, $n \in \mathbb{N}^+$. By convention, we set $0! = 1$. Obviously we can write a recurrence relation for the factorial function, $(n + 1)! = (n + 1) \cdot n!$.

Interestingly, the discrete definition of the factorial can be extended to a smooth, real function with the property that

$$\Gamma(x + 1) = x\Gamma(x), \quad x \in (0, \infty). \tag{2.101}$$

Euler constructed such a function in 1729 with the normalization $\Gamma(1) = 1$ by using the convergent integral

$$\Gamma(x) \equiv \int_0^\infty e^{-t} t^{x-1} dt. \tag{2.102}$$

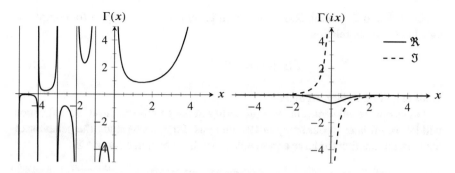

Fig. 2.5 The plot of the gamma function $\Gamma(x)$ for real (left) and purely imaginary (right) x. Note that for real arguments the gamma function is real and the thin vertical lines on the left panel denote asymptotes of the function

Starting from this integral representation, it is straightforward to show that the functional equation (2.101) holds. Using partial integration we find

$$\Gamma(z+1) = \int_0^\infty e^{-x} x^z dx = -x^z e^{-x}\Big|_0^\infty + z \int_0^\infty e^{-x} x^{z-1} dx$$

$$= z \int_0^\infty e^{-x} x^{z-1} dx = z\Gamma(z). \tag{2.103}$$

Furthermore, for $z \equiv n \in \mathbb{N}$, using $\Gamma(1) = 1$ and Eq. (2.101), it follows that the gamma function takes the values $\Gamma(n) = (n - 1)!$. Thus the gamma function is an extension of the factorial function to numbers which are not integers. Figure 2.5 shows the plot of the gamma function for real and purely imaginary x. Notice in particular the singularities at non-positive integers $x = 0, -1, -2, \ldots$.

The gamma function was extended to complex variables by Gauss (1811) using the infinite product

$$\Gamma(z) = \lim_{n \to \infty} \frac{n! n^z}{z(z+1) \cdots (z+n)}, \tag{2.104}$$

which defines a meromorphic function in z. From this representation we can see explicitly that the gamma function exhibits singularities in the complex plane, namely, $z \in C \setminus \{0, -1, -2, \ldots\}$. However, there are several ways of defining the complex gamma function besides the above product representation [33, 34]. For example, the integral definition of the gamma function for complex z is of course a generalization of the definition of Eq. (2.102),

$$\Gamma(z) \equiv \int_0^\infty e^{-x} x^{z-1} dx, \tag{2.105}$$

which is due to Bernoulli. It is clear that in general $\Gamma(z)$ is complex for complex z and that the basic relation

$$\Gamma(z+1) = z\Gamma(z), \quad z \in \mathbb{C}. \tag{2.106}$$

holds also for complex arguments.

Let us now recall some classical properties of the gamma function [34, 35] which will be useful later in our study of MB integrals. First, let us study the behavior of the gamma function for large arguments. It can be shown that for $n \in \mathbb{Z}$

$$\Gamma(n+1) = n! \stackrel{n \to \infty}{\sim} \sqrt{2\pi n} n^n e^{-n}, \tag{2.107}$$

while for $z \in \mathbb{C}$ we have

$$\Gamma(z) \stackrel{|z| \to \infty}{\sim} \sqrt{\frac{2\pi}{z}} z^z e^{-z}. \tag{2.108}$$

The second equation is the (leading term of the) Stirling formula. From the above two equations, we could naively deduce, due to the different square root factors, that for large values Eqs. (2.107) and (2.108) are in contradiction. However, this is not the case; see Problem 2.4. From Eq. (2.108) it is clear that the gamma function is strictly increasing for positive real arguments, as expected. It is also very easy to see that the rate of increase is super-exponential, i.e., $\lim_{x \to \infty} \Gamma(x)/e^x = \infty$. It is less obvious, but nevertheless true, that for complex arguments $z = x + iy$, $x, y \in \mathbb{R}$, both the real and imaginary parts of the gamma function fall off faster than exponentially for large imaginary parts, i.e., as $|y| \to \infty$; see Problem 2.5. In order to get a feeling for the magnitudes of gamma function values, below we give some representative numerical and analytical values obtained with Mathematica.

$$\Gamma(1) = \Gamma(2) = 1, \tag{2.109}$$

$$\Gamma(10.1) = 4.54761 \cdot 10^5, \tag{2.110}$$

$$\Gamma(-10.1) = -2.21342 \cdot 10^{-6}, \tag{2.111}$$

$$\Gamma(100.1) = 1.47845 \cdot 10^{156}, \tag{2.112}$$

$$\Gamma(-100.1) = -6.86951 \cdot 10^{-158}, \tag{2.113}$$

$$\Gamma(-1 \pm i) = -0.171533 + 0.326483 i \tag{2.114}$$

$$\Gamma(-1 \pm 10 i) = -4.99740 \cdot 10^{-9} \pm 1.07847 \cdot 10^{-8} i \tag{2.115}$$

$$\Gamma(-1 \pm 100 i) = 1.51438 \cdot 10^{-71} \pm 1.27644 \cdot 10^{-73} i \tag{2.116}$$

$$\Gamma\left(\frac{1}{2}\right) = \sqrt{\pi}, \tag{2.117}$$

$$\frac{1}{\Gamma(-1)} = \frac{1}{\Gamma(0)} = 0. \tag{2.118}$$

We note that the last equation is to be understood in the sense of a limit.

Next, consider Euler's reflection (or complement) formula, which states that for $0 < z < 1$

$$\Gamma(z)\Gamma(1 - z) = \frac{\pi}{\sin \pi z}. \tag{2.119}$$

Similarly

$$\Gamma\left(\frac{1}{2} + z\right)\Gamma\left(\frac{1}{2} - z\right) = \frac{\pi}{\cos \pi z}. \tag{2.120}$$

These so-called complement relations can sometimes be useful for manipulating MB integrals, and we will also use them when we discuss the analytic solution of one-dimensional MB integrals in Chap. 5. Notice moreover that Eq. (2.119) allows us to determine the behavior of the gamma function around $z = 0$. Indeed, using $\sin(\pi z) = \pi z + O(z^3)$ and $\Gamma(1) = 1$, we find

$$\Gamma(z) \xrightarrow{z \to 0} \frac{1}{z}. \tag{2.121}$$

In fact, the above calculation can be generalized and the behavior of the gamma function determined around any non-positive integer. To do so, let us consider $z \to -n + z$ in Eq. (2.119), where $n \in \mathbb{N}$. Then we have

$$\Gamma(-n + z) = \frac{1}{\Gamma(1 + n - z)} \frac{\pi}{\sin \pi(-n + z)}. \tag{2.122}$$

However $\sin \pi(-n + z) = (-1)^n \sin(\pi z)$ for integers, while $\Gamma(1 + n - z) = n!$ for $z \to 0$; hence finally

$$\Gamma(-n + z) \xrightarrow{z \to 0} \frac{(-1)^n}{n!} \frac{1}{z}. \tag{2.123}$$

Of course, the same conclusion can be reached by applying the recursion relation in Eq. (2.106) to reduce $\Gamma(-n + z)$ to $\Gamma(z)$ and using Eq. (2.121); see Problem 2.6.

Thus, we see that the gamma function has poles at all non-positive integers, a fact that will become significant shortly, when we discuss residues and Cauchy's theorem. In Fig. 2.6 we show the location of these poles for some gamma functions with specific arguments which will be considered in Chap. 4; see Fig. 4.1.

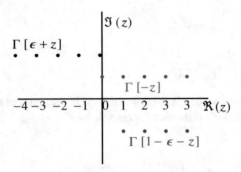

Fig. 2.6 Positions of the singularities on the real axis (horizontal) for three gamma functions. To avoid visual overlapping of dots in the figure, poles of single gammas are drawn at different vertical levels. The ϵ parameter is assumed to be real and positive, $\epsilon > 0$. The singular behavior of the gamma function at the poles can be seen in Fig. 2.5

Going beyond the above results, the expansion of the gamma function around integer numbers to higher orders is often needed in explicit computations. It is common that the small parameter of the expansion is the ϵ variable that appears in dimensional regularization where the space-time dimension is set to $d = 4 - 2\epsilon$. For example, the expansions of $\Gamma(z)$ around $z = 0$ and $z = 1$ in the small parameter ϵ read

$$\Gamma(\epsilon) = \frac{1}{\epsilon} - \gamma + \frac{1}{2}\epsilon\left[\zeta(2) + \gamma^2\right] + \frac{1}{6}\left[-2\zeta(3) - 3\gamma\zeta(2) - \gamma^3\right]\epsilon^2 + O\left(\epsilon^3\right)$$

$$\equiv \frac{1}{\epsilon} - \gamma + g_1\epsilon + g_2\epsilon^2 + O\left(\epsilon^3\right), \tag{2.124}$$

$$\Gamma(1 + \epsilon) = 1 - \gamma\epsilon + \frac{1}{2}\epsilon^2\left[\zeta(2) + \gamma^2\right] + \ldots \equiv 1 - \gamma\epsilon + g_1\epsilon^2 + \ldots. \tag{2.125}$$

We see that in the expansions a couple of irrational constants have appeared. These constants are either related to the harmonic series[10] or are values of the zeta function at positive integers,

$$\gamma = 0.577\,215\,664\,901\,532\ldots \quad \text{Euler's constant,} \tag{2.126}$$

$$\zeta_2 \equiv \zeta(2) = \mathrm{Li}_2(1) = \frac{\pi^2}{6} = 1.644\,93\ldots, \tag{2.127}$$

$$\zeta(3) = \mathrm{Li}_3(1) = 1.202\,056\,903\,159\,594\ldots. \tag{2.128}$$

---

[10] For the origin of Euler's gamma constant, see Chapter "Gamma's Birthplace" in [11]. It was for long not clear if γ is irrational, and G. H. Hardy offered to vacate his Savilian Chair at Oxford to anyone who could prove gamma to be irrational. The emergence of the zeta function is discussed in this book also.

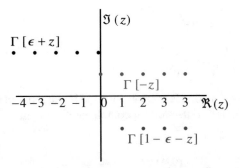

Fig. 2.6 Positions of the singularities on the real axis (horizontal) for three gamma functions. To avoid visual overlapping of dots in the figure, poles of single gammas are drawn at different vertical levels. The ϵ parameter is assumed to be real and positive, $\epsilon > 0$. The singular behavior of the gamma function at the poles can be seen in Fig. 2.5

Going beyond the above results, the expansion of the gamma function around integer numbers to higher orders is often needed in explicit computations. It is common that the small parameter of the expansion is the ϵ variable that appears in dimensional regularization where the space-time dimension is set to $d = 4 - 2\epsilon$. For example, the expansions of $\Gamma(z)$ around $z = 0$ and $z = 1$ in the small parameter ϵ read

$$\Gamma(\epsilon) = \frac{1}{\epsilon} - \gamma + \frac{1}{2}\epsilon\left[\zeta(2) + \gamma^2\right] + \frac{1}{6}\left[-2\zeta(3) - 3\gamma\zeta(2) - \gamma^3\right]\epsilon^2 + O\left(\epsilon^3\right)$$

$$\equiv \frac{1}{\epsilon} - \gamma + g_1\epsilon + g_2\epsilon^2 + O\left(\epsilon^3\right), \tag{2.124}$$

$$\Gamma(1 + \epsilon) = 1 - \gamma\epsilon + \frac{1}{2}\epsilon^2\left[\zeta(2) + \gamma^2\right] + \ldots \equiv 1 - \gamma\epsilon + g_1\epsilon^2 + \ldots. \tag{2.125}$$

We see that in the expansions a couple of irrational constants have appeared. These constants are either related to the harmonic series[10] or are values of the zeta function at positive integers,

$$\gamma = 0.577\,215\,664\,901\,532\ldots \quad \text{Euler's constant,} \tag{2.126}$$

$$\zeta_2 \equiv \zeta(2) = \text{Li}_2(1) = \frac{\pi^2}{6} = 1.644\,93\ldots, \tag{2.127}$$

$$\zeta(3) = \text{Li}_3(1) = 1.202\,056\,903\,159\,594\ldots. \tag{2.128}$$

---

[10] For the origin of Euler's gamma constant, see Chapter "Gamma's Birthplace" in [11]. It was for long not clear if γ is irrational, and G. H. Hardy offered to vacate his Savilian Chair at Oxford to anyone who could prove gamma to be irrational. The emergence of the zeta function is discussed in this book also.

$$\Gamma\left(\frac{1}{2}\right) = \sqrt{\pi}, \tag{2.117}$$

$$\frac{1}{\Gamma(-1)} = \frac{1}{\Gamma(0)} = 0. \tag{2.118}$$

We note that the last equation is to be understood in the sense of a limit.

Next, consider Euler's reflection (or complement) formula, which states that for $0 < z < 1$

$$\Gamma(z)\Gamma(1 - z) = \frac{\pi}{\sin \pi z}. \tag{2.119}$$

Similarly

$$\Gamma\left(\frac{1}{2} + z\right)\Gamma\left(\frac{1}{2} - z\right) = \frac{\pi}{\cos \pi z}. \tag{2.120}$$

These so-called complement relations can sometimes be useful for manipulating MB integrals, and we will also use them when we discuss the analytic solution of one-dimensional MB integrals in Chap. 5. Notice moreover that Eq. (2.119) allows us to determine the behavior of the gamma function around $z = 0$. Indeed, using $\sin(\pi z) = \pi z + O(z^3)$ and $\Gamma(1) = 1$, we find

$$\Gamma(z) \xrightarrow{z \to 0} \frac{1}{z}. \tag{2.121}$$

In fact, the above calculation can be generalized and the behavior of the gamma function determined around any non-positive integer. To do so, let us consider $z \to -n + z$ in Eq. (2.119), where $n \in \mathbb{N}$. Then we have

$$\Gamma(-n + z) = \frac{1}{\Gamma(1 + n - z)} \frac{\pi}{\sin \pi(-n + z)}. \tag{2.122}$$

However $\sin \pi(-n + z) = (-1)^n \sin(\pi z)$ for integers, while $\Gamma(1 + n - z) = n!$ for $z \to 0$; hence finally

$$\Gamma(-n + z) \xrightarrow{z \to 0} \frac{(-1)^n}{n!} \frac{1}{z}. \tag{2.123}$$

Of course, the same conclusion can be reached by applying the recursion relation in Eq. (2.106) to reduce $\Gamma(-n + z)$ to $\Gamma(z)$ and using Eq. (2.121); see Problem 2.6.

Thus, we see that the gamma function has poles at all non-positive integers, a fact that will become significant shortly, when we discuss residues and Cauchy's theorem. In Fig. 2.6 we show the location of these poles for some gamma functions with specific arguments which will be considered in Chap. 4; see Fig. 4.1.

In order to explain these relations, let us define the *generalized harmonic number of order a of N* as the finite sum

$$H_{N,a} \equiv \sum_{k=1}^{N} \frac{1}{k^a}.$$ (2.129)

This sum is also often denoted as $S_a(N)$. With this definition, the zeta function[11] at a is simply the harmonic number of order a of infinity,

$$\zeta(a) = \sum_{k=1}^{\infty} \frac{1}{k^a} = H_{\infty,a}.$$ (2.130)

Then Euler's constant is the limit at infinity of the difference of the harmonic number $H_{N,1}$ and the natural logarithm of N,

$$\gamma = \lim_{N \to \infty} \left[\sum_{k=1}^{N} \frac{1}{k} - \ln(N) \right],$$ (2.131)

while $\zeta(2)$, $\zeta(3)$, and so on are simply the values of the convergent sums

$$\zeta(2) = \sum_{k=1}^{\infty} \frac{1}{k^2}, \quad \zeta(3) = \sum_{k=1}^{\infty} \frac{1}{k^3}.$$ (2.132)

Incidentally, using the notion of harmonic numbers, it is possible to write a compact formula for the all-order expansion of the gamma function around positive integers n in terms of $\Gamma(1 + \epsilon)$,

$$\frac{\Gamma(n + \epsilon)}{\Gamma(n)} = \Gamma(1 + \epsilon) \exp \left[-\sum_{k=1}^{\infty} \frac{(-\epsilon)^k}{k} S_k(n - 1) \right].$$

Another useful relation which will be used in Chap. 5 is the Legendre duplication formula,

$$\Gamma(2z) = \frac{2^{2z-1}}{\sqrt{\pi}} \Gamma(z) \Gamma \left(z + \frac{1}{2} \right),$$ (2.133)

---

[11] For a history of the zeta function and logarithm, see [36]. Physical applications of spectral zeta function are discussed in Lecture Notes [37].

which is a special case of the so-called product theorem for gamma functions,

$$\Gamma(nz) = (2\pi)^{\frac{1-n}{2}} n^{nz-\frac{1}{2}} \prod_{k=0}^{n-1} \Gamma\left(z + \frac{k}{n}\right).$$ (2.134)

As we will see, one also often encounters the derivatives of the gamma function in actual calculations; see, e.g., Eq. (2.162) or the file MB_miscellaneous_Springer.nb in the auxiliary material in [5] (polygammas can be seen in examples of ϵ expanded MB integrals). Hence, we recall the definition of the *digamma* function $\psi(z)$ of a (in general complex) variable z,

$$\psi(z) \equiv \frac{d}{dz} \ln \Gamma(z) = \frac{1}{\Gamma(z)} \frac{d}{dz} \Gamma(z).$$ (2.135)

The polygamma function of order m is then the m-th derivative of the digamma function

$$\psi^{(m)}(z) \equiv \frac{d^m}{dz^m} \psi(z).$$ (2.136)

Thus, the digamma function is simply the polygamma function of order zero, $\psi(z) = \psi^{(0)}(z)$. *It can be shown that polygamma functions have poles of order $m + 1$ at all non-positive integers*; see Problem 2.7.

For arguments whose real part is positive, $\Re(z) > 0$, the digamma function can be represented also as a one-dimensional integral over a real integration variable,

$$\psi(z) = \int_0^\infty [(1+t)^{-1} - (1+t)^{-z}]\frac{dt}{t} - \gamma, \qquad \Re(z) > 0,$$ (2.137)

or alternatively

$$\psi(z) = \int_0^1 \frac{t^{z-1} - 1}{t - 1} dt - \gamma, \qquad \Re(z) > 0.$$ (2.138)

This relation is employed when solving MB integrals via reduction to Euler-type integrals, a method that will be explored in Sect. 5.3. We note that we can easily derive similar integral representations for polygamma functions by simply taking derivatives of the above formulae with respect to z. (For $\Re(z) > 0$ the order of differentiation and integration can be interchanged.)

Among the many relations that hold for polygamma functions, we recall here only those that are the most relevant for manipulating MB integrals. First, using Eq. (2.105), it is quite straightforward to show that

$$\psi(z + 1) = \psi(z) + \frac{1}{z}$$ (2.139)

In order to explain these relations, let us define the *generalized harmonic number of order a of N* as the finite sum

$$H_{N,a} \equiv \sum_{k=1}^{N} \frac{1}{k^a}.$$ (2.129)

This sum is also often denoted as $S_a(N)$. With this definition, the zeta function[11] at a is simply the harmonic number of order a of infinity,

$$\zeta(a) = \sum_{k=1}^{\infty} \frac{1}{k^a} = H_{\infty,a}.$$ (2.130)

Then Euler's constant is the limit at infinity of the difference of the harmonic number $H_{N,1}$ and the natural logarithm of N,

$$\gamma = \lim_{N \to \infty} \left[\sum_{k=1}^{N} \frac{1}{k} - \ln(N) \right],$$ (2.131)

while $\zeta(2)$, $\zeta(3)$, and so on are simply the values of the convergent sums

$$\zeta(2) = \sum_{k=1}^{\infty} \frac{1}{k^2}, \quad \zeta(3) = \sum_{k=1}^{\infty} \frac{1}{k^3}.$$ (2.132)

Incidentally, using the notion of harmonic numbers, it is possible to write a compact formula for the all-order expansion of the gamma function around positive integers n in terms of $\Gamma(1 + \epsilon)$,

$$\frac{\Gamma(n + \epsilon)}{\Gamma(n)} = \Gamma(1 + \epsilon) \exp \left[-\sum_{k=1}^{\infty} \frac{(-\epsilon)^k}{k} S_k(n - 1) \right].$$

Another useful relation which will be used in Chap. 5 is the Legendre duplication formula,

$$\Gamma(2z) = \frac{2^{2z-1}}{\sqrt{\pi}} \Gamma(z) \Gamma\left(z + \frac{1}{2}\right),$$ (2.133)

---

[11] For a history of the zeta function and logarithm, see [36]. Physical applications of spectral zeta function are discussed in Lecture Notes [37].

which is a special case of the so-called product theorem for gamma functions,

$$\Gamma(nz) = (2\pi)^{\frac{1-n}{2}} n^{nz-\frac{1}{2}} \prod_{k=0}^{n-1} \Gamma\left(z + \frac{k}{n}\right).$$

(2.134)

As we will see, one also often encounters the derivatives of the gamma function in actual calculations; see, e.g., Eq. (2.162) or the file MB_miscellaneous_Springer.nb in the auxiliary material in [5] (polygammas can be seen in examples of ϵ expanded MB integrals). Hence, we recall the definition of the *digamma* function $\psi(z)$ of a (in general complex) variable z,

$$\psi(z) \equiv \frac{d}{dz} \ln \Gamma(z) = \frac{1}{\Gamma(z)} \frac{d}{dz} \Gamma(z).$$

(2.135)

The polygamma function of order m is then the m-th derivative of the digamma function

$$\psi^{(m)}(z) \equiv \frac{d^m}{dz^m} \psi(z).$$

(2.136)

Thus, the digamma function is simply the polygamma function of order zero, $\psi(z) = \psi^{(0)}(z)$. *It can be shown that polygamma functions have poles of order $m + 1$ at all non-positive integers*; see Problem 2.7.

For arguments whose real part is positive, $\Re(z) > 0$, the digamma function can be represented also as a one-dimensional integral over a real integration variable,

$$\psi(z) = \int_0^\infty [(1+t)^{-1} - (1+t)^{-z}] \frac{dt}{t} - \gamma, \qquad \Re(z) > 0,$$

(2.137)

or alternatively

$$\psi(z) = \int_0^1 \frac{t^{z-1} - 1}{t - 1} dt - \gamma, \qquad \Re(z) > 0.$$

(2.138)

This relation is employed when solving MB integrals via reduction to Euler-type integrals, a method that will be explored in Sect. 5.3. We note that we can easily derive similar integral representations for polygamma functions by simply taking derivatives of the above formulae with respect to z. (For $\Re(z) > 0$ the order of differentiation and integration can be interchanged.)

Among the many relations that hold for polygamma functions, we recall here only those that are the most relevant for manipulating MB integrals. First, using Eq. (2.105), it is quite straightforward to show that

$$\psi(z + 1) = \psi(z) + \frac{1}{z}$$

(2.139)

and in the general case

$$\psi^{(m)}(z+1) = \psi^{(m)}(x) + (-1)^m m! z^{-m-1}. \tag{2.140}$$

These relations allow us to reduce the values of the polygamma functions at non-negative integers to values at one, where we have

$$\psi(1) = -\gamma \quad \text{and} \quad \psi^{(m)}(1) = -(-1)^m m! \zeta(m+1), \quad m \in \mathbb{N}^+. \tag{2.141}$$

Thus, in terms of the generalized harmonic numbers introduced above, we find for $N \in \mathbb{N}^+$

$$\psi(N+1) = \sum_{k=1}^{N} \frac{1}{k} - \gamma = S_1(N) - \gamma, \tag{2.142}$$

and

$$\psi^{(m)}(N+1) = (-1)^m m! \sum_{k=1}^{m} \frac{1}{k^{m+1}} - (-1)^m m! \zeta(m+1) \tag{2.143}$$

$$= (-1)^m m! \left[S_{m+1}(N) - \zeta(m+1) \right]. \tag{2.144}$$

Thus, for example, $\psi(1) = -\gamma$, $\psi(2) = 1 - \gamma$, $\psi(3) = 3/2 - \gamma$, and so on.

One more relation that we will make use of in Chap. 5 is the relation

$$\psi^{(m_q)}(1-z) = (-1)^{m_q} \left[\psi^{(m_q)}(z) + \pi \frac{\partial^{m_q}}{\partial z^{m_q}} \cot(\pi z) \right], \tag{2.145}$$

which can be used to convert all $\psi^{(m)}(1-z)$ functions to $\psi^{(m)}(z)$ functions.

We note finally that expanding polygamma functions around integer parameters, we again encounter the Euler γ constant and zeta values, e.g.,

$$\psi(1+\epsilon) = -\gamma + \zeta(2)\epsilon - \zeta(3)\epsilon^2 + O(\epsilon^3), \tag{2.146}$$

$$\psi^{(1)}(1+\epsilon) = \zeta(2) - 2\zeta(3)\epsilon + 3\zeta(4)\epsilon^2 + O(\epsilon^3), \tag{2.147}$$

$$\psi^{(2)}(1+\epsilon) = -2\zeta(3) + 6\zeta(4)\epsilon - 12\zeta(5)\epsilon^2 + O(\epsilon^3). \tag{2.148}$$

2.6   Residues and Cauchy's Residue Theorem

The Cauchy integral theorem (also known as the Cauchy-Goursat theorem) states that if a function is analytic in a domain (without singularities); then for any closed contour in the domain, that contour integral is zero. If singularities exist, the integral over an anticlockwise[12] directed closed path C_+ is (Cauchy's residue theorem)

$$\oint_{C_+} F(z)dz = 2\pi i \sum_{z=z_i} \text{Res}[F(z)] \tag{2.149}$$

where the residues $\text{Res}[F(z)]|_{z=z_i}$ are coefficients a^i_{-1} of the Laurent series of $F(z)$ around z_i:

$$F(z) = \sum_{n=-N}^{\infty} a^i_n (z - z_i)^n = \frac{a^i_{-N}}{(z - z_i)^N} + \cdots + \frac{a^i_{-1}}{(z - z_i)} + a^i_0 + \cdots \tag{2.150}$$

$$\boxed{\text{Res}[F(z)]|_{z=z_i} = a^i_{-1}} \tag{2.151}$$

If $G(z)$ has a Taylor expansion around z_0, then it is

$$\boxed{\text{Res}[G(z)\, F(z)]|_{z=z_i} = \sum_{n=1}^{N} \frac{a^i_{-n}}{k!} \frac{d^n}{dz^n} G(z)|_{z=z_i}} \tag{2.152}$$

Due to this property, we need for applications not only $\Gamma(z)$ but also its derivatives.

The basic equation (2.149) says that calculation of integrals in the complex plane can be done by summing residues enclosed by the contour of integration.

This is a simple trick which can help to solve many otherwise difficult integrals in a simple way.

Residues for Integrals over Real and Complex Variables

Using Cauchy's theorem evaluate

$$I_1 = \oint_{\Omega} \frac{dz}{z^4 + 1}, \tag{2.153}$$

---

[12] The positive direction of the curve is defined as anticlockwise for which the winding number is 1; see Problem 2.8.

Fig. 2.7 Contours Ω and Ω_1 applied in Eqs. (2.153) and (2.154), respectively

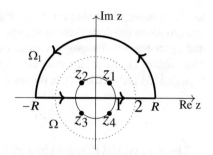

where Ω is any closed contour[13] around $z = 0$ with radius $R > 1$ (see Fig. 2.7), and the integral

$$I_2 = \int_{-\infty}^{\infty} \frac{dx}{x^4 + 1}, \tag{2.154}$$

for which the contour Ω_1 in Fig. 2.7 can be applied.

We have four solutions for $z^4 + 1 = 0$, $z_1 = e^{\pi i/4}$, $z_2 = e^{3\pi i/4}$, $z_3 = e^{5\pi i/4}$, $z_4 = e^{7\pi i/4}$. All solutions $z_1, \ldots, z_4$ lie inside Ω; see Fig. 2.7.

Thus we have

$$\text{Res}[I_1]|_{z=z_1} = -\frac{1}{4\sqrt{2}}(1+i), \quad \text{Res}[I_1]|_{z=z_2} = -\frac{1}{4\sqrt{2}}(-1+i), \tag{2.155}$$

$$\text{Res}[I_1]|_{z=z_3} = -\frac{1}{4\sqrt{2}}(-1-i), \quad \text{Res}[I_1]|_{z=z_4} = -\frac{1}{4\sqrt{2}}(1-i), \tag{2.156}$$

and the sum

$$I_1 = \sum_{i=1}^{4} \text{Res}[I_1]|_{z=z_i} = 0. \tag{2.157}$$

In Mathematica, a residue calculated in Eq. (2.155) would have the following form:

```
In[11]:= Residue[1/(1+z^4),{z,Exp[I*Pi/4]}]
```

$$\tag{2.158}$$

In contrast to the I_1 integral, I_2 is a real integrand over real contour of integration. Nonetheless, we can apply Cauchy's theorem by extending the path of integration

---

[13] Contours of this kind are known as the Hankel contours [32].

to the complex plane (the thick Ω_1 contour in Fig. 2.7). This is possible since the integral along the semicircle in the upper half-plane vanishes in the $R \to \infty$ limit, and so we are not changing the value of the integral by including this contribution. Then inside Ω_1 there are only two singularities at z_1 and z_2, solved already in Eqs. (2.155) and (2.155); then the result is

$$I_2 = 2\pi i \cdot \left(-\frac{1}{4\sqrt{2}} \cdot 2i \right) = \frac{\pi}{2\sqrt{2}}. \tag{2.159}$$

These are typical examples how integrals can be solved using Cauchy's theorem. For more examples, see, for instance, [6].

Residues of the Gamma Function

What are residues for $\Gamma(z)$ and $\Gamma^2(z)$ function at points $z = i, 1 + i, -1, 0, 1$?

Singularities of the gamma function sit on the real axis and are negative integers; see Fig. 2.5 and Eq. (2.118). Series at points $z = i, i + 1$ Laurent expanded as in Eq. (2.150) have no a^i_{-1} coefficients, e.g.,

```
In[12]:= Series[Gamma[z],{z,I,1}]
Out[12]:= Gamma[i]+Gamma[i]PolyGamma[0,i](z-i)+O[z-i]^2
```

In general,

$$\operatorname{Res}[\Gamma[z]]_{z=-n} = \frac{(-1)^n}{n!} \tag{2.160}$$

$$\operatorname{Res}[G[z]\Gamma[z]]_{z=-n} = \frac{(-1)^n}{n!} G[-n] \tag{2.161}$$

$$\operatorname{Res}[G[z]\Gamma^2[z]]_{z=-n} = \frac{2\,\psi^{(0)}[n+1]G[-n] + G'[-n]}{(n!)^2}, \tag{2.162}$$

where $G[z]$ is an analytic function of z. Equation (2.160) was proved in Eq. (2.123), while the relation in Eq. (2.161) is a straightforward consequence of Eq. (2.160). Note the appearance of the polygamma function in Eq. (2.162); see Problem 2.9.

Summing Gamma Poles and Integration over the Bromwich Contour

Let us calculate the sum of residues of $\Gamma[-z]$ when the contour of integration goes through $\Re[z] = \pm 1/2$ for the first 10 or 100 poles, see Fig. 2.8.

In Mathematica, a residue calculated for $\Gamma[-z]$ would have the following form:

Fig. 2.8 Integration contours Ω_1 and Ω_2. Contours of integration are at fixed $\Re(z) = -\frac{1}{2}, \frac{1}{2}$ which correspond to the sums starting at $n = 0, 1$ at In[13] and In[14], respectively

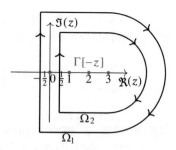

```
In[13]:= n1=Sum[-Residue[Gamma[-z],{z,n}],{n,0,100}];
In[14]:= n2=Sum[-Residue[Gamma[-z],{z,n}],{n,1,100}];
In[15]:= n1-n2;;
```

The minus signs in the first and second In commands are due to the direction of integration, which is clockwise; see Fig. 2.8. The result of the sum in the first command which starts from $n = 0$ is 0.367879. In fact, taking only ten terms gives the same accuracy. The result of the sum in the second command which starts from $n = 1$ is -0.632121. The difference is $n1 - n2 = 1$, which is the value of the residue for $\Gamma[-z]$ at $z = 0$.

In Chap. 6 we will discuss how to calculate MB integrals numerically. For the above case, we can make it directly in Mathematica. The contour is over a straight line parallel to the vertical axis. Such a contour is called the Bromwich contour [32]

```
In[16]:= NIntegrate[1/(2 Pi) Gamma[-z]/.z->-1/2+I
  ↪  y,{y,-10,10}];
```

$$(2.163)$$

The integral gives the same value as the sum of residues In[13] above. Please note a change of variables in Eq. (2.163) and that to get the same accuracy as for sums, integration can be restricted to the region $(-10, 10)$, as the gamma function falls down quickly with increasing $|z|$ (Fig. 2.5).

Choosing Right Direction of the Integration Contour

Calculate the integral of $\Gamma[z]$ along a line parallel to the imaginary axis at $\Re(z) = -1/2$.

The idea is to approximate an integral over a closed path by an integral over a part of the path only, as shown in Eq. (2.163). We can write

$$\oint_{-1/3-i\infty}^{-1/3+i\infty} dz\Gamma[z] \approx \oint_{-1/3-9i}^{-1/3+9i} dz\Gamma[z] = (-i)\,3.97173. \qquad (2.164)$$

This integral is assumed to be part of a closed path, closing to the left or to the right, resulting in different sums of residues representing the integral.

Closing to the left (anticlockwise),

$$\oint_{-1/2-i}^{-1/2+i} dz\Gamma[z] = 2\pi i \sum_{n=1}^{\infty} \frac{(-1)^n}{n!} = 2\pi i\,\frac{1-e}{e} = -3.97173\,i \qquad (2.165)$$

while closing contour on the right,

$$(-1)\,2\pi i \sum_{n=0}^{0} \frac{(-1)^n}{n!} = -2\pi i \neq -3.97173\,i. \qquad (2.166)$$

While both the approximation in Eq. (2.164) and the first sum in Eq. (2.165) represent the integral, the second one, Eq. (2.166), does not. The reason is simple: from the section on the gamma function, we know that the integrand in Eq. (2.164) becomes small when closing the path to the left. But when we close the integration path to the right, there is a non-negligible contribution from the positive real axis at infinity, and the sum of residues fails to represent the integral in Eq. (2.164).

QFT Example: FI as the MB Integral and the Series

Let us consider the diagram with two internal propagators as drawn below.

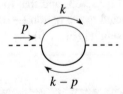

The corresponding integral will be considered in Sect. 3.4; its MB form is (see Eq. (3.50))

$$G(1)_{\text{SE2l2m}} = \int_{-i\infty-1/8}^{+i\infty-1/8} dz_1\,(-s)^{-\epsilon} \left(-\frac{m^2}{s}\right)^{z_1} \frac{\Gamma[1-\epsilon-z_1]^2\Gamma(-z_1)\Gamma(\epsilon+z_1)}{\Gamma(2-2\epsilon-z_1)}.$$

$$(2.167)$$

Its singular structure due to ϵ poles of gamma functions involved will be shown in the chapter on the resolution of singularities in Fig. 4.1. Expanding to order ϵ^{-1} and summing residues we get

$$\sum_{n=0}^{\infty} \frac{s^n}{n!} \frac{\Gamma^3(n+1)}{\Gamma(2+2n)} = \frac{4 \arcsin[\sqrt{s}/2]}{\sqrt{4-s}\sqrt{s}}. \tag{2.168}$$

The relation in Eq. (2.168) is derived in the file `MB_SE212m_Springer.nb` in the auxiliary material in [5].

2.7 Gamma and Hypergeometric Functions, Hypergeometric Integrals

Due to their sophistication, the hypergeometric functions are more often than we think present in solutions to the exact physical problems. For instance, in classical mechanics an exact solution for the pendulum period is

$$T = 2\pi \sqrt{\frac{l}{g}} \times {}_2F_1\left[\frac{1}{2}, \frac{1}{2}; 1; \sin^2\theta\right], \tag{2.169}$$

and in the limit of small θ, the classical formula emerges $T = 2\pi\sqrt{\frac{l}{g}}$. For more examples, see [32].

To prove Barnes lemmas which give useful relations among gamma functions applied in the next chapters, relations between the MB integrals and hypergeometric functions are worth studying. Hypergeometric series or (ordinary) Gauss functions come from the extension of the ordinary geometric series

$$1 + x + x^2 + x^3 + \ldots \tag{2.170}$$

to the series of the form

$$1 + \frac{ab}{c}\frac{z}{1!} + \frac{a(\mathbf{a}+\mathbf{1})b(\mathbf{b}+\mathbf{1})}{c(\mathbf{c}+\mathbf{1})}\frac{z^2}{2!} + \ldots \tag{2.171}$$

where in each next term of the sum, we increase by one the number of terms in which parameters a, b, c are involved and simultaneously each added parameter is increased by one comparing to the previously added factor (boldface in Eq. (2.171)). Similarly with each next term in the sum, the power of variable z is increased by one, normalized by the factorial increasing accordingly by one (denoted in boldface in Eq. (2.171)).

Usually the series in Eq. (2.171) is written in the following way:

$$_2F_1[a, b; c; z].\tag{2.172}$$

This is the Gauss function. The two semicolons in Eq. (2.172) separate parameters in numerators, denominators, and variable (here z), respectively. Most of the elementary functions which appear in mathematical physics can be expressed in terms of the Gauss function, for instance,

$$e^z = \sum_{n=0}^{\infty} \frac{z^n}{n!} = \lim_{b \to 0}\{_2F_1[1, b; ; 1; z/b]\},\tag{2.173}$$

$$\log(1 + z) = \sum_{n=0}^{\infty} \frac{(-z)^n}{1 + n} = {}_2F_1[1, 1; 2; -z].\tag{2.174}$$

Other functions worth mentioning which can be derived from the general function in Eq. (2.172) are Bessel and Airy functions or the Hermite polynomials.

The definition of the Gauss function in Eq. (2.172) can be extended to the generalized Gauss (hypergeometric) function written in a compact way as follows:

$$1 + \frac{a_1 a_2 \ldots a_A}{b_1 b_2 \ldots b_B} \frac{z}{1!} + \frac{a_1(a_1 + 1)a_2(a_2 + 1) \ldots a_A(a_A + 1)}{b_1(b_1 + 1)b_2(b_2 + 1) \ldots b_B(b_B + 1)} \frac{z^2}{2!} + \cdots$$

$$\equiv \sum_{n=0}^{\infty} \frac{(a_1)_n (a_2)_n \ldots (a_A)_n}{(b_1)_n (b_2)_n \ldots (b_B)_n} \frac{z^n}{n!}.\tag{2.175}$$

Here $(a)_n \equiv a(a+1)\cdots(a+n-1) = \frac{\Gamma[a+n]}{\Gamma[a]}$ is Pochhammer's symbol. If the sum of this series with $A + B$ parameters exists, the function can be denoted as [33, 38]

$$_A F_B[a_1, a_2, \ldots, a_A; b_1, b_2, \ldots, b_B; z].\tag{2.176}$$

Some shorter notation is also used:

$$\sum_{n=0}^{\infty} \frac{((a)_A)_n}{((b)_B)_n} \frac{z^n}{n!} = {}_A F_B[(a); (b); z].\tag{2.177}$$

Truly, it can hardly be shorter. Also, the products of several gamma functions as they appear in the MB integrals can be written as

$$\frac{\Gamma(a_1)\Gamma(a_2)\ldots\Gamma(a_A)}{\Gamma(b_1)\Gamma(b_1)\ldots\Gamma(b_1)} = \Gamma\begin{bmatrix} a_1, a_2, \ldots, a_A \\ b_1, b_2, \ldots, b_B \end{bmatrix} \equiv \Gamma[(a); (b)].\tag{2.178}$$

Using the above notation, the so-called first Barnes lemma (1BL) can be written symbolically in the following way:

$$\frac{1}{2\pi i} \int_{-i\infty}^{+i\infty} \Gamma\left(a+z, b+z, c-z, d-z\right) dz = \Gamma\left[\begin{matrix} a+c, a+d, b+c, b+d \\ a+b+c+d \end{matrix}\right]$$

(2.179)

with the condition $\Re(a+b+c+d) < 1$. We will write this equation in a manifest form in Eq. (3.46) when considering construction of MB representations for FI.

MB Integrals and Hypergeometric Functions

Using the relation in Eq. (2.161), we can cast the MB contour formula in Eq. (1.43) in the following form for $|z| < 1$ (by closing the contour to the right):

$$\int_C \frac{dz}{2\pi i} \frac{\Gamma(a+z)\Gamma(b+z)\Gamma(-z)}{\Gamma(c+z)}(-s)^z = \sum_{n=0}^{\infty} \frac{\Gamma(a+n)\Gamma(b+n)}{\Gamma(c+n)} \frac{s^n}{n!}$$

$$= \frac{\Gamma(a)\Gamma(b)}{\Gamma(c)} \, {}_2F_1(a, b; c; s).$$

(2.180)

The proof for Eq. (2.180) is given in [34]; see Problem 2.10. Putting $b = c$, we can show that from ${}_2F_1(a, b; b; s)$ the MB master formula in Eq. (1.44) follows.

The continuation of the hypergeometric series for $|s| > 1$ is performed using the intermediate formula

$$F(s) = \sum_{n=0}^{\infty} \frac{\Gamma(a+n)\Gamma(1-c+a+n)\sin[(c-a-n)\pi]}{\Gamma(1+n)\Gamma(1-a+b+n)\cos(n\pi)\sin[(b-a-n)\pi]}(-s)^{-a-n}$$

$$+ \sum_{n=0}^{\infty} \frac{\Gamma(b+n)\Gamma(1-c+b+n)\sin[(c-b-n)\pi]}{\Gamma(1+n)\Gamma(1-a+b+n)\cos(n\pi)\sin[(a-b-n)\pi]}(-s)^{-b-n}$$

and yields

$$\frac{\Gamma(a)\Gamma(b)}{\Gamma(c)} \, {}_2F_1(a, b; c; s) = \frac{\Gamma(a)\Gamma(b-a)}{\Gamma(c-a)}(-s)^{-a} \, {}_2F_1\left(a, 1-c+a; 1-b+a; \frac{1}{s}\right)$$

$$+ \frac{\Gamma(b)\Gamma(a-b)}{\Gamma(c-b)}(-s)^{-b} \, {}_2F_1\left(b, 1-c+b; 1-a+b; \frac{1}{s}\right)$$

(2.181)

Fig. 2.9 Integration contour and singularities in the complex z plane connected with Eq. (2.179)

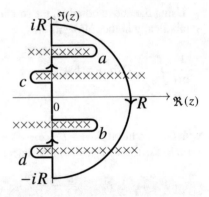

Proof of 1BL

Armed with complex analysis, we can now understand how the above condition $\mathfrak{R}(a+b+c+d) < 1$ in Eq. (2.179) emerges. In Fig. 2.9 poles of gamma functions of the integrand in Eq. (2.179) are given.

Due to Cauchy's residue theorem discussed in Sect. 2.6, we are interested in the situation of vanishing integral round the semi-circle when approaching infinity, $R \to \infty$. Such integral can be parametrized with $z = Re^{it}$ as

$$I_{\text{semi-circle}} = \frac{-1}{2\pi i} \int_{-\pi/2}^{\pi/2} \Gamma(a + Re^{it}, b + Re^{it}, c - Re^{it}, d - Re^{it}) Re^{it} i\, dt.$$

$$(2.182)$$

Then, taking asymptotic behavior of the gamma function (2.108), we have $|I_{\text{semi-circle}}| < \text{const} \times R^{-a-b+1-c+1-d}$, which vanishes with $R \to \infty$ when $\mathfrak{R}[a+b+c+d] < 1$.

To show that the integral over the remaining part of the closed contour in Fig. 2.9 which is parallel to the imaginary axis from $-iR$ to $+iR$ equals to the right-hand side of Eq. (2.179), we need to sum the residues at $z = c + n$ and $z = d + n$, $n = 0, 1, 2, \ldots \in \mathbb{Z}$

$$I_A = \sum_{n=0}^{[R]} \Gamma(a+c+n, b+c+n, d-c-n) \frac{(-1)^{n-1}}{n!}$$

$$+ \sum_{n=0}^{[R]} \Gamma(a+d+n, b+d+n, c-d-n) \frac{(-1)^{n-1}}{n!}. \qquad (2.183)$$

In the limit of $[R] \to \infty$, we get the two sums which involve the hypergeometric function $_2F_1$:

$$I_A = \Gamma(a+c, b+c, d-c) \, _2F_1[a+c, b+c; 1+c-d; 1]$$
$$+ \Gamma(a+d, b+d, c-d) \, _2F_1[a+d, b+d; 1+d-c; 1]. \qquad (2.184)$$

If we use Gauss's theorem [33],

$$_2F_1[a, b; c; 1] = \begin{bmatrix} c, c-a-b \\ c-a, c-b \end{bmatrix}, \qquad (2.185)$$

we can merge the two terms into the right-hand side of Eq. (2.179), q.e.d.

The so-called second Barnes lemma reads

$$\frac{1}{2\pi i} \int_{-i\infty}^{+i\infty} \Gamma \begin{bmatrix} a+z, b+z, b+c, c+z, 1-d-z, -z \\ e+z \end{bmatrix} dz$$
$$= \Gamma \begin{bmatrix} a, b, c, 1-d+a, 1-d+b, 1-d+c \\ e-a, e-b, e-c \end{bmatrix}, \qquad (2.186)$$

with Saalschutzian condition $1 + a + b + c = d + e$.

We will write this equation in a manifest form when considering physics applications in Eq. (3.48).

Barnes Lemmas Eqs. (2.179) and (2.186) can be generalized with the help of Gauss summation theorem for the hypergeometric function $_2F_1$ into a general theorem which includes all the Barnes-type integrals [33]

$$\frac{1}{2\pi i} \int_{-i\infty}^{+i\infty} \Gamma \begin{bmatrix} (a)+z, (b)-z \\ (c)+z, (d)-z \end{bmatrix} dz \qquad (2.187)$$

$$= \sum_{\mu=1}^{A} \Gamma \begin{bmatrix} (a)-a_\mu, (b)+a_\mu \\ (c)-a_\mu, (d)+a_\mu \end{bmatrix} {}_{B+C}F_{A+D-1} \begin{bmatrix} (b)+a_\mu, 1+a_\mu-(c); \\ 1+a_\mu-(a)', (d)+a_\mu; \end{bmatrix} (-1)^{A+C} \end{bmatrix}$$

$$= \sum_{\nu=1}^{B} \Gamma \begin{bmatrix} (a)+b_\nu, (b)-b_\nu \\ (c)+b_\nu, (d)-b_\nu \end{bmatrix} {}_{A+D}F_{B+C-1} \begin{bmatrix} (a)+b_\nu, 1+b_\nu-(d); \\ (c)+b_\nu-(a), 1+b_\nu-(b)'; \end{bmatrix} (-1)^{B+D} \end{bmatrix}.$$

where

$$\mathfrak{R}(c + d - a - b) > 0, \quad A - C = B - D \geq 0. \tag{2.188}$$

Here prime in the notation $(\ldots)'$ indicates that denominators in Eq. (2.187) give factorial elements in each term of the series.

In principle, using Eq. (2.187) one can perform the integration over any free z variable. However, utility of such integration is limited due to the complexity of an emerging representation for generalized hypergeometric functions $_AF_B(; \pm 1)$. So far in practice, only the first and second Barnes lemmas are in use; see next chapters.

More efficient way to decrease dimensionality of MB integrals is to perform a transformation of variables in such a way that Barnes lemmas can be applied. Such an approach has been implemented in the package barnesroutines.m [39] for MB integrals with fixed contours which appear after expansion of MB representation in ϵ. It performs a search for suitable transformations with restrictions to absolute values of matrix elements for this transformation. See Sect. 3.6 for details.

To finish this section, let us mention some multivariate generalizations of hypergeometric functions. This is a vast topic and here we are only scratching the surface [40]. Let us begin with generalizations to two variables. Perhaps the most well-known of these are the Appell functions F_1, F_2, F_3, and F_4. These can be defined by the following double sums:

$$F_1(\alpha, \beta, \beta', \gamma, x, y) = \sum_{m=0}^{\infty} \sum_{n=0}^{\infty} \frac{(\alpha)_{m+n}(\beta)_m(\beta')_n}{(\gamma)_{m+n}m!n!} x^m y^n, \qquad |x| < 1, \ |y| < 1, \tag{2.189}$$

$$F_2(\alpha, \beta, \beta', \gamma, \gamma', x, y) = \sum_{m=0}^{\infty} \sum_{n=0}^{\infty} \frac{(\alpha)_{m+n}(\beta)_m(\beta')_n}{(\gamma)_m(\gamma')_n m!n!} x^m y^n, \qquad |x| + |y| < 1, \tag{2.190}$$

$$F_3(\alpha, \alpha', \beta, \beta', \gamma, x, y) = \sum_{m=0}^{\infty} \sum_{n=0}^{\infty} \frac{(\alpha)_m(\alpha')_n(\beta)_m(\beta')_n}{(\gamma)_{m+n}m!n!} x^m y^n, \qquad |x| < 1, \ |y| < 1, \tag{2.191}$$

$$F_4(\alpha, \beta, \gamma, \gamma', x, y) = \sum_{m=0}^{\infty} \sum_{n=0}^{\infty} \frac{(\alpha)_{m+n}(\beta)_{m+n}}{(\gamma)_m(\gamma')_n m!n!} x^m y^n, \qquad |\sqrt{x}| + |\sqrt{y}| < 1, \tag{2.192}$$

where we have indicated the region of convergence for each sum. These double sums are obviously generalizations of the hypergeometric series for the Gauss function but extended to two variables. Other generalizations can be defined along the same lines, and in fact, the Appell functions are just the first 4 in the set of 34 so-called Horn functions. More elaborate generalizations, still of two variables, such as the Kampé de Fériet function have also been considered. Indeed, all four Appell functions are just special cases of the more general Kampé de Fériet function.

Returning to the Appell functions, we mention that they also admit representations as two-dimensional real Euler-type integrals, as well as MB integrals. Here we limit ourselves to presenting these for the first Appell function F_1 only, which we will encounter later in Sect. 5.1.6. We have

$$
F_1(\alpha, \beta, \beta', \gamma, x, y) = \frac{\Gamma(\gamma)}{\Gamma(\beta)\Gamma(\beta')\Gamma(\gamma - \beta - \beta')}
$$
$$
\times \int_0^1 du \int_0^{1-u} dv \, u^{\beta-1} v^{\beta'-1} (1 - u - v)^{\gamma-\beta-\beta'-1} (1 - ux - vy)^{-\alpha},
$$
$$
(2.193)
$$

but the F_1 function (though not the other Appell functions) actually admits a simple one-dimensional real integral representation as well,

$$
F_1(\alpha, \beta, \beta', \gamma, x, y) = \frac{\Gamma(\gamma)}{\Gamma(\alpha)\Gamma(\gamma - \alpha)} \int_0^1 du \, u^{\alpha-1} (1 - u)^{\gamma-\alpha-1} (1 - ux)^{-\beta} (1 - uy)^{-\beta'}.
$$
$$
(2.194)
$$

The F_1 function can also be written as a two-dimensional MB integral,

$$
F_1(\alpha, \beta, \beta', \gamma, x, y) = \frac{\Gamma(\gamma)}{\Gamma(\alpha)\Gamma(\beta)\Gamma(\beta')} \int_{-i\infty}^{+i\infty} \frac{dz_1}{2\pi i} \int_{-i\infty}^{+i\infty} \frac{dz_2}{2\pi i}
$$
$$
\times \frac{\Gamma(\alpha + z_1 + z_2)\Gamma(\beta + z_1)\Gamma(\beta' + z_2)}{\Gamma(\gamma + z_1 + z_2)} \Gamma(-z_1)\Gamma(-z_2)(-x)^{z_1}(-y)^{z_2}.
$$
$$
(2.195)
$$

Turning to multivariable generalizations, we mention here the H-function of several variables. This function has been discussed in various forms by several authors in the literature and here we adopt the definition of [41].[14] In the most general case, the H-function of N variables is defined as follows:

$$
H[x, (\alpha, A), (\beta, B); L_s] = (2\pi i)^N \int_{L_s} \Theta(s) x^s ds,
$$
$$
(2.196)
$$

where

$$
\Theta(s) = \frac{\prod_{j=1}^m \Gamma\left(\alpha_j + \sum_{k=1}^N a_{j,k} s_k\right)}{\prod_{j=1}^n \Gamma\left(\beta_j + \sum_{k=1}^N b_{j,k} s_k\right)}.
$$
$$
(2.197)
$$

---

[14] Our definition given in Eq. (2.196) is different from the H-function considered by [41] only in the replacement of x^{-s} by x^s. We have made this replacement in order to unify this definition with previous expressions for special functions represented as MB integrals.

Here $s = (s_1, \ldots, s_N)$, $x = (x_1, \ldots, s_N)$, $\alpha = (\alpha_1, \ldots, \alpha_m)$ and $\beta = (\beta_1, \ldots, \beta_n)$ denote vectors of complex numbers, while

$$A = (a_{j,k})_{m \times N} \quad \text{and} \quad B = (b_{j,k})_{n \times N} \tag{2.198}$$

are matrices of real numbers. Finally

$$x^s = \prod_{k=1}^{N} (x_k)^{s_k}; \qquad ds = \prod_{k=1}^{N} ds_k; \qquad L_s = L_{s_1} \times \cdots \times L_{s_N}, \tag{2.199}$$

where L_{s_k} is an infinite contour in the complex s_k-plane running from $-i\infty$ to $+i\infty$ such that $\Theta(s)$ has no singularities for $s \in L_s$.

The H-function defined in Eq. (2.196) above generalizes nearly all known special functions of N variables, e.g., Lauricella functions $F_A^{(N)}$, $F_B^{(N)}$, $F_C^{(N)}$, and $F_D^{(N)}$; the G-function of N variables; the special H-function of N variables; etc. For the specific cases of $N = 1$ and 2, it essentially reduces to the known Fox's H-function of one variable and the H-function of two variables defined by various authors scattered in the literature.

Problems

Problem 2.1 As we have discussed, the logarithm of the product of two complex numbers w and z can be written as the sum of the logarithms plus an additional phase.

$$\ln(wz) = \ln w + \ln z + \eta(w, z). \tag{2.200}$$

Show that for *generic* complex numbers w and z (i.e., assuming that both the real and imaginary parts of w and z are non-zero), this phase can be written in the following form:

$$\eta(w, z) = 2\pi i \Big[\theta(-Im\, w)\theta(-Im\, z)\theta(+Im\, wz) - \theta(+Im\, w)\theta(+Im\, z)\theta(-Im\, wz)\Big].$$

<u>Hint</u>: The argument of the left-hand side of Eq. (2.200), $\ln(wz)$, lies in the interval $(-\pi, \pi]$ by definition. Consider the circumstances under which the argument of the right-hand side falls outside this interval and how it can be mapped to this interval by adding an integer multiple of $2\pi i$.

Is the expression for $\eta(w, z)$ valid also for nongeneric complex numbers? If not, why not?

<u>Hint</u>: Consider the case when w or z (or both) is a negative real number and the case when both w and z are purely imaginary.

Problem 2.2 Consider the general dilogarithmic integral discussed in Sect. 2.3.2:

$$\int \frac{dx}{mx + b} \ln(nx + a). \tag{2.201}$$

We gave two solutions for this integral, one in Eq. (2.56) and one as the direct output of integration with `Mathematica`. Show that the real parts of the two solutions always coincide up to a constant that is independent of x, while the imaginary parts may differ by a function of the form $C \cdot \ln(mx + b)$.

<u>Hint</u>: Use Eq. (2.66) to relate the dilogarithms in the two expressions; then examine their difference. Be careful to keep in mind the proper functional identities for logarithms of complex variables such as Eqs. (2.31) and (2.32).

Problem 2.3 Derive the shuffle product of $G(a; z)G(b; z)$ starting from the representations of these functions as integrals.

<u>Hint</u>: Consider the integral representation

$$G(a; z)G(b; z) = \int_0^z \frac{dt_1}{t_1 - a} \int_0^z \frac{dt_2}{t_2 - b} \tag{2.202}$$

and split the two-dimensional domain of integration (the square with corners $(0, 0)$, $(0, z)$, $(z, 0)$, and (z, z)) along the diagonal. See, e.g., [24] for details.

Problem 2.4 Prove the asymptotic behavior of the gamma function for integer and complex variables; Eqs. (2.107) and (2.108). Try to understand the origin of different behavior of the square root factors for $n \to \infty$ and $z \to \infty$ in two formulas.

Hint: Relation in Eq. (2.107) is the so-called Stirling formula; the proof can be found among others in many statistics textbooks. More terms in the expansion are given in Eq. (6.3).

Problem 2.5 Consider the behavior of the gamma function along the imaginary axis, i.e., $\Gamma(iy)$, with $y \in \mathbb{R}$. Show that for $|y| \to \infty$, both the real and imaginary parts of $\Gamma(iy)$ fall off as $|y|^{-1/2}e^{-i\pi|y|/2}$.

Hint: Use $z^z = e^{z \ln z}$ and the properties of the exponential function to manipulate the asymptotic formula in Eq. (2.108) to the desired form.

Problem 2.6 Prove Eq. (2.123).

Hint: Use the recursive property of the gamma function and Eq. (2.121).

Problem 2.7 Argue that polygamma functions have poles of order $m + 1$ at all non-positive integers.

Hint: Consider the behavior of the gamma function around non-positive integers as given in Eq. (2.123). Then use the definition of the polygamma functions (Eqs. 2.135 and 2.136) to deduce the behavior of these functions around non-positive integers.

Problem 2.8 This is a problem bridging mathematical and physics intuition. The integral in Eq. (2.149) in Sect. 2.6 depends on the direction of curve C. In the mirror world, the direction of the curve is opposite (left- and right-handedness are reversed). Changing C_+ to C_- in Eq. (2.149) changes the sign of the result. Does this mean that Cauchy's integral can be determined only up to the sign and it is not well-defined if parity is conserved? How is the problem fixed in mathematics?

Hint: Consider the Jordan curve, notions of the interior and exterior; see [42].

Problem 2.9 Prove Eq. (2.162). Assume that $G(z)$ is non-singular at $z = -n$, $n \in \mathbb{N}$.

Hint: By definition, the residue is the coefficient of $(z - (-n))^{-1}$ in the Laurent expansion of the function. One way of computing this expansion is to use Eq. (2.122) to express $\Gamma(z)$ in terms of $\Gamma(1-z)$ and $\frac{\pi}{\sin \pi z}$. Then, $G(z)$ and $\Gamma(1-z)$ are regular at $z = -n$ and can be simply Taylor-expanded (note that the derivative of the gamma function, $\psi^{(0)}$, will appear). On the other hand, $\left(\frac{\pi}{\sin \pi z}\right)^2$ is obviously singular as $\frac{1}{(z+n)^2}$ around $z = -n$, but the Laurent expansion is easy to construct, e.g., starting from the series representation of the sine. Alternatively, you can consider the Taylor expansion of the function $\left[\frac{(z+n)\pi}{\sin \pi z}\right]^2$. Then, it is straightforward to obtain the coefficient of $(z + n)^{-1}$.

Problem 2.10 Show that from $_2F_1(a, b; c; z)$ the basic MB formula in Eq. (1.44) follows:

$$\frac{1}{(A + B)^\lambda} = \frac{1}{\Gamma(\lambda)} \frac{1}{2\pi i} \int\limits_{c-i\infty}^{c+i\infty} dz \Gamma(\lambda + z)\Gamma(-z) A^z B^{-\lambda-z}.$$

Hint: Use a series in Eq. (2.180) with $b = c$ and apply relations given just below Eq. (1.44).

References

1. T. Needham, A. Needham, *Visual Complex Analysis* (Oxford University Press, 1997)
2. E. Wegert, Phase plots of complex functions: A journey in illustration. arXiv:1007.2295
3. J. Vollinga, S. Weinzierl, Numerical evaluation of multiple polylogarithms. Comput. Phys. Commun. **167**, 177 (2005). arXiv:hep-ph/0410259, https://doi.org/10.1016/j.cpc.2004.12.009
4. E. Wegert, *Visual Complex Functions* (Springer Basel AG, 2012)
5. https://github.com/idubovyk/mbspringer, http://jgluza.us.edu.pl/mbspringer
6. I.M. Roussos, *Improper Riemann Integrals* (CRC, Taylor & Francis Group, 2014)
7. K. Rudolph, G.M. Poore, Minted - highlighted source code for *LaTeX*. https://github.com/gpoore/minted
8. J.A. Scott, In praise of the Catalan constant. Math. Gazette **86**(505), 102–103 (2002). https://doi.org/10.2307/3621589
9. G. 't Hooft, M. Veltman, Scalar one loop integrals. Nucl. Phys. **B153**, 365–401 (1979)
10. S. Actis, M. Czakon, J. Gluza, T. Riemann, Virtual hadronic and heavy-Fermion $O(\alpha^2)$ corrections to Bhabha scattering. Phys. Rev. **D78**, 085019 (2008). arXiv:0807.4691, https://doi.org/10.1103/PhysRevD.78.085019
11. J. Havil, *Gamma: Exploring Euler's Constant* (Princeton University Press, 2019)
12. G. Passarino, M. Veltman, One loop corrections for e^+e^- annihilation into $\mu^+\mu^-$ in the Weinberg model. Nucl. Phys. **B160**, 151 (1979). https://doi.org/10.1016/0550-3213(79)90234-7
13. A. Devoto, D.W. Duke, Table of integrals and formulae for Feynman diagram calculations. Riv. Nuovo Cim. **7N6**, 1–39 (1984). https://doi.org/10.1007/BF02724330
14. D. Maitre, HPL, a mathematica implementation of the harmonic polylogarithms. Comput. Phys. Commun. **174**, 222–240 (2006). arXiv:hep-ph/0507152, https://doi.org/10.1016/j.cpc.2005.10.008

15. L. Naterop, A. Signer, Y. Ulrich, handyG —Rapid numerical evaluation of generalised polylogarithms in Fortran. Comput. Phys. Commun. **253**, 107165 (2020). arXiv:1909.01656, https://doi.org/10.1016/j.cpc.2020.107165

16. T. Gehrmann, E. Remiddi, Numerical evaluation of harmonic polylogarithms. Comput. Phys. Commun. **141**, 296–312 (2001). arXiv:hep-ph/0107173, https://doi.org/10.1016/S0010-4655(01)00411-8

17. S. Buehler, C. Duhr, CHAPLIN - complex harmonic polylogarithms in Fortran. Comput. Phys. Commun. **185**, 2703–2713 (2014). arXiv:1106.5739, https://doi.org/10.1016/j.cpc.2014.05.022

18. A.B. Goncharov, Multiple polylogarithms, cyclotomy and modular complexes. Math. Res. Lett. **5**, 497–516 (1998). arXiv:1105.2076, https://doi.org/10.4310/MRL.1998.v5.n4.a7

19. A.B. Goncharov, Multiple polylogarithms and mixed Tate motives. arXiv:math/0103059

20. J.A. Lappo-Danilevskij, Mémoires sur la théorie des systémes des équations différentielles linéaires. Vol. II, Travaux Inst. Physico-Math. Stekloff **7**, 5–210 (1935)

21. K.T. Chen, Iterated path integrals. Bull. Am. Math. Soc. **83**(5), 831–879 (1977)

22. C. Duhr, F. Dulat, PolyLogTools — polylogs for the masses. JHEP **08**, 135 (2019). arXiv:1904.07279, https://doi.org/10.1007/JHEP08(2019)135

23. A.B. Goncharov, Galois symmetries of fundamental groupoids and noncommutative geometry. Duke Math. J. **128**, 209 (2005). arXiv:math/0208144, https://doi.org/10.1215/S0012-7094-04-12822-2

24. C. Duhr, Mathematical aspects of scattering amplitudes, in *Theoretical Advanced Study Institute in Elementary Particle Physics: Journeys Through the Precision Frontier: Amplitudes for Colliders* pp. 419–476 (2015). arXiv:1411.7538, https://doi.org/10.1142/9789814678766_0010

25. S. Weinzierl, Feynman Integrals. arXiv:2201.03593

26. E. Remiddi, J. Vermaseren, Harmonic polylogarithms. Int. J. Mod. Phys. **A15**, 725–754 (2000). arXiv:hep-ph/9905237, https://doi.org/10.1142/S0217751X00000367

27. E. Remiddi, Differential equations for Feynman graph amplitudes. Nuovo Cim. **A110**, 1435–1452 (1997). arXiv:hep-th/9711188

28. T.-F. Feng, C.-H. Chang, J.-B. Chen, H.-B. Zhang, GKZ-hypergeometric systems for Feynman integrals. Nucl. Phys. B **953**, 114952 (2020). arXiv:1912.01726, https://doi.org/10.1016/j.nuclphysb.2020.114952

29. L. de la Cruz, Feynman integrals as A-hypergeometric functions. JHEP **12**, 123 (2019). arXiv:1907.00507, https://doi.org/10.1007/JHEP12(2019)123

30. I. Gel'fand, A. Zelevinskii, M. Kapranov, GKZ-hypergeometric systems for Feynman integrals. Hypergeometric Funct. Toral Manif. **23**, 94 (1989). https://doi.org/10.1007/BF01078777doi:10.1007/BF01078777

31. J.L. Bourjaily, et al., Functions beyond multiple polylogarithms for precision collider physics (2022). arXiv:2203.07088

32. E. de Oliveira, *Solved Exercises in Fractional Calculus*. Studies in Systems, Decision and Control (Springer International Publishing, 2019). https://doi.org/10.1007/978-3-030-20524-9

33. L.J. Slater, *Generalized Hypergeometric Functions* (Cambridge University Press, 1966)

34. E. Whittaker, G. Watson, *A Course of Modern Analysis* (Cambridge University Press, 1965)

35. R. Remmert, *Classical Topics in Complex Function Theory*, vol. 172 (Springer, 1998). https://doi.org/10.1007/978-1-4757-2956-6

36. J. Havil, *John Napier: Life, Logarithms, and Legacy* (Princeton University Press, 2014)

37. E. Elizalde, *Ten Physical Applications of Spectral Zeta Functions* (Springer, Berlin, Heidelberg, 2012). https://doi.org/10.1007/978-3-642-29405-1

38. E.W. Barnes, A new development of the theory of the hypergeometric functions. Proc. Lond. Math. Soc. **s2-6**(1), 141–177 (1908). https://doi.org/10.1112/plms/s2-6.1.141

39. D. Kosower, Mathematica program barnesroutines.m version 1.1.1 (Jul 2009), available at the MB Tools webpage. http://projects.hepforge.org/mbtools/

40. S. Abreu, R. Britto, C. Duhr, E. Gardi, J. Matthew, From positive geometries to a coaction on hypergeometric functions. JHEP **02**, 122 (2020). arXiv:1910.08358, https://doi.org/10.1007/JHEP02(2020)122

41. N. Hai, H. Srivastava, The convergence problem of certain multiple mellin-barnes contour integrals representing h-functions in several variables. Comput. Math. Appl. **29**(6), 17–25 (1995)

42. E. Zeidler, *Quantum Field Theory II: Quantum Electrodynamics: A Bridge between Mathematicians and Physicists* (Springer Science & Business Media, 2009). https://doi.org/10.1007/978-3-540-85377-0

Mellin-Barnes Representations for Feynman Integrals

3

Abstract

Starting from Feynman integrals defined in momentum space, Feynman and Schwinger parametrizations are introduced and their equivalence is shown. Further, the properties of Feynman graphs and the corresponding F and U Symanzik polynomials are discussed. The MB master formula and a suite of useful programs for constructing and evaluating MB integrals are introduced. Peculiarities of the construction of an MB representation due to the planarity of a given diagram are examined, and the notion of the loop-by-loop (LA) and global (GA) approaches are introduced. The Cheng-Wu theorem is proved and applied to the derivation of MB representations for non-planar diagrams. The first and second Barnes lemmas are used in order to maximally simplify the structure and a number of complex variables in MB representations. Then, specific one-loop to three-loop examples of MB integral constructions are shown. An alternative "method of brackets" for constructing MB representations is also introduced. Finally, the application of the MB method to the computation of real phase space integrals is discussed.

3.1 Representations of Feynman Integrals

As discussed in Sect. 1.2, our starting point is the integral in Eq. (1.22). Omitting the prefactors $e^{\epsilon \gamma L}$ and $(2\pi \mu)^{(4-d)L}$, let us focus on scalar integrals with $T(k) = 1$,

$$G_L[1] = \frac{1}{(i\pi^{d/2})^L} \int \frac{d^d k_1 \ldots d^d k_L}{(q_1^2 - m_1^2)^{n_1} \ldots (q_i^2 - m_i^2)^{n_j} \ldots (q_N^2 - m_N^2)^{n_N}}. \tag{3.1}$$

© The Author(s), under exclusive license to Springer Nature Switzerland AG 2022
I. Dubovyk et al., *Mellin-Barnes Integrals*, Lecture Notes in Physics 1008,
https://doi.org/10.1007/978-3-031-14272-7_3

A single Feynman propagator D_i is of the form

$$D_i = q_i^2 - m_i^2 + i\delta = \left[\sum_{l=1}^{L} c_i^l k_l + \sum_{e=1}^{E} d_i^e p_e\right]^2 - m_i^2 + i\delta, \tag{3.2}$$

where k_l and p_e are internal and external momenta, respectively. The c_i^l, $d_i^e \in$ $[-1, 1]$ are integer coefficients and depend on a particular topology. Here we also added a small imaginary part $i\delta$ to the propagator as was discussed in Sect. 1.2.

To proceed further we introduce a generalized Feynman parameter representation

$$\frac{1}{D_1^{n_1} D_2^{n_2} \dots D_N^{n_N}} = \frac{\Gamma(n_1 + \dots + n_N)}{\Gamma(n_1) \dots \Gamma(n_N)} \tag{3.3}$$

$$\times \int_0^1 dx_1 \dots \int_0^1 dx_N \frac{x_1^{n_1-1} \dots x_N^{n_N-1} \delta(1 - x_1 - \dots - x_m)}{(x_1 D_1 + \dots + x_N D_N)^{N_\nu}}$$

with $N_\nu = n_1 + \dots + n_N$.

An alternative is the so-called Schwinger or α-parameter representation, which comes from the generalized identity (Problem 3.1)

$$\frac{1}{D_1^{n_1} D_2^{n_2} \dots D_N^{n_N}} = \frac{i^{-N_\nu}}{\Gamma(n_1) \dots \Gamma(n_N)} \tag{3.4}$$

$$\times \int_0^\infty d\alpha_1 \dots \int_0^\infty d\alpha_N \alpha_1^{n_1-1} \dots \alpha_N^{n_N-1} e^{i[\alpha_1 D_1 + \dots + \alpha_N D_N]}.$$

Here it is useful to show the connection between these two representations; we will use it later in Sect. 3.8 discussing the Cheng-Wu theorem. Using the identity

$$1 = \int_0^\infty \frac{d\lambda}{\lambda} \delta\left(1 - \frac{1}{\lambda} \sum_{i=1}^{N} \alpha_i\right) \tag{3.5}$$

and changing variables from α_i to $\alpha_i = \lambda x_i$, one can find

$$\frac{1}{D_1^{n_1} D_2^{n_2} \dots D_N^{n_N}} = \frac{i^{-N_\nu}}{\Gamma(n_1) \dots \Gamma(n_N)} \int_0^\infty dx_1 \dots \int_0^\infty dx_N x_1^{n_1-1} \dots x_N^{n_N-1} \tag{3.6}$$

$$\times \int_0^\infty d\lambda \lambda^{N_\nu-1} \delta\left(1 - \sum_{i=1}^{N} x_i\right) e^{i\lambda \sum_{i=1}^{N} x_i D_i}.$$

Integrating over λ we come to the Feynman parameter representation of Eq. (3.4). Note that all x_i are positive while the sum of x_i must be unity. Therefore the integration region can be limited:

$$0 < x_i < 1 \quad \Leftrightarrow \quad 0 < x_i < \infty.$$

Let us now consider the momentum-dependent function

$$m^2(\mathbf{x}) = x_1 D_1 + \ldots + x_i D_i + \ldots + x_N D_N = k_i M_{ij} k_j - 2Q_j k_j + J \qquad (3.7)$$

where M is an $(L \times L)$-matrix, $Q = Q(x_i, p_e)$ an L-vector, and $J = J(x_i x_j, m_i^2, p_{e_i} \cdot p_{e_j})$.

Before integration over loop momenta, one has to perform several preparatory steps:

- Shift momenta in order to remove linear terms in k:

$$k \to k + M^{-1}Q \Rightarrow m^2 = kMk - QM^{-1}Q + J. \qquad (3.8)$$

  Shifts over internal momenta leave the integrals unchanged.
- Wick rotations—transform Minkowskian space into the Euclidean for all loop momenta:

$$k_0 \to ik_0; \quad k_j \to k_j (1 \leqslant j \leqslant d - 1) \Rightarrow k^2 \to -k^2; \quad d^d k \to i d^d k.$$

- Diagonalization of the matrix M:

$$k^{\dagger} M k = (V(x)k)^{\dagger} V(x) M V(x)^{-1} V(x)k; \quad k(x) = V(x)k;$$

$$V M V^{-1} = M_{\text{diag}}; \quad (V^{\dagger} = V^{-1})$$

$$kMk \Rightarrow k(x) M_{\text{diag}} k(x) = \sum_{i=1}^{L} \alpha_i k_i^2(x).$$

The operation leaves integrals unchanged. After such manipulations the function m^2 has the following form:

$$m^2 = -\sum_{i=1}^{L} \alpha_i k_i^2 - QM^{-1}Q + J.$$

- Rescale k_i:

$$k_i \to \sqrt{\alpha_i} k_i \Rightarrow d^d k_i \to (\alpha_i)^{-d/2} d^d k_i \quad \text{and} \quad \prod_{i=1}^{L} \alpha_i = \det M.$$

Finally, we obtain

$$G_L[1] = (-1)^{N_\nu} (i)^L (\det M)^{-d/2} \frac{\Gamma(N_\nu)}{\prod\limits_{i=1}^{N} \Gamma(n_i)} \int dx_1 \ldots dx_N \int \frac{Dk_1 \ldots Dk_L}{\left(\sum\limits_{i=1}^{L} k_i^2 + QM^{-1}Q - J \right)^{N_\nu}}$$

(3.9)

or

$$G_L[1] = \frac{(i)^{L-N_\nu} (\det M)^{-d/2}}{\prod\limits_{i=1}^{N} \Gamma(n_i)} \int d\alpha_1 \ldots d\alpha_N \int Dk_1 \ldots Dk_L \, e^{-i\left(\sum\limits_{i=1}^{L} k_i^2 + QM^{-1}Q - J \right)},$$

(3.10)

with $Dk = \dfrac{d^d k}{i\pi^{d/2}}$.

Now the integration over loop momenta can be done in a simple way (see Problem 3.18):

$$i^L \int \frac{Dk_1 \ldots Dk_L}{\left(\sum\limits_{i=1}^{L} k_i^2 + \mu^2(x) \right)^{N_\nu}} = \frac{\Gamma\left(N_\nu - \frac{d}{2} L \right)}{\Gamma(N_\nu)} \frac{1}{(\mu^2(x))^{N_\nu - \frac{dL}{2}}},$$

(3.11)

$$\int Dk_1 \ldots Dk_L e^{-i\left(\sum\limits_{i=1}^{L} k_i^2 + \mu^2(\alpha) \right)} = (-i)^{-Ld/2} e^{-i\mu^2(\alpha)},$$

(3.12)

with $\mu^2(x) = QM^{-1}Q - J$. The final result (Feynman parametrization) is

$$\boxed{G_L[1] = \frac{(-1)^{N_\nu} \Gamma\left(N_\nu - \frac{d}{2} L \right)}{\prod\limits_{i=1}^{N} \Gamma(n_i)} \int \prod_{j=1}^{N} dx_j \, x_j^{n_j - 1} \delta\left(1 - \sum_{i=1}^{N} x_i \right) \frac{U(\mathbf{x})^{N_\nu - d(L+1)/2}}{F(\mathbf{x})^{N_\nu - dL/2}}}$$

(3.13)

where we introduced two Feynman graph polynomials U and F

$$m^2 = kMk - 2Qk + J \Leftrightarrow U = \det M,$$

(3.14)

$$F = -\det M\, J + Q M^T Q. \tag{3.15}$$

In Schwinger or alpha representation, we have

$$G_L[1] = \frac{(i)^{L-N_v}(det M)^{-d/2}}{\displaystyle\prod_{i=1}^{N}\Gamma(n_i)} \int \prod_{j=1}^{N} d\alpha_j\, \alpha_j^{n_j-1}\, e^{-i\frac{F(\alpha)}{U(\alpha)}}, \tag{3.16}$$

The relations for U and F in Eqs. (3.14) and (3.15) are crucial for further studies. They are called Symanzik polynomials[1] [2]. The Schwinger representation has been used in the context of the method of brackets (see Sect. 3.13) and multifold sums [3]. Here we framed the Feynman parametrization of Eq. (3.13) as we will explore it further in the context of MB integrals.

Other Useful FI Representations

Other parametrizations used in Feynman integral studies in connection with expansions by regions, unitarity, or differential approaches are [4]:

1. Lee-Pomeransky representation [5]

$$I(v) = \frac{(-1)^{N_v}\Gamma(D/2)}{\Gamma((L+1)D/2 - N_v)\prod_j \Gamma(v_j)} \int_0^\infty \left(\prod_{j=1}^{N} dz_j\, z_j^{v_j-1}\right)(U+F)^{-D/2}. \tag{3.17}$$

2. Baikov representation [6]

$$G(v_1 \ldots v_N) = \frac{\pi^{(L-N)/2} S_E^{(E+1-D)/2}}{[\Gamma((D-E-L+1)/2)]_L}$$

$$\times \int \left(\prod_{i=1}^{L}\prod_{j=i}^{L+E} ds_{ij}\right) S^{(D-E-L-1)/2} \prod_{j=1}^{N} D_j^{-v_j}, \tag{3.18}$$

where

$$[\Gamma(x)]_L \equiv \Gamma(x)\,\Gamma(x-1)\ldots\Gamma(x-L).$$

---

[1] The three independent proofs of the Feynman diagram parametrization due to Symanzik (not given in the original work) were present by Nakanishi, Shimamoto, and Kinoshita. For references, see [1].

The quantities S and S_E come from the Jacobian of the variable transformation and have the form

$$S = \det\left\{ s_{ij}\big|_{i,j=1...L+E} \right\}, \quad S_E = \det\left\{ s_{ij}\big|_{i,j=L+1...L+E} \right\}.$$

Here we have loop momenta k_i and external momenta $p_1, \ldots, p_E$ with [7]

$$s_{ij} = s_{ji} = k_i \cdot q_j; \quad i = 1, \ldots, L; \quad j = 1, \ldots, K, \tag{3.19}$$

where $q_1, \ldots, q_L = k_1, \ldots, k_L$, $q_{L+1}, \ldots, q_{L+E} = p_1, \ldots, p_E$, and $K = L + E$.

The functions D_j are linear functions of the variables s_{ij}, so that $\prod_{i=1}^{L} \prod_{j=i}^{L+E} ds_{ij} \propto dD_1 \ldots dD_N$. Thus, we have

$$G(\boldsymbol{\nu}) \propto \int \left(\prod_{j=1}^{N} D_j^{-\nu_j} dD_j \right) P^{(D-E-L-1)/2},$$

where $P(D_1, \ldots D_N)$ is *Baikov polynomial* obtained from S by expressing s_{ij} via $D_1, \ldots D_N$.

3.2 Topological Structure of Feynman Diagrams, Graph Polynomials

The functions U and F are called graph polynomials. They are polynomials in the Feynman parameters and have the following properties:

- They are homogeneous in the Feynman parameters; U is of degree L; F is of degree $L + 1$.
- U is linear in each Feynman parameter. If all internal masses are zero, then also F is linear in each Feynman parameter.
- In expanded form each monomial of U has a coefficient $+1$.

U and F are the first and the second Symanzik polynomials of the graph, respectively. These polynomials can be also derived from the topology of the underlying graph.

To explain the graphical method of graph polynomial construction, one needs to introduce some basic definitions first:

- Spanning tree T of the graph G is a subgraph with the following properties:
  - T contains all the vertices of G.

- The number of loops in T is zero.
- T is connected.

  T can be obtained from G by deleting L edges (L—number of loops in G).
- Spanning k-forest $\mathcal{T}_k$ for the graph G has the same properties as T, but it is not required that a spanning forest is connected. Instead we require that it should have exactly k connected components.

  F can be obtained from G by deleting $L + k - 1$ edges.

If $\mathcal{T}$ is the set of all spanning forests of G and $\mathcal{T}_k$ is set of all spanning k-forests of G then

$$\mathcal{T} = \bigcup_{k=1}^{r} \mathcal{T}_k \quad (r - \text{number of vertices}).$$

Each element of $\mathcal{T}_k$ has k connected components $(T_1, \ldots, T_k)$. With P_{T_i} we denote a set of external momenta attached to T_i for a given k-forest. Depending on the "direction" of external momenta (whether they are incoming or outgoing), they enter the P_{T_i} with a different relative sign.

The graph polynomials U and F can be obtained from the spanning trees and the spanning 2-forests of a graph G as follows:

$$U = \sum_{T \in \mathcal{T}_1} \prod_{e_i \notin T} x_i, \tag{3.20}$$

$$F = - \sum_{(T_1, T_2) \in \mathcal{T}_2} \left(\prod_{e_i \notin (T_1, T_2)} x_i \right) \left(\sum_{p_i \in P_{T_1}} p_i \right) \left(\sum_{p_j \in P_{T_2}} p_j \right) + U \sum_{i=1}^{n} x_i m_i^2 \tag{3.21}$$

$$\equiv F_0 + U \sum_{i=1}^{n} x_i m_i^2. \tag{3.22}$$

A simple one-loop example how to find Symanzik polynomials is given in Fig. 3.1.

Cuts of internal lines (lines removed in Fig. 3.1) are made according to Eqs. (3.20) and (3.22) such that:

- U: (i) every vertex is still connected to every other vertex by a sequence of uncut lines; (ii) no further cuts are made without violating (i).
- F: (iii) the cuts divide the graph into two disjointed parts such that within each part (i) and (ii) are valid and at least one external momentum line is connected to each part.

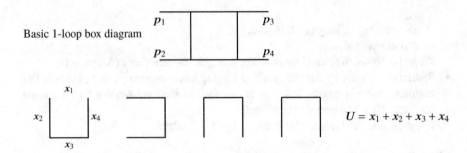

Trees contributing to the U polynomial for the 1-loop box diagram are drawn above.

$$F = t \cdot x_1 x_3 + s \cdot x_2 x_4$$

2 - trees contributing to the F polynomial for the 1-loop box diagram are drawn above.

Fig. 3.1 Graphical construction of F and U Symanzik polynomials. Kinematic variables t and s are defined as $t = (p_1 - p_3)^2$ and $s = (p_1 + p_2)^2$. External particles are considered massless

Regarding Fig. 3.1, let us note that the delta function $\delta(1 - \sum_{i=1}^{N} x_i)$ in Eq. (3.13) in any one-loop diagram goes over all variables x_i, so $U = 1$. This feature will be used in Sect. 3.7.

As a next example, for a general vertex in Fig. 3.2 which can also be a part of a multiloop diagram, we have

$$U = x_1 + x_2 + x_3 \equiv 1, \tag{3.23}$$

$$F_0 = -(q_2 + q_3)^2 x_1 x_2 - q_2^2 x_1 x_3 - q_3^2 x_2 x_3, \tag{3.24}$$

$$F = F_0 + U(x_1 m_1^2 + x_2 m_2^2 + x_3 m_3^2). \tag{3.25}$$

A more complicated example for a non-planar two-loop vertex is shown in Fig. 3.3 where single terms for trees and forests are shown (see Problem 3.3), and further examples are used in the next sections.

For practical purposes, F and U polynomials can be determined algebraically using the definitions of Eqs. (3.14) and (3.15). This is encoded, for instance, in the package MB.m [8]; see Problem 3.4.

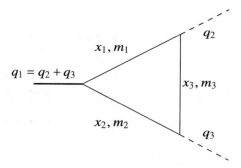

Fig. 3.2 A general one-loop vertex, which can be a part of a multiloop diagram

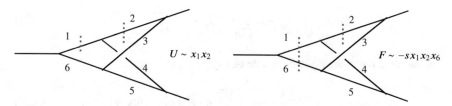

Fig. 3.3 Construction of the U and F polynomials for the non-planar massless vertex V6l0m; single terms are shown. See Problem 3.3 for a complete solution

3.3   Master Mellin-Barnes Formula: Prescription for the Contour, Proof

The backbone of the procedure to build up Mellin-Barnes representations is the following relation:

$$\frac{1}{(A+B)^\lambda} = \frac{1}{\Gamma(\lambda)} \frac{1}{2\pi i} \int_{c-i\infty}^{c+i\infty} dz\, \Gamma(\lambda+z)\Gamma(-z) A^z B^{-\lambda-z}, \tag{3.26}$$

where

- The integration contour separates the poles of $\Gamma(-z)$ from those of $\Gamma(\lambda+z)$,
- A and B are complex numbers such that $|arg(A) - arg(B)| < \pi$.

Proof of the Generic MB Formula

The proof based on the series and summing residues can be found in [9]. Here we will briefly outline it. We start from a relation

$$\frac{1}{(A+B)^\lambda} = \frac{1}{A^\lambda} \frac{1}{(1+B/A)^\lambda} \equiv \frac{1}{A^\lambda} \frac{1}{(1+\tilde{B})^\lambda} \tag{3.27}$$

Fig. 3.4 Integration contour
for Eq. (3.29)

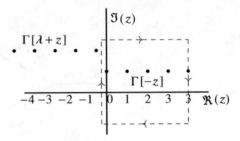

which we expand as Taylor series

$$\text{LHS} = \frac{1}{(1 + \tilde{B})^\lambda} = \sum_{n=0}^{\infty} (-1)^n \frac{\lambda \dots (\lambda + n - 1)}{n!} \tilde{B}^n. \tag{3.28}$$

On the other hand, the right-hand side of Eq. (3.26) is

$$\text{RHS} = \frac{1}{(1 + \tilde{B})^\lambda} = \frac{1}{\Gamma(\nu)} \frac{1}{2\pi i} \int_{-i\infty}^{+i\infty} dz \; \tilde{B}^z \Gamma(\lambda + z) \Gamma(-z). \tag{3.29}$$

According to Cauchy's residue theorem, Eq. (2.149), $\int_C f(z) dz = 2\pi i \sum_i Res_{z_i} f$, by closing the integration contour to the right (see Fig. 3.4), and taking a series of residues (with a minus sign) at points $z = 0, 1, 2, \dots$ (Problem 3.5), we obtain:

$$\text{RHS} = \frac{1}{\Gamma(\lambda)} \frac{1}{2\pi i} \int_{-i\infty}^{+i\infty} dz \; 2\pi i \sum_{n=0}^{\infty} \frac{(-1)^n}{n!} \Gamma(\lambda + n) \tilde{B}^n \tag{3.30}$$

By putting $\Gamma(\lambda + n) = \lambda \dots (\lambda + n - 1) \Gamma(\lambda)$, we can see that LHS = RHS.

The Mellin-Barnes relation in Eq. (3.26) can be iterated and easily extended as a sum of several terms:

$$\frac{1}{(A_1 + \dots + A_n)^\lambda} = \frac{1}{\Gamma(\lambda)} \frac{1}{(2\pi i)^{n-1}} \int_{c-i\infty}^{c+i\infty} dz_2 \dots \int_{c-i\infty}^{c+i\infty} dz_n \prod_{i=2}^{n} A_i^{z_i}$$

$$\times A_1^{-\lambda - z_2 - \dots - z_n} \Gamma(\lambda + z_2 + \dots + z_n) \prod_{i=2}^{n} \Gamma(-z_i). \tag{3.31}$$

This formula can be applied to the Feynman parameter representation in Eq. (3.13). Different strategies for this procedure will be discussed in the next sections, but in the end the graph polynomials are split into pieces, and the

integration over Feynman parameters can be done using the relation

$$\int_0^1 \prod_{i=1}^N dx_i \, x_i^{n_i-1} \, \delta(1 - x_1 - \ldots - x_N) = \frac{\Gamma(n_1)\ldots\Gamma(n_N)}{\Gamma(n_1 + \ldots + n_N)}, \tag{3.32}$$

which is a generalization of the Euler formula

$$\int_0^1 dx \, x^{\alpha_1-1}(1-x)^{\alpha_2-1} = \frac{\Gamma(\alpha_1)\Gamma(\alpha_2)}{\Gamma(\alpha_1 + \alpha_2)}. \tag{3.33}$$

In the most general form, an MB representation can be written as the I_{MB} integral of the form

$$I_{\text{MB}} = \frac{1}{(2\pi i)^r} \int_{-i\infty}^{+i\infty} \cdots \int_{-i\infty}^{+i\infty} \prod_i^r dz_i \, \mathbf{F}(Z, S, \epsilon) \frac{\prod_{j=1}^{N_n} \Gamma(\Lambda_j)}{\prod_{k=1}^{N_d} \Gamma(\Lambda_k)}. \tag{3.34}$$

$\mathbf{F}$ depends on: Z – set of integration variables whose length is usually smaller than r

S – set of kinematic parameters and masses

$$\Lambda_i : \text{ is a linear combination of } z_i \text{ and } \epsilon, \text{ e.g., } \Lambda_i = \sum_l \alpha_{il} z_l + \gamma_i + \delta_i \epsilon. \tag{3.35}$$

In practice $\mathbf{F}$ is a product of S elements raised to some power as shown below:

$$\mathbf{F} \sim \prod_k X_k^{\sum_i (\alpha_{ki} z_i + \gamma_k + \delta_k \epsilon)}, \tag{3.36}$$

where $\alpha_{ij}, \gamma_i, \delta_i \in \mathbb{Z}$ and X_k are the ratios of kinematical invariants and masses, e.g.,

$$X = \left\{ -\frac{s}{m_1^2}, \frac{m_1^2}{m_2^2}, \frac{s}{t}, \ldots \right\}. \tag{3.37}$$

The procedure where the sum of terms in the Symanzik polynomials of Eqs. (3.20) and (3.22) is transformed into the MB integrals is performed automatically in the AMBRE project [10–13]; see also the Appendix and the web page [14] where auxiliary packages with examples related to MB calculations can be found. In the next section, we show the first simple example of the construction of the MB representation.

3.4   Construction of MB Representations for Feynman Integrals: An Example with Basic Steps

Let us start from the one-loop virtual self-energy case (Fig. 3.5).
    The scalar part of the diagram is

$$G(1)_{\text{SE2l2m}} = \int \frac{d^d k}{(k^2 - m^2 + i\delta)^{\nu_1}((k+p)^2 - m^2 + i\delta)^{\nu_2}}, \tag{3.38}$$

where all the typical constant factors were omitted. Index SE2l2m stands for self-energy (SE) one-loop with two massive (2l2m) internal lines. We will use analogous nomenclature later on. After calculating U and F polynomials (see Sect. 3.2), we have:

$$U = x_1 + x_2, \qquad F = m^2(x_1 + x_2)^2 - sx_1x_2 - i\delta. \tag{3.39}$$

Feynman parametrization for this diagram reads

$$G(1)_{\text{SE2l2m}} = \frac{\Gamma(\nu_1 + \nu_2 - \frac{d}{2})}{\Gamma(\nu_1)\Gamma(\nu_2)} \int_0^1 \prod_{j=1}^2 dx_j x_j^{\nu_j - 1} \delta\left(1 - \sum_{i=1}^2 x_i\right)$$

$$\times \frac{(x_1 + x_2)^{\nu_1 + \nu_2 - d}}{(m^2(x_1 + x_2)^2 - sx_1x_2 - i\delta)^{\nu_1 + \nu_2 - d/2}}. \tag{3.40}$$

We see that if we apply the Dirac δ function to the U polynomial, which is simply a sum of Feynman parameters, we get 1. In general, every one-loop n-point diagram has the U polynomial of the form $x_1 + \ldots + x_n$, so $U = 1$ for all one-loop cases.

Fig. 3.5 Two-point (self-energy) diagram

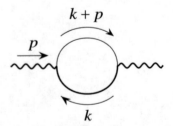

At this point Eq. (3.26) can be used to start constructing a Mellin-Barnes representation. We use it to replace a sum of Feynman parameters in the F polynomial into its product with additional integration over the complex space:

$$\frac{1}{F^\lambda} = \frac{1}{(m^2(x_1 + x_2)^2 - sx_1x_2 - i\delta)^\lambda}$$

$$= \frac{1}{\Gamma(\lambda)} \frac{1}{2\pi i} \int_{c-i\infty}^{c+i\infty} dz_1 \Gamma(\lambda + z_1)\Gamma(-z_1)(m^2 - i\delta)^{z_1}(-s - i\delta)^{-\lambda-z_1}$$

$$\times (x_1x_2)^{-\lambda-z_1}[x_1 + x_2]^{2z_1}, \tag{3.41}$$

where $\lambda = \nu_1 + \nu_2 - d/2$. The term $[x_1 + x_2]^{2z_1}$ in this case can be dropped since we already have the δ function. But here we show this term explicitly for generality. The term again can be changed according to Eq. (3.26) resulting in

$$\frac{1}{F^\lambda} = \frac{1}{\Gamma(\lambda)} \frac{1}{(2\pi i)^2} \int_{c-i\infty}^{c+i\infty} dz_1 \frac{1}{\Gamma(-2z_1)} \int_{c-i\infty}^{c+i\infty} dz_2 \Gamma(\lambda + z_1)\Gamma(-z_1)\Gamma(-2z_1 + z_2)$$

$$\times \Gamma(-z_2)(m^2 - i\delta)^{z_1}(-s - i\delta)^{-\lambda-z_1} x_1^{-\lambda-z_1+z_2} x_2^{-\lambda+z_1-z_2}. \tag{3.42}$$

Next step is to insert Eq. (3.42) back into Eq. (3.40) and collect powers of Feynman parameters, which in our case are

$$x_1^{a_1-1} = x_1^{(-\lambda-z_1+z_2+\nu_1)-1},$$

$$x_2^{a_2-1} = x_2^{(-\lambda+z_1-z_2+\nu_2)-1}. \tag{3.43}$$

The integration over Feynman parameters is performed using the following formula:

$$\int_0^1 \prod_{i=1}^n dx_j x_j^{a_j-1} \delta\left(1 - \sum_{i=1}^n x_i\right) = \frac{\Gamma(a_1)\dots\Gamma(a_n)}{\Gamma(a_1 + \dots + a_n)}, \tag{3.44}$$

which in practice is restricted to collecting relevant powers of Feynman parameters. The final Mellin-Barnes representation for the self-energy diagram of Eq. (3.38) is

$$G(1)_{\text{SE2l2m}} = \frac{(-1)^{\nu_1+\nu_2}}{\Gamma(\nu_1)\Gamma(\nu_2)} \int_{c-i\infty}^{c+i\infty} \frac{dz_1}{2\pi i} \int_{c-i\infty}^{c+i\infty} \frac{dz_2}{2\pi i} (m^2 - i\delta)^{z_1}(-s - i\delta)^{d/2-\nu_1-\nu_2-z_1}$$

$$\times \Gamma(-d/2 + \nu_1 + \nu_2 + z_1)\Gamma(-z_1)\Gamma(-2z_1 + z_2)\Gamma(-z_2)$$

$$\times \frac{\Gamma(d/2 - \nu_1 + z_1 - z_2)\Gamma(d/2 - \nu_2 - z_1 + z_2)}{\Gamma(-2z_1)\Gamma(d - \nu_1 - \nu_2)}. \tag{3.45}$$

For more complicated cases, the F polynomial often contains more than two terms. In such cases it is straightforward to use the general formula in Eq. (3.31).

Usually in Mellin-Barnes representations, infinitesimal complex part $i\delta$ is omitted. This doesn't mean that $i\delta$ is irrelevant. Even if $i\delta$ is omitted at the beginning, it must be restored later to make the analytic continuation to the physical domain possible.

In general, the description of how to restore $i\delta$ from the propagator structure in Eq. (3.38) is as follows: one needs to add $-i\delta$ to each negative invariant, e.g., $(-s)^{z_1}$ should be replaced by $(-s - i\delta)^{z_1}$, if $s > 0$. This will choose the correct branch of the complex functions.

The small imaginary part is not needed for positive invariants since the rising of positive numbers to a complex power doesn't lead to any ambiguities. This recipe clearly follows from Feynman parametrization steps shown in Sect. 3.1.

Analytic continuation can be summarized nicely using the notion of F and U polynomials, as discussed in [15].

The structure of the F polynomial affects the final MB representation form. Obviously, this is due to the fact that Eq. (3.31) changes the sum of n terms raised to a certain power to a $(n-1)$-dimensional integral over the complex space. This observation is helpful when one wants to estimate the dimensionality of the final MB representation only by looking at the F polynomial.

At this stage, we can further simplify the two-dimensional MB integral in Eq. (3.45) by applying the following Barnes lemmas:

- First Barnes lemma (1BL)

$$\int_{z_0 - i\infty}^{z_0 + i\infty} dz\, \Gamma(a + z)\Gamma(b + z)\Gamma(c - z)\Gamma(d - z)$$
$$= \frac{\Gamma(a + c)\Gamma(a + d)\Gamma(b + c)\Gamma(b + d)}{\Gamma(a + b + c + d)},$$

(3.46)

where

$$a + b + c + d < 1, \quad a, b, c, d \in \mathbb{R}.$$

(3.47)

- Second Barnes lemma (2BL)

$$\int_{z_0 - i\infty}^{z_0 + i\infty} dz\, \frac{\Gamma(a + z)\Gamma(b + z)\Gamma(c + z)\Gamma(d - z)\Gamma(e - z)}{\Gamma(f + z)}$$
$$= \frac{\Gamma(a + d)\Gamma(a + e)\Gamma(b + d)\Gamma(b + e)\Gamma(c + d)\Gamma(c + e)}{\Gamma(a + b + d + e)\Gamma(a + c + d + e)\Gamma(b + c + d + e)},$$

(3.48)

where

$$a + b + c + d + e = f. \tag{3.49}$$

The condition in Eq. (3.47) is fulfilled automatically as arguments of gamma functions in MB representations depend on ϵ, which is not fixed.

These lemmas can be proved using a notion of hypergeometric functions as it was made for a case of 1BL in the previous chapter (Sect. 2.7); for an alternative proof, see Problem 5.2.

We can now apply 1BL to Eq. (3.45), and simultaneously substituting powers of propagators equal one $\nu_1 = \nu_2 = 1$ and $d = 4 - 2\epsilon$, we get the final, compact result:

$$G_1(1)_{\text{SE2l2m}} = \int_{c-i\infty}^{c+i\infty} \frac{dz_1}{2\pi i} (m^2)^{z_1} (-s)^{-\epsilon-z_1} \frac{\Gamma(1 - \epsilon - z_1)^2 \Gamma(-z_1) \Gamma(\epsilon + z_1)}{\Gamma(2 - 2\epsilon - 2z_1)}. \tag{3.50}$$

The above steps leading to Eq. (3.50) can be found in the file `MB_SE2l2m_Springer.nb` in the auxiliary material in [16]. We will discuss this integral, solving it analytically and numerically in the following sections.

3.5 Simplifying MB Representations

In the previous section, we have shown how to get the MB representation for the one-loop self-energy integral (with the same mass for the two propagators). If we go beyond the one-loop level, the situation is more involved, especially when we consider multileg multiloop scalar and tensor virtual integrals. Beyond one-loop:

- $U(\mathbf{x}) \neq 1$,
- Dimensionality of the MB representations starts to depend on the $U(\mathbf{x})$ structure,
- Nontrivial simplifications of the graph polynomials are necessary.

It appears that two different strategies can be developed, which depend on the planarity of Feynman diagrams. Usually for planar diagrams, a so-called loop-by-loop approach (LA) is the most efficient. In this approach the basic Mellin-Barnes relation in Eq. (3.31) and Feynman parameter integrations change sums of terms in the Symanzik polynomials into integrals over complex space. These transformations are successively made for each internal momenta of the Feynman integrals. In a planar case, there is no mismatch, and after each step an effective (planar) diagram appears. In addition, the LA has an advantage because at each step we consider one-loop subloops, so $U = 1$ follows automatically, and we have to care only about F polynomial, which should be factorized using the most efficient method in order to achieve the smallest dimension of the constructed MB representation.

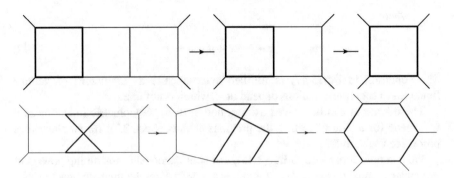

Fig. 3.6 Limitations of the LA approach. In the planar case (the first row), momentum flow is conserved after each step in all vertices, which is not true in the non-planar case (the second row)

In non-planar cases a well-defined connected diagram cannot always be drawn after each step. As a consequence, not all vertices conserve momenta [17]; see Fig. 3.6 and Problem 3.7.

That is why in this case we undertake a more natural approach where F and U polynomials are transformed to the MB representation using Eq. (3.31) in one step. However, this global approach (GA) has the price that F and U polynomials may result in more complicated MB representations. Still, the GA approach to F and U polynomials has an important property: *The polynomials are homogeneous functions of the Feynman parameters of some given degree, depending on the underlying topology.* So, we can systematically rescale Feynman variables and, in turn, change Dirac delta function under the integral, which defines the region of integration. In this case, the Dirac delta function can be used on a restricted subset of Feynman parameters [18]. This trick is in fact the Cheng-Wu theorem [19]. We will discuss it in Sect. 3.8.

> The general problem is for which choice of Feynman variables (e.g., rescaling) the factorization of the polynomials will be the most efficient, leading to the smallest number of terms and, accordingly, to the lowest dimensionality of the MB representations.

In practice, a choice between the LA and GA strategies for a given integral aiming at the lowest MB dimensionality of integrals seems not to be unique. For instance, in a massless case of the non-planar two-loop diagram, a minimal fourfold MB representation was derived starting from the global Feynman parameter representation [20]. On the other hand, in the massive case, it appears that the LA is more efficient; an eightfold MB representation can be obtained [21]. There is no known MB representation smaller than tenfold for GA [21].Moreover, a structure of MB integrals may depend also on the kinematic point and mass thresholds, which will be discussed in Sect. 6.2.4. Thus, it is not easy to get a general and efficient program for the construction of a broad class of MB representations.

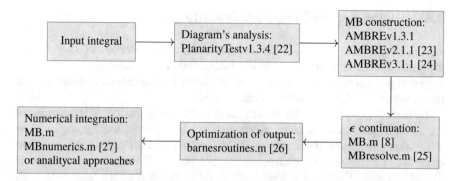

Fig. 3.7 The operational sequence of the MB-suite. This flowchart shows the main steps and the corresponding software to produce a Mellin-Barnes representation of a Feynman integral and perform its numerical or analytic solution

Figure 3.7 shows which method is used to which type of the diagram and up to which multileg-multiloop integrals the MB-suite of packages work efficiently. See [8, 22–27] for details on the dedicated packages.

The MB-suite comprises several tools, and the calculational procedure goes for dimensionally regulated Feynman integrals in the momentum space like this:

(i) Start from integrals expressed by Feynman parameters; see Eq. (3.13) and the example in Eq. (3.40).

(ii) Transform Feynman integrals into the MB representations; see Sect. 3.4 with the AMBRE package [10–13] controlled for automation procedures by the PlanarityTest.m package (choice between LA and GA) [17, 28].

(iii) Perform an analytic continuation in ϵ with MB.m or MBresolve.m [8, 29] for original integrals of dimension d, $d = 4 - 2\epsilon$, with integration paths parallel to the imaginary axis.

(iv) Expand the Mellin-Barnes integrals as series in small ϵ with MB.m or MBresolve.m.

(v) Perform simplifications using Barnes lemmas and barnesroutines.m [14].

After step (v) the original representation of the Feynman integral in terms of MB integrals expanded in ϵ is formulated. One may now start to calculate them, either analytically or numerically, or in a mixed approach—see next sections and chapters.

3.6   Using Barnes Lemmas Efficiently

As already shown in Sect. 3.4, the dimension of the representation can be decreased using the Barnes lemmas by analytical integration over z-variables which do not appear in powers of kinematic variables, so Barnes lemmas are an essential part of the construction of MB representations.

In Table 3.1, we show different ways to decrease the number of MB integrations for a simple polynomial which nevertheless can be a part of more complex U or F functions. (i) Direct application of the MB formula of Eq. (3.31) with integration using Eq. (3.32) gives a three-dimensional MB integral. (ii) From the second row of the table, one can see that a factorization and application of MB relation to each factorized term reduces the number of MB integrations to two. (iii) Now let us consider the simplest polynomial containing only two linear terms. As shown in the following equation, after obvious integration steps and omitting coefficients, we get a combination of gamma functions that exactly fits the 1BL

$$(x_1 + x_2)^p \to \int dz_1 x_1^{z_1} x_2^{p-z_1} \Gamma(-z_1) \Gamma(-p + z_1) \delta(1 - x_1 - x_2)$$

$$\to \int dz_1 \Gamma(-z_1) \Gamma(-p + z_1) \Gamma(z_1 + 1) \Gamma(p - z_1 + 1) / \Gamma(p + 2). \qquad (3.51)$$

In this way, factorization together with 1BL gives no additional MB integrations automatically—the same situation we have had in Sect. 3.4 where we expressly did not simplify the term $(x_1 + x_2)^2$. One can recursively prove this property for a linear combination of any length. (iv) As an alternative, one can perform a rescaling of integration variables with an additional delta function, as briefly shown in the fourth row of Table 3.1. Such procedure leads to a very efficient simplification that also gives no extra MB integrations. This approach is used intensively for the construction of MB representations within the GA approach; see Sect. 3.9.

Another way to achieve the same result without factorizing linear terms is to remove as many as possible z-variables out of exponents of invariants and masses and then to check for 1BL. Such a procedure can be done by a simple linear shift of one of the z-variables. Usually, we have a combination $\sum_{j=i_1}^{i_N} \alpha_j z_j$ in the exponent

Table 3.1 Simplification of graph polynomials by factorization, Barnes lemmas, and rescaling. Corresponding manipulations can be found at the file MB_Simpl_Springer.nb in the auxiliary material in [16]

(i)	$x_1 x_3 + x_1 x_4 + x_2 x_3 + x_2 x_4$	3-dim representation
(ii)	$(x_1 + x_2)(x_3 + x_4)$	2-dim representation
(iii)	$(x_1 + x_2)(x_3 + x_4) + \mathrm{BL}$	0-dim representation
(iv)	$(x_1 + x_2)(x_3 + x_4) \to [x_1 \to v_1 \xi_{11}, x_2 \to v_1 \xi_{12}, \\ \delta(1 - \xi_{11} - \xi_{12}); x_3 \to v_2 \xi_{21}, \ldots] \to v_1 v_2$	0-dim representation

of one of the invariants and the same combination in the exponent of another one, and such combination doesn't appear in other exponents. In this situation one can make a shift $z_{i_1} \to z_{i_1} - \sum_{j=i_2}^{i_N} \alpha_j z_j$ and check Barnes' first lemma for all variables $z_{i_2} \ldots z_{i_N}$. This algorithm is implemented in AMBRE and works very well for the LA approach where the F polynomial has only degree 2. In other words, it allows to catch all linear combinations for one-loop sub-diagrams due to a relative simplicity of the F polynomial; for details see the next section.

Going beyond one-loop, graph polynomials become more and more complicated and factorization of linear sub-expressions becomes not so obvious. In this case based on experience with shifts of variables described above, we can go further and try find some suitable linear transformation of integration variables which will allow us to apply one of the Barnes lemmas. Such search has been implemented in the package barnesroutines.m [26] for MB integrals with fixed contours which appear after the expansion of the MB representation in ϵ.

However, we can also apply the Barnes lemmas to the original MB integral before ϵ expansion with a fixed contour of integration. This approach is interesting for two reasons. First, the application of Barnes lemmas to MB integrals with fixed contours may lead to a large number of MB integrals. Second, starting with the ϵ expansion from a high-dimensional MB representation, we get in general a bigger cascade of integrals than starting from a representation where dimensionality was already decreased by Barnes lemmas.

Let us discuss then a general strategy how to determine suitable z transformations for applying the Barnes lemmas to the MB integrals before ϵ expansion. This strategy gives the minimal number of dimensions for MB integrals.

We start by encoding the z-dependence of gamma functions of the MB integral in Eq. (3.34) in a matrix form

$$M_\Gamma Z = \begin{bmatrix} \alpha_{ij}(\text{numerator}) \\ \cdots\cdots \\ \alpha_{ij}(\text{denominator}) \end{bmatrix} \begin{pmatrix} z_1 \\ \vdots \\ z_r \end{pmatrix}. \tag{3.52}$$

M_Γ is a rectangular $(N_n + N_d) \times r$ matrix whose upper part contains α-coefficients in Eq. (3.35) from the numerator in Eq. (3.34) and whose bottom part contains the analogous α-coefficients from the denominator in Eq. (3.34). Z is an r-vector of integration variables z_i. Now, any linear variable transformation can be represented as

$$M_\Gamma Z = M_\Gamma U U^{-1} Z = M_\Gamma U Z', \quad Z' = U^{-1} Z, \tag{3.53}$$

with a non-singular $r \times r$ transformation matrix U. M_Γ encodes a new z structure of gamma functions. Barnes lemmas can be applied if columns in $M_\Gamma U$ have specific structure.

- For the first Barnes lemma, elements in a column from $N_n + 1$ to $N_n + N_d$ must be equal to 0, and elements from 1 to N_n must contain the set $\{1, 1, -1, -1\}$, while all others must be also equal to 0. This can be formulated in terms of overdetermined systems of linear equations

$$M_\Gamma X = \{B_1\}. \tag{3.54}$$

X is an unknown r-vector representing a column in U and $\{B_1\}$ is a set of all possible right-hand sides. For general N_n one has $\dfrac{3 N_n!}{4!(N_n - 4)!}$ different r.h.s.

Because systems are overdetermined, they may have no solutions at all. This means that there are no transformations leading to Barnes' first lemma. If some number of solutions n_s is found, one can proceed further with the construction of the matrix U. Each solution represents a column in the matrix and can be placed on any position starting from the diagonal U. The procedure of replacing columns in a diagonal matrix by our solutions continues until the matrix becomes singular. After all these transformations, the maximal amount of added columns before the matrix becomes singular gives the number of integrations which can be done with the help of Barnes' first lemma.

- For a discussion concerning efficient application of the second Barnes lemma to the construction of MB integrals, we refer the reader to [30]. For 2BL the condition on arguments of gamma functions in Eq. (3.49) is important.

The search for transformations for both lemmas can be performed recursively until no solution can be found. The resulting dimensionality may also depend on the order in which the search is performed. Searching for a transformation for the second lemma can be successful after a search and application of the first lemma and vice versa. The efficiency of this procedure will be shown in the next sections.

3.7   Loop-by-Loop (LA) Approach

In the LA approach, each subloop of a multiloop diagram is treated separately in an iterative way. Within this approach, the overall delta function applied to any one-loop diagram results in $U = 1$, and we have to care only about the F polynomial. Aspects of deriving MB representations, such as the order of integration, simplifications of a result via Barnes lemmas, etc., using the LA are discussed in [10, 11], and a lot of instructive examples can be found in [13]. For planar cases the automatic construction of MB representations by the LA seems to be optimal; see, for example, QED ladder diagrams shown in Fig. 3.8 and Table 3.2.

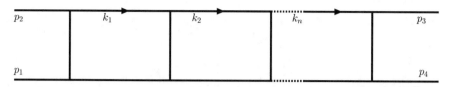

Fig. 3.8 n-loop ladder diagram with $k_1, \ldots, k_n$ internal and $p_1, \ldots, p_4$ external momenta

Table 3.2 Optimal results for ladder diagrams defined in Fig. 3.8. It has been found in [18] that Dim(massive case) = Dim(massless case) + #loops

Dimensions of planar ladder MB representations	Massless cases				Massive (QED) cases			
Number of loops (L)	1	2	3	4	1	2	3	4
No Barnes' first lemma (BL1)	1	4	7	10	3	8	13	18
With BL1	1	4	7	10	2 (1+1)	6 (4+2)	10 (7+3)	14 (10+4)

3.7.1 LA Approach, Planar Example

Let us see how it works for the two-loop vertex in Fig. 3.9. The steps given below can be found at the file MB_V6l3m1M_Springer.nb in the auxiliary material in [16]. Such a vertex appears in the two-loop Z-boson decay calculations [12,31]. We consider the integral with three massive Z-boson and one Higgs boson propagators, $m_1 = m_3 = m_6 = M_Z$, $m_2 = M_H$, and $m_4 = m_5 = 0$. The integral representation of this diagram is the following:

$$I_{\text{V6l3m1M}} = \int d^d k_1 d^d k_2 \frac{1}{(k_1^2 - M_H^2)((k_1 - k_2)^2 - M_Z^2)(k_2^2 - M_Z^2)}$$

$$\times \frac{1}{(k_1 + p_1)^2 (k_2 + p_1)^2 ((k_1 + p_1 + p_2)^2 - M_Z^2)} \quad (3.55)$$

with $p_1^2 = p_2^2 = 0$ and $(p_1 + p_2)^2 = s$.

The minimal MB dimensionality for this integral can be obtained if one first integrates over the subloop triangle marked by dashed lines in Fig. 3.10, and then over remaining propagators, so the order of integration is $k_2 \rightarrow k_1$.

For the first iteration, we have

$$I_{\text{V6l3m1M}}^{(1)} = \int d^d k_2 \frac{1}{((k_1 - k_2)^2 - M_Z^2)(k_2^2 - M_Z^2)(k_2 + p_1)^2}. \quad (3.56)$$

The F polynomial corresponding to this integral is

$$F^{(1)}(\mathbf{x}) = M_Z^2 x_1 + M_Z^2 x_2 - k_1^2 x_1 x_2 - (k_1 + p_1)^2 x_1 x_3. \quad (3.57)$$

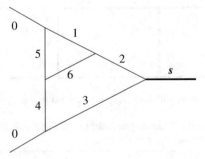

Fig. 3.9 Two-loop vertex diagram for which LA is applied. Different mass configurations for propagators 1–6 lead to different dimensions of MB integrals; see example $I^{MB}_{V6l3m1M}$ and Eq. (3.62)

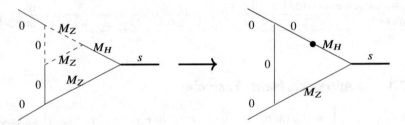

Fig. 3.10 Example of the LA: first iteration, triangle subloop (dashed lines); second, four propagators effectively combined into a vertex

The linear dependence on Feynman parameters for M_Z^2 terms takes place due to the fact that $U = 1$. The MB representation for this subloop can be written as

$$I^{MB\,(1)}_{V6l3m1M} = \dots (-1)^{2-\epsilon-z_1-z_2}(M_Z^2)^{z_1+z_2}(k_1^2)^{-z_3}((k_1+p_1)^2)^{-1-\epsilon-z_1-z_2-z_3}$$

$$\times \Gamma(-z_1)\Gamma(-\epsilon-z_2)\Gamma(-z_2)\Gamma(-\epsilon-z_1-z_2-z_3)\Gamma(-z_3)$$

$$\times \Gamma(1+z_2+z_3)\Gamma(1+\epsilon+z_1+z_2+z_3)/\Gamma(1-2\epsilon-z_1-z_2).$$

$$(3.58)$$

Integrations over z-variables in a complex plane are assumed.

In the next stage, we integrate over the remaining loop momentum k_1

$$I^{(2)}_{V6l3m1M} = \int \frac{d^d k_1}{(k_1^2)^{z_3}(k_1^2 - M_H^2)((k_1+p_1)^2)^{2+\epsilon+z_1+z_2+z_3}((k_1+p_1+p_2)^2 - M_Z^2)}.$$

$$(3.59)$$

This integral has four propagators. Note that now the powers of k_1^2 and $(k_1 + p_1)^2$ are complex and depend on z_i variables. One notices that during the first step, it is possible to modify F by the term $\pm M_H^2 x_1 x_2$ and get in the next iteration only one propagator $k_1^2 - M_H^2$ with the complex power. However, in this case the MB representation for that modified F function will have a term $(-M_H^2)^z$ which forces the integration to be always of the "Minkowskian" type; details will be explained in Chap. 6. Nonetheless, such a construction is according to the LA strategy where the propagators in Eq. (3.59) can be represented as a vertex diagram where propagators k_1^2 and $k_1^2 - M_H^2$ are combined into one line as shown in Fig. 3.10 and this doesn't bring any complication for further construction.

The F polynomial for the final iteration is

$$F^{(2)}(\mathbf{x}) = M_H^2 x_2 + M_Z^2 x_4 - s x_1 x_4 - s x_2 x_4, \tag{3.60}$$

and we end up with a six-dimensional MB representation:

$$I_{\mathrm{V6l3m1M}}^{\mathrm{MB}} = (\ldots)(M_H^2)^{z_4}(M_Z^2)^{z_{125}}(-s)^{-2-2\epsilon-z_{1245}} \tag{3.61}$$

$$\times \Gamma(-z_1)\Gamma(-\epsilon - z_2)\Gamma(-z_2)\Gamma(-\epsilon - z_{123})\Gamma(1 + z_{23})$$

$$\times \Gamma(1 + \epsilon + z_{123})\Gamma(-1 - 2\epsilon - z_{124})\Gamma(-z_4)\Gamma(-z_5)\Gamma(-1 - 2\epsilon - z_{1256})$$

$$\times \Gamma(-z_6)\Gamma(-z_3 + z_6)\Gamma(2 + 2\epsilon + z_{12456})/\Gamma(1 - 2\epsilon - z_{12})\Gamma(-3\epsilon - z_{1245}).$$

where we abbreviated $z_{123} = z_1 + z_2 + z_3$, etc.

We can further simplify this representation. Already at the stage of F polynomials, one can see that, for example, $F^{(1)}(\mathbf{x})$ can be rewritten in the form

$$F^{(1)}(\mathbf{x}) = M_Z^2(x_1 + x_2) - k_1^2 x_1 x_2 - (k_1 + p_1)^2 x_1 x_3 \tag{3.62}$$

and $x_1 + x_2$ can be considered as one term and the MB formula can be applied recursively, first to the F polynomial as a whole and then to the term $x_1 + x_2$. This always allows to apply BL1 to a z-variable used in the MB transformation of the sum $x_1 + x_2$ as explained in Sect. 3.6. The similar trick can be also done for the term $-s x_4 (x_1 + x_2)$ in $F^{(2)}(\mathbf{x})$.

However, for automation in AMBRE package, another strategy was chosen (see previous section). After the shift $z_1 \rightarrow z_1 - z_2 - z_5$, Barnes' first lemma is applied for variables z_2 and z_6 giving a four-dimensional final representation.

3.7.2 LA Approach, Non-planar Example

Next example is a non-planar two-loop vertex shown in Fig. 3.11.

There is only one massive propagator, $m_1 = m_2 = m_3 = m_4 = m_5 = 0$ and $m_6 = M_Z$. This integral is especially demanding for the SD method; for a

Fig. 3.11 The two-loop non-planar vertex diagram for which LA is also applied

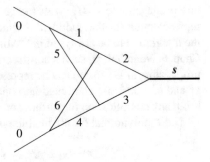

discussion see [32] and [33]. The integral representation in this case is (see the file `MB_V611MZ_Springer.nb` in the auxiliary material in [16])

$$I_{V611m}^{(1)} = \int \int \frac{d^d k_1 d^d k_2}{k_1^2 (k_1 - k_2)^2 k_2^2 ((k_1 - k_2 + p_1)^2 - M_Z^2)(k_2 + p_2)^2 (k_1 + p_1 + p_2)^2}.$$
(3.63)

The main purpose of the example is to show difficulties related to the application of the LA method to non-planar diagrams.

We start with a one-loop sub-diagram and the order of integrations k_2, k_1. There are four propagators which include k_2 (box integral)

$$I_{V611m}^{(1)} = \int d^d k_2 \frac{1}{(k_1 - k_2)^2 (k_2)^2 ((k_1 - k_2 + p_1)^2 - M_Z^2)(k_2 + p_2)^2},$$
(3.64)

which has a more complicated $F^{(1)}(\mathbf{x})$ function:

$$F^{(1)}(\mathbf{x}) = -k_1^2 x_1 x_2 - \left((k_1 + p_1)^2 - M_Z^2 \right) x_2 x_3 - (k_1 + p_2)^2 x_1 x_4$$

$$- (k_1 + p_1 + p_2)^2 x_3 x_4 + M_Z^2 x_3 (x_1 + x_3 + x_4).$$
(3.65)

More problems appear when we go to the integration over loop momenta k_1

$$I_{V611m}^{(2)} = \int d^d k_1 \frac{1}{((k_1)^2)^{1-z_1} ((k_1 + p_1)^2 - M_Z^2)^{-z_3} ((k_1 + p_2)^2)^{-z_5}}$$

$$\times \frac{1}{((k_1 + p_1 + p_2)^2)^{3+\epsilon+z_{123456}}}.$$
(3.66)

In Eq. (3.66) the propagators do not form a diagram with conserved momentum flow. Propagators $(k_1)^2 ((k_1 + p_1)^2 - M_Z^2)(k_1 + p_1 + p_2)^2$ correspond to a one-loop vertex diagram, but $(k_1 + p_2)^2$ must be considered as an artificial numerator; in non-planar case the exponent of such numerator is not an integer but a complex number.

After the k_1 integration, $F^{(2)}(\mathbf{x})$ takes the form

$$F^{(2)}(\mathbf{x}) = M_Z^2 x_2 + 2s x_2 x_3 - 2s x_1 x_4, \tag{3.67}$$

and the final six-dimensional representation can be written in the form[2]

$$I_{V6l1m}^{MB} = \frac{1}{(-2s)^{2+2\epsilon}} \int\limits_{-i\infty}^{i\infty} \frac{dz_1}{2\pi i} \cdots \frac{dz_8}{2\pi i} (-1)^{z_8} \left(\frac{M_Z^2}{-2s}\right)^{z_2} \Gamma(2 + \epsilon + z_{1235} - z_7)\Gamma(-z_7)$$

$$\times \Gamma(-z_1)\Gamma(-1 - 2\epsilon - z_{13})\Gamma(1 + z_{13})\Gamma(-1 - \epsilon - z_{15})\Gamma(1 + z_{15})$$

$$\times \Gamma(-1 - 2\epsilon - z_{128})\Gamma(1 - \epsilon + z_{135} - z_{78})\Gamma(-z_8)\Gamma(2 + 2\epsilon + z_{28})\Gamma(-z_5 + z_8)$$

$$\times \Gamma(-z_3 + z_{78})\Gamma(-z_2 + z_7)\Gamma(-1 - \epsilon - z_{123} + z_7)/\Gamma(-2\epsilon)\Gamma(1 - z_1)$$

$$\times \Gamma(-1 - 2\epsilon - z_{123} + z_7)\Gamma(-3\epsilon - z_2)\Gamma(3 + \epsilon + z_{1235} - z_7). \tag{3.68}$$

The main disadvantage of this representation is the factor $(-1)^{z_8}$ with a complex variable z_8 which forces the integral to be always of the Minkowskian type.

By construction this representation is correct and in principle can be used for the evaluation of the integral, but due to its Minkowskian form, it cannot be evaluated numerically by MB.m which works for Euclidean kinematic. In the next sections, we will describe another approach to construct MB representations for all types of non-planar diagrams without $(-1)^z$-type factors just discussed. To achieve it, we consider first the powerful Cheng-Wu theorem.

3.8 Cheng-Wu Theorem

The theorem (CW) has been considered by Cheng and Wu in [19].

Theorem 3.1 *For the Feynman parameter representation in Eq. (3.13), the Cheng-Wu (CW) theorem states that the same formula holds with the delta function $\delta(1 - \sum_{i=1}^{N} x_i)$ replaced by*

$$\delta\left(\sum_{i \in \Omega} x_i - 1\right), \tag{3.69}$$

where Ω is an arbitrary subset of the lines $1, \ldots, L$, when the integration over the rest of the variables, i.e., for $i \notin \Omega$, is extended to the integration from zero to infinity.

---

[2] After ϵ expansion only maximally five-dimensional MB integrals remain, up to the constant in ϵ.

Proof For the proof we use the identity in Eq. (3.5) and restrict it to the subset Ω, namely,

$$1 = \int_0^\infty \frac{d\lambda}{\lambda} \delta \left(1 - \frac{1}{\lambda} \sum_{i=1}^N \alpha_i \right) \Leftrightarrow 1 = \int_0^\infty \frac{d\lambda}{\lambda} \delta \left(1 - \frac{1}{\lambda} \sum_{i \in \Omega} \alpha_i \right). \tag{3.70}$$

Then we change variables from α_i to $\alpha_i = \lambda x_i$, as applied for obtaining the representation in Eq. (3.6). The key point in the calculation is that, as noticed in [19], the terms $\prod_{j=1}^N dx_j \, x_j^{n_j-1}$, $U(x)^{N_\nu - d(L+1)/2}$, and $F(x)^{N_\nu - dL/2}$ in Eq. (3.13) are homogeneous in x of order 0. Indeed, $\prod_{j=1}^N dx_j \, x_j^{n_j-1} \sim N_\nu \equiv A$, $U(x)^{N_\nu - d(L+1)/2} \sim (N_\nu - d(L+1)/2)L \equiv B$, $F(x)^{N_\nu - dL/2} \sim (N_\nu - dL/2)(L+1) \equiv C$, as U and F are of order L and $L+1$, respectively, so $A+B+C = 0$. So, we can freely rescale any subset of x parameters in (3.70).   □

In [34] an alternative proof of the CW theorem is given using a notion of sector decomposition [35] where independently of the choice of the coefficients a_i in the general integral $\int_0^\infty dx \, f(x) \delta \left(1 - \sum_i a_i x_i \right)$, the considered integrand does not change and $f(x)$ must be the homogenous function in variables x. We propose this proof as the Problem 3.9. Another proof of CW is given in [36].

3.9   Global (GA) Approach

The second possibility to construct an MB representation for a given Feynman integral is to integrate simultaneously over all loop momenta. In this case $U(x)$ is not equal to 1 anymore, but we can avoid highly oscillating factors of the form $(-1)^z$ as given in the previous example; see Eq. (3.68). A naive way to construct a representation is to apply the MB master formula of Eq. (3.31) to both $U(x)$ and $F(x)$ polynomials and then try to simplify the result using Barnes lemmas. However, in this way, one faces several problems. First, after integration over Feynman parameters, one always gets $\Gamma(0)$ in the denominator. This happens precisely because of the homogeneity of the original Feynman parameter representation. We can see it in the following way:

GA and the $\Gamma(0)$ Problem

Let us consider the sunset diagram in Fig. 3.12.
    The Symanzik polynomials for the corresponding FI are

$$U = x_1 x_2 + x_2 x_3 + x_1 x_3, \tag{3.71}$$

$$F = -p^2 x_1 x_2 x_3 + m^2 x_1 U, \tag{3.72}$$

Fig. 3.12 The two-loop sunset diagram with one non-zero mass and Feynman parameters x_1, x_2, x_3

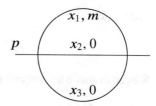

for which (see Eq. (3.13)) the relevant part of the integration with Feynman parameters is

$$I \propto \int dx_1 dx_2 dx_3 \delta(\ldots) \frac{U^{N_\nu - d(L+1)/2}}{F^{N_\nu - dL/2}}. \qquad (3.73)$$

Substituting Eq. (3.72) into Eq. (3.73) and using the MB relation of Eq. (3.26) iteratively, we have

$$I \propto (-p^2 x_1 x_2 x_3)^{z_1} (m^2 x_1 U)^{-(N_\nu - dL/2) - z_1} U^{N_\nu - d(L+1)/2} \qquad (3.74)$$

$$\propto (x_1 x_2 x_3)^{z_1} x_1^{-N_\nu + dL/2 - z_1} U^{-d/2 - z_1} \qquad (3.75)$$

$$\propto (x_1 x_2 x_3)^{z_1} x_1^{-N_\nu + dL/2 - z_1} (x_1 x_2)^{z_2} (x_2 x_3)^{z_3} (x_1 x_3)^{-d/2 - z_2 - z_3 - z_1}. \qquad (3.76)$$

Now we perform the x-integration where the sum of exponents of Feynman parameters enters, see the denominator of Eq. (3.44). Gathering Feynman parameter factors together we get

$$x_1^{-N_\nu + dL/2 - d/2 - z_3 - z_1} x_2^{z_1 + z_2 + z_3} x_3^{-d/2 - z_2}. \qquad (3.77)$$

Hence, keeping in mind the overall prefactor $\prod_i x_i^{n_i - 1}$, the sum of exponents gives $d(L-2)/2$ which for $L = 2$ is 0. According to Eq. (3.44), the result translates to the zero argument of the gamma function in the denominator. From another side, we know that MB representation for this integral does not vanish, so the gamma function in the denominator should be cancelled after suitable integration with the help of BL. We propose this as Problem 3.11.

To overcome this problem, we apply CW theorem by starting from the expression in Eq. (3.75):

$$(x_1 x_2 x_3)_1^z x_1^{-N_\nu + dL/2 - z_1} (x_1 x_2 + x_3(x_1 + x_2))^{-d/2 - z_1}. \qquad (3.78)$$

Now we use CW to the Ω subset in Eq. (3.78)

$$\int_0^1 dx_1 dx_2 dx_3 \delta(1 - x_1 - x_2 - x_3) \to \int_0^\infty dx_3 \int_0^1 dx_1 dx_2 \delta(1 - x_1 - x_2) \qquad (3.79)$$

and we get

$$x_1^{-N_v+dL/2}x_2^{z_1}x_3^{z_1}(x_3+x_1x_2)^{-d/2-z_1}.$$ (3.80)

We can evaluate the integral over x_3

$$\int_0^\infty dx_2 x_3^{z_1}(x_3+x_1x_2)^{-d/2-z_1} = x_1^{1-\frac{d}{2}}x_2^{1-\frac{d}{2}}\frac{\Gamma[1+z_1]\Gamma[-1+d/2]}{\Gamma[d/2+z_1]},$$ (3.81)

and adding integrations over x_2 and x_3, we get

$$\int_0^1 dx_1 \int_0^1 dx_2 \delta(1-x_1-x_2)(x_1x_2)^{-N_v+dL/2+1-\frac{d}{2}}x_2^{z_1+1-\frac{d}{2}}$$ (3.82)

$$= \frac{\Gamma[-N_v+1+d(L-1)/2]\Gamma[z_1+1-d/2]}{\Gamma[-N_v+2+z_1+d(L-2)/2]}.$$ (3.83)

This time, coming back to Eq. (3.44), the argument of gamma in the denominator is non-zero, and we can construct MB representation. It can be shown that in general the result as given in Eq. (3.77) gives "zero" exponents, independently of the number of loops considered.

An option to regulate this singularity is to shift the exponent of one of the Feynman parameters by arbitrary δ and then take the limit $\delta \to 0$, before doing the analytic continuation in the dimensional parameter $\epsilon \to 0$. One should stress that shifting the original exponents of propagators $n_i \to n_i + \eta$, similar to what we will do in Example 4.2.3, does not help because $\Gamma(0)$ appears automatically and it does not depend on powers of propagators n_i. Second, even at the two-loop level, the amount of terms in $U(\mathbf{x})$ and $F(\mathbf{x})$ is quite large; the number of MB integrations is equal to the number of terms in a polynomial, minus one; see Eq. (3.31). It leads to a very high-dimensional MB representation for which it is difficult to catch all possible simplifications by Barnes lemmas.

3.9.1 GA Approach and CW Theorem, the Non-planar Double Box

To show how to construct the MB representations properly, we start with a detailed explanation of one of the first successful implementations of the global approach to a nontrivial Feynman integral, namely, the non-planar massless double box diagram [20] given in Fig. 3.13, with on-shell external legs ($p_i^2 = 0$).

Fig. 3.13 The non-planar double box topology

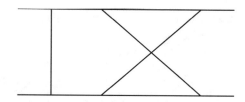

We define an integral for this diagram as (all external momenta p_i are incoming in cyclic notation)

$$B_7^{NP} = \iint d^d k_1 d^d k_2 \frac{1}{[(k_1 + k_2 + p_1 + p_2)^2]^{n_1} [(k_1 + k_2 + p_2)^2]^{n_2} [(k_1 + k_2)^2]^{n_3}}$$
$$\frac{1}{[(k_1 - p_3)^2]^{n_4} [(k_1)^2]^{n_5} [(k_2 - p_4)^2]^{n_6} [(k_2)^2]^{n_7}}. \tag{3.84}$$

An explicit form of the Symanzik polynomials for this integral is (see Problem 3.4)

$$U(x) = x_1 x_2 + x_1 x_4 + x_2 x_4 + x_1 x_5 + x_2 x_5 + x_2 x_6 + x_4 x_6 + x_5 x_6 \tag{3.85}$$
$$+ x_1 x_7 + x_4 x_7 + x_5 x_7 + x_6 x_7,$$

$$F(x) = -s\, x_1 x_2 x_5 - s\, x_1 x_3 x_5 - s\, x_2 x_3 x_5 - u\, x_2 x_4 x_6$$
$$-s\, x_3 x_5 x_6 - t\, x_1 x_4 x_7 - s\, x_3 x_5 x_7 - s\, x_3 x_6 x_7. \tag{3.86}$$

Changing sums of monomials in x into products using the MB master formula (Eq. (3.31)) leads to an 18-fold MB integral (11 and 7 complex variables come from U and F, respectively). Certainly, it can be factorized in a better way. The factorization proposed in [20] looks as follows:

$$U(x) = (x_1 + x_6)(x_2 + x_7) + (x_3 + x_4 + x_5)(x_1 + x_2 + x_6 + x_7), \tag{3.87}$$

$$F(x) = -t\, x_1 x_4 x_7 - u\, x_2 x_4 x_6 - s\, x_1 x_2 x_5 - s\, x_3 x_6 x_7 - s\, x_3 x_5 (x_1 + x_2 + x_6 + x_7), \tag{3.88}$$

where the longest factorized term has four Feynman parameters $x_1 + x_2 + x_6 + x_7$. The factorizations in U and F are connected with a proper collection of Feynman parameters in front of the variables k_1, k_2, as shown schematically in Fig. 3.14.

Now we can apply the CW Theorem 3.1 with Eq. (3.69), keeping in the δ-function the longest factorized subset of Feynman parameters, so it can be dropped out from the Symanzik polynomials, and the integral becomes

$$B_7^{NP} = \frac{(-1)^{N_v} \Gamma(N_v - d)}{\Gamma(n_1) \dots \Gamma(n_N)} \int_0^\infty dx_3 dx_4 dx_5 \int_0^1 dx_1 dx_2 dx_6 dx_7 \delta(1 - (x_1 + x_2 + x_6 + x_7))$$

$$\frac{((x_1 + x_6)(x_2 + x_7) + x_3 + x_4 + x_5)^{N_v - \frac{3d}{2}}}{(-t\, x_1 x_4 x_7 - u\, x_2 x_4 x_6 - s\, x_1 x_2 x_5 - s\, x_3 x_6 x_7 - s\, x_3 x_5)^{N_v - d}}. \tag{3.89}$$

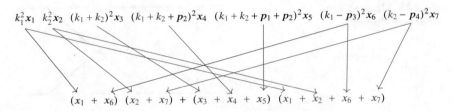

Fig. 3.14 Efficient factorization scheme for the U polynomial in Eq. (3.87)

In the next step, we apply the MB relation to U and F not expanding the term $(x_1 + x_6)(x_2 + x_7)$:

$$
B_7^{NP} = \frac{(-1)^{N_\nu}}{\Gamma(n_1)\ldots\Gamma(n_N)} \int\limits_{-i\infty}^{i\infty} \frac{dz_1}{2\pi i} \cdots \frac{dz_4}{2\pi i} \int dx_1 \ldots dx_7 \, (-s)^{-N_\nu + d - z_2 - z_3}(-t)^{z_2}(-u)^{z_3}
$$

$$
\times \, \Gamma(-z_1)\Gamma(-z_2)\Gamma(-z_3)\Gamma(-z_4)\Gamma(N_\nu - d + z_1 + z_2 + z_3 + z_4)
$$

$$
\times \, x_1^{-N_\nu + d - z_1 - z_2 - z_3} x_2^{z_2 + z_3} x_3^{-N_\nu + d - z_2 - z_3 - z_4} x_4^{z_1 + z_3} x_5^{z_2 + z_4} x_6^{z_1 + z_2} x_7^{z_3 + z_4}
$$

$$
\times \, (x_3 + x_4 + x_5 + (x_1 + x_6)(x_2 + x_7))^{N_\nu - \frac{3d}{2}}. \tag{3.90}
$$

To perform the integration over CW variables, we can use iteratively the following integration formula:

$$
\int\limits_0^\infty dx \, x^{z_1}(x + y)^{z_2} = \frac{y^{1 + z_1 + z_2}\Gamma(1 + z_1)\Gamma(-1 - z_1 - z_2)}{\Gamma(-z_2)}. \tag{3.91}
$$

In the last step, we apply the MB master relation of Eq. (3.31) to the terms $(x_1 + x_6)$ and $(x_2 + x_7)$. As we know from the previous section, the corresponding z-variables can be removed using first Barnes lemma. Finally, we end up with a four-dimensional MB representation,

$$
B_7^{NP} = \frac{(-1)^{N_\nu}}{\Gamma(n_1)\ldots\Gamma(n_7)} \int\limits_{-i\infty}^{i\infty} \frac{dz_1}{2\pi i} \cdots \frac{dz_4}{2\pi i} (-s)^{4 - 2\epsilon - N_\nu - z_{23}}(-t)^{z_3}(-u)^{z_2}
$$

$$
\frac{\Gamma(-z_1)\Gamma(-z_2)\Gamma(-z_3)\Gamma(-z_4)\Gamma(2 - \epsilon - n_{45})\Gamma(2 - \epsilon - n_{67})}{\Gamma(4 - 2\epsilon - n_{4567})\Gamma(n_{45} + z_{1234})\Gamma(n_{67} + z_{1234})\Gamma(6 - 3\epsilon - N_\nu)}
$$

$$
\Gamma(n_2 + z_{23})\Gamma(n_4 + z_{24})\Gamma(n_5 + z_{13})\Gamma(n_6 + z_{34})\Gamma(n_7 + z_{12})\Gamma^3(-2 + \epsilon + n_{4567} + z_{1234})
$$

$$
\Gamma(4 - 2\epsilon - n_{124567} - z_{123})\Gamma(4 - 2\epsilon - n_{234567} - z_{234})\Gamma(-4 + 2\epsilon + N_\nu + z_{1234}),
$$

$$
\tag{3.92}
$$

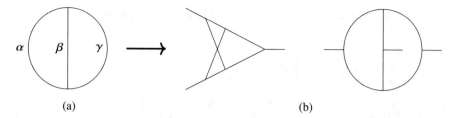

α β γ $\longrightarrow$

(a)   (b)

Fig. 3.15 The two-loop skeleton diagram (**a**) and creation of the two-loop non-planar vertex diagram (**b**). In (**b**) both diagrams are topologically equivalent

with notations $z_{i...j...k} = z_i + \ldots + z_j + \ldots + z_k$ and $n_{i...j...k} = n_i + \ldots + n_j + \ldots + n_k$.

3.9.2 General Two-Loop Skeleton Diagrams

As discussed in [37], *to get any two-loop diagram, it is sufficient to attach external lines/legs to the lines and/or vertices of the skeleton diagram given in* Fig. 3.15. Lines α, β, and γ in the skeleton diagram are also called the chains.

Non-planar Vertex from the Skeleton Diagram

To get a non-planar two-loop vertex diagram in Fig. 3.15b, one has to attach one external leg to each line on the skeleton diagram (a), while for the two-loop planar box, we would have to attach two external legs to the line α and two external legs to line γ. For the non-planar double box, we would have to attach two legs to line α and one leg to each of lines β and γ.

Let us consider the two-loop skeleton diagram in Fig. 3.15. A possible choice of the loop momenta k_1 and k_2 in the chains contains the following combinations of loop momenta: k_1, k_2 and $k_1 + k_2$ or $k_1 - k_2$ (depending on the orientation of momentum flow in the chains), plus corresponding external momenta, depending on the topology. Let us now introduce the following transformation of the Feynman parameters belonging to each chain:

$$\{\mathbf{x}\}_i : \quad x_k \to v_i \xi_{ik} \ \times \delta \left(1 - \sum_{k=1}^{\eta_i} \xi_{ik} \right), \tag{3.93}$$

where i denotes the chain index and $k \in [1, \eta_i]$, with η_i - the number of propagators in the chain. The δ-function keeps the number of variables unchanged.

Now the integration over Feynman parameters looks as follows:

$$\int_0^1 \prod_{i=1}^N dx_i \; \delta\left(1 - \sum_{i=1}^N x_i\right) \rightarrow$$

$$\int_0^1 \prod_{i=1}^3 dv_i v_i^{\eta_i - 1} \; \delta\left(1 - \sum_{i=1}^3 v_i\right) \int_0^1 \prod_{i=1}^3 \prod_{k=1}^{\eta_i} d\xi_{ik} \prod_{i=1}^3 \delta\left(1 - \sum_{k=1}^{\eta_i} \xi_{ik}\right). \qquad (3.94)$$

This transformation dramatically simplifies graph polynomials.

Change of Variables for CW Theorem

Let us consider a two-loop non-planar completely massless vertex, whose U and F are represented in Fig. 3.3; see the file `MB_V610m_Springer.nb` in the auxiliary material in [16] for detailed derivations. We can see that applying the MB formula to the constructed U and F terms in Fig. 3.3 would result in multi-dimensional MB representations, 11- and 5-dimensional, respectively. Let us see how the transformation of Eq. (3.93) will change the situation. The starting Feynman integral is

$$\int \frac{d^d k_1 d^d k_2}{[k_1^2]^{n_1}[(p_1 - k_1)^2]^{n_2}[(p_1 - k_1 - k_2)^2]^{n_3}[(p_2 + k_1 + k_2)^2]^{n_4}[(p_2 + k_2)^2]^{n_5}[k_2^2]^{n_6}}. \qquad (3.95)$$

The momentum-dependent function $m^2(\mathbf{x})$ (see Eq. (3.7)) and the parameter transformations for the integral are the following:

$$
\begin{aligned}
m^2(\mathbf{x}) &= x_1(p_1 - k_1 - k_2)^2 & x_1 &\rightarrow v_1 \xi_{11} \\
&+ x_2(p_2 + k_1 + k_2)^2 & x_2 &\rightarrow v_1 \xi_{12} \\
&+ x_3(k_1)^2 & x_3 &\rightarrow v_2 \xi_{21} \\
&+ x_4(p_1 - k_1)^2 & x_4 &\rightarrow v_2 \xi_{22} \\
&+ x_5(p_2 + k_2)^2 & x_5 &\rightarrow v_3 \xi_{31} \\
&+ x_6(k_2)^2 & x_6 &\rightarrow v_3 \xi_{32}.
\end{aligned}
\qquad (3.96)
$$

Utilizing the δ-functions $\prod_{i=1}^3 \delta\left(1 - \sum_{k=1}^{\eta_i} \xi_{ik}\right)$ in Eq. (3.94), the first Symanzik polynomial U for any two-loop diagram becomes

$$U(\mathbf{x}) \rightarrow U(\mathbf{v}) = v_1 v_2 + v_2 v_3 + v_1 v_3. \qquad (3.97)$$

Due to dependence on kinematic variables, the second Symanzik polynomial F has more complicated structure and depends on a definite topology. In our case, after the transformation in Eq. (3.96), we get

$$F = -s\xi_{11}\xi_{22}\xi_{31}v_1v_2v_3 - s\xi_{12}\xi_{21}\xi_{32}v_1v_2v_3 - s\xi_{31}\xi_{32}v_1v_3^2 - s\xi_{31}\xi_{32}v_2v_3^2. \tag{3.98}$$

In summary, using Eq. (3.31) after rescaling in Eq. (3.97), for *any* two-loop diagram, the U polynomial can generate maximally a two-dimensional MB integral. Let us see now how applying the CW Theorem 3.1 we can simplify further U and F structures, and so decrease the dimensionality of MB representations.

Applying the CW Theorem

Looking on the homogeneity of variables v and ξ in Eq. (3.98), we can see that the CW Theorem 3.1 can be applied only to the v-variables. Thus, according to [37], we can generalize this fact of the homogeneity of the v-variables: *For any two-loop diagram, the graph polynomials U and F can be represented formally in the form corresponding to the so-called sunrise diagram* in Fig. 3.16, where p^2 and m_i^2 depend on ξ_{ik} and the kinematic invariants (S) of the initial diagram:

$$G(X) \sim \int \prod d\xi_{ik}\delta\left(1 - \sum_k \xi_{ik}\right)$$

$$\int \frac{d^d k_1 d^d k_2}{[k_1^2 - m_1^2(S, \xi_{ik})]^{n_1}[k_2^2 - m_2^2(S, \xi_{ik})]^{n_2}[(p + k_1 + k_2)^2 - m_3^2(S, \xi_{ik})]^{n_3}}. \tag{3.99}$$

In case of the considered massless non-planar vertex in Eq. (3.95), we have specifically

$$p^2 = -s(\xi_{12}\xi_{22}\xi_{31} + \xi_{12}\xi_{21}\xi_{32} - \xi_{31}\xi_{32}), \tag{3.100}$$

Fig. 3.16 The general sunrise diagram

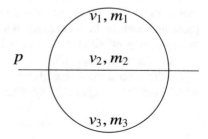

$$m_1^2 = -s\xi_{31}\xi_{32}, \quad m_2^2 = m_3^2 = 0. \tag{3.101}$$

In order to use the CW Theorem 3.1, there are only three possibilities to choose a subset in $\delta\left(1 - \sum_{i=1}^{3} v_i\right)$ during the transformation of Eq. (3.93). This choice depends on the structure of F; from $v_1 + v_2$, $v_2 + v_3$, or $v_1 + v_3$, one should take the combination which makes F as simple as possible. In case of Eq. (3.98), it is $v_1 + v_2$, so we have

$$U = v_3 + v_1 v_2, \tag{3.102}$$

$$F = -s\xi_{11}\xi_{22}\xi_{31}v_1 v_2 v_3 - s\xi_{12}\xi_{21}\xi_{32}v_1 v_2 v_3 - s\xi_{31}\xi_{32}v_1 v_3^2. \tag{3.103}$$

Applying the MB master formula to F and integrating over v_3, we get a twofold MB integral

$$
I_{V6l0m}^{MB} = (-1)^{N_v}(-s)^{4-2\epsilon-N_v} \int_{-i\infty}^{i\infty} \int_{-i\infty}^{i\infty} \frac{dz_1}{2\pi i} \frac{dz_2}{2\pi i}
$$
$$
\times \frac{\Gamma(2-\epsilon-n_{12})\Gamma(2-\epsilon-n_{56})\Gamma(4-2\epsilon-n_{12356}-z_1)\Gamma(-z_1)\Gamma(n_1+z_1)}{\Gamma(n_1)\Gamma(n_2)\Gamma(n_3)\Gamma(n_4)\Gamma(n_5)\Gamma(4-2\epsilon-n_{1256})\Gamma(n_6)\Gamma(n_5+z_1)}
$$
$$
\times \frac{\Gamma(-z_2)\Gamma(n_2+z_2)\Gamma(n_6+z_2)\Gamma(-4+2\epsilon+N_v+z_{12})}{\Gamma(8-4\epsilon-2n_{1256}-n_{34}-z_{12})\Gamma(n_{12}+z_{12})\Gamma(n_{56}+z_{12})\Gamma(4-2\epsilon-n_{12456}-z_2)}. \tag{3.104}
$$

> In case of massive diagrams, namely, when some of the propagators have masses m_i^2, the best option is to not expand in the first place the term $U \sum x_i m_i^2$, and instead to apply Eq. (3.31) to F *considering U as an independent variable.*

Adding a mass parameter to one of the propagators in the non-planar vertex results only in one additional MB integration. However, this is not a general way, for example, in case of the non-planar QED double box in Fig. 3.13 with $p_i^2 = m^2$ and masses m^2 in propagators with powers n_1, n_3, n_4, and n_7 in Eq. (3.84), the above procedure leads to the appearance of a "pseudo" Minkowskian factor $(-m^2)^{z_i}$ in the MB representation, similar to the case discussed already in Sect. 3.7 (see Eq. (3.68)), which restricts this approach.

On the other hand, using LA, one gets an eight-dimensional representation [21] which points to the applicability of LA also to some non-planar diagrams.

3.9.3 Generalization to Three-Loop Integrals

At the three-loop level, there are two generating topologies indicated in Fig. 3.17 with different properties [37]. The diagram on the left side of Fig. 3.17 and all its derivatives can be cut to two disconnected one-loop pieces. Propagators corresponding to the three-loop diagrams can have two combinations of two loop momenta (if three different loop momenta in one propagator are present, one loop momentum can be always eliminated). In the three-loop case, the number of v-variables in the general transformation rule in Eq. (3.93) is five. Typical examples of these types of diagrams for the box topologies are depicted in Fig. 3.18.

For the left skeleton diagram in Fig. 3.17, after a transformation of variables, all diagrams up to replacing of indices have a U polynomial in the following form (Problem 3.10):

$$U = v_1 v_2 v_3 + v_1 v_2 v_4 + v_2 v_3 v_4 + v_1 v_2 v_5 + v_1 v_3 v_5 + v_2 v_3 v_5 + v_1 v_4 v_5 + v_3 v_4 v_5.$$
(3.105)

Many diagrams of this type have planar subloops and can be treated in a different way; see the next section.

The diagram on the right side in Fig. 3.17 cannot be divided into two one-loop pieces. Its propagators have all three different combinations of two loop momenta and number of v-variables is six. This skeleton generates the most complicated non-planar topologies; see, for example, Fig. 3.19 and Problem 3.13.

All child diagrams have now again the same U function in the form

$$U = v_1 v_2 v_3 + v_1 v_2 v_4 + v_1 v_3 v_4 + v_1 v_2 v_5 + v_1 v_3 v_5 + v_2 v_3 v_5 + v_2 v_4 v_5 + v_3 v_4 v_5$$
$$+ v_1 v_2 v_6 + v_2 v_3 v_6 + v_1 v_4 v_6 + v_2 v_4 v_6 + v_3 v_4 v_6 + v_1 v_5 v_6 + v_3 v_5 v_6 + v_4 v_5 v_6.$$
(3.106)

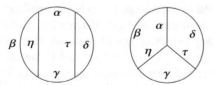

Fig. 3.17 Basic skeleton generating diagrams with propagators $\alpha, \ldots, \tau$ for the three-loop topologies discussed in [37], so-called `Ladder` (on left) and `Mercedes` (on right) topologies

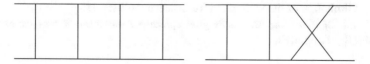

Fig. 3.18 The three-loop box topologies obtained from the `Ladder` skeleton diagram in Fig. 3.17. Diagrams originated from this skeleton have the same U polynomial structure

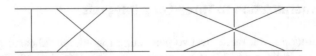

Fig. 3.19 The three-loop box topologies obtained from the right skeleton diagram in Fig. 3.17 and that have another U polynomial structure

As in the two-loop case, further simplification can be done with the help of the CW Theorem 3.1. The shortest form of U can be obtained if we choose three CW variables. An analysis of the form of U shows that there are four different possibilities to choose CW variables. For Eq. (3.106) one of them is

$$\int_0^\infty dv_2 dv_3 dv_4 \int_0^1 dv_1 dv_5 dv_6 \delta(1 - v_1 - v_5 - v_6),$$ (3.107)

which gives

$$U_{CW} = v_2 v_3 + v_2 v_4 + v_3 v_4 + v_1 v_2 v_5 + v_1 v_3 v_5 + v_1 v_2 v_6 + v_1 v_4 v_6 + v_1 v_5 v_6 + v_3 v_5 v_6 + v_4 v_5 v_6$$ (3.108)

and reduces the number of terms in U from 16 to 10.

One of possible ways to make the integration over suitable CW variables is to use the following factorization trick:

$$U_{CW} = v_2(v_3 + v_4 + v_1 v_5) + v_3(v_4 + v_1 v_5) + v_1 v_6(v_2 + v_5) + v_4 v_6(v_1 + v_5) + v_3 v_5 v_6.$$ (3.109)

There are six different possibilities to get four terms in U and altogether we have 24 variants to choose CW variables and factorize U. A final choice is based, first, on a minimization of the amount of terms in F and, second, on the presence in F of the same factorization patterns as in Eq. (3.109) for U.

Now, to construct the MB representation as in the two-loop case, we do not have to necessarily expand mass terms in F (see the previous section). Similarly, it is not necessary to expand any factorized combination which corresponds to the pattern in Eq. (3.109). After applying Eq. (3.31) to U and F, we can integrate recursively the polynomials over v_3 and then over v_4 using Eq. (3.91). The integration over v_2 can be also done using Eq. (3.91). Finally, we apply again Eq. (3.31) to the term $v_1 + v_5$. In the end this MB integration can be removed using 1BL. As one can see within this algorithm, the U polynomial gives already 4 MB integrations. F is usually more complex and the final MB representation can be very high dimensional. This is a natural limitation of the MB method for massive multileg FI.

As a rule of thumb, the GA usually gives optimal representations if from the beginning Length[U] $\leq$ Length[F].

Fig. 3.20 The three-loop planar self-energy diagram I_{SE610m} defined in Eq. (3.110)

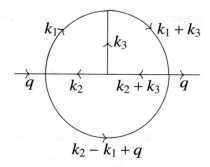

This example shows an application of GA and 1BL to a planar three-loop self-energy massless diagram, resulting in the same MB integral dimension as by LA. The integral is

$$I_{\text{SE610m}} = \int \frac{dk_1 dk_2 dk_3}{[k_1^2]^{n_1} [k_2^2]^{n_2} [k_3^2]^{n_3} [(k_1 + k_3)^2]^{n_4} [(k_2 + k_3)^2]^{n_5} [(k_2 - k_1 + q)^2]^{n_6}}. \tag{3.110}$$

The diagram is given in Fig. 3.20 where the corresponding MB representation can be found in the file MB_SE610m_Springer.nb in the auxiliary material in [16].

The LA method gives here four-dimensional MB representation; see Section E4 in [38] and Problem 3.14. This diagram corresponds to the right skeleton diagram in Fig. 3.17 with two external legs attached to vertices on the circle, so variable transformation of the type in Eq. (3.93) is not needed. The U polynomial has the same structure as in Eq. (3.109); corresponding F polynomial can be simplified to the following form:

$$F = -sv_3v_4v_6(v_1 + v_5) - sv_2v_3(v_4 + v_1v_5) - sv_1v_3v_6(v_2 + v_5). \tag{3.111}$$

This means three additional MB integrations plus four integrations which come from the U polynomial. The resulting seven-dimensional output from AMBREv3.2 is given in the file MB_SE610m_Springer.nb in the auxiliary material in [16] (see also the next page for the MB representation in the framed Eq. (3.114)). The integration over z7 corresponds to the term $v_1 + v_5$ and can be immediately removed by applying 1BL and making the shift z6 -> z6 - z3 - z2; one can apply 1BL to the variable z2 and 2BL to the variable z3, obtaining a four-dimensional representation as in case of the LA. The shift in momenta was found using the algorithm described in Sect. 3.6.

To complete this section and the discussion of the GA method for three-loop cases, we present a strategy of choosing CW variables and a factorization for the first type of three-loop diagrams in Fig. 3.18. Because here we have only five

v-variables, the optimal choice is to consider two CW variables. For Eq. (3.11) this leads to

$$\int_0^\infty dv_1 dv_2 \int_0^1 dv_3 dv_4 dv_5 \delta(1 - v_3 - v_4 - v_5) \tag{3.112}$$

and

$$U_{CW} = v_1 v_2 + v_2 v_3 v_4 + v_1 v_3 v_5 + v_2 v_3 v_5 + v_1 v_4 v_5 + v_3 v_4 v_5. \tag{3.113}$$

```
{(((-s)^(-3 eps))
   Gamma[-z1] Gamma[-z2] Gamma[3 eps + z1 + z2]
   Gamma[-z3] Gamma[-z4] Gamma[1 + z1 + z4]
   Gamma[-z5] Gamma[1 - 3 eps + z3 + z5]
   Gamma[1 - eps + z1 + z3 + z4 + z5]
   Gamma[-2 + 3 eps - z1 - z2 - z3 - z4 - 2 z5 - z6]
   Gamma[-1 + eps - z2 - z5 - z6]
   Gamma[-1 + 3 eps - z3 - z5 - z6]
   Gamma[1 + z2 + z6] Gamma[2 - 4 eps + z3 + z4 + z5 + z6]
   Gamma[2 - 2 eps + z1 + z2 + z3 + z4 + z5 + z6 - z7]
   Gamma[-z7] Gamma[-z1 - z4 + z7] Gamma[1 - 2 eps + z5 + z7])/
   (Gamma[2 - 4 eps] Gamma[-z1 - z4]
   Gamma[2 - 3 eps + z1 + z3 + z4 + z5]
   Gamma[-1 + 3 eps - z2 - z3 - 2 z5 - z6]
   Gamma[2 - eps + z1 + z2 + z3 + z4 + z5 + z6])}
```

$$\tag{3.114}$$

This operation reduces the number of terms in U from 8 to 6. As in the previous case, there are also four possibilities to choose CW variables. A factorization scheme to integrate over CW variables can be the following:

$$U_{CW} = v_2(v_1 + v_3 v_4) + v_3 v_5(v_2 + v_4) + v_1 v_5(v_3 + v_4). \tag{3.115}$$

Due to the smaller number of variables and terms, there are now only four possibilities to factorize U. All other steps go the same way, as for Eq. (3.109). After applying Eq. (3.31) to U and F, we integrate over v_1 and v_2 using Eq. (3.91). The combination $v_3 + v_4$ again can be removed using 1BL. In this case, U polynomial results in two additional MB integrations.

3.10   Hybrid (HA) Approach

At the three-loop, we can construct MB representations by combining the LA and GA approaches. This depends on whether or not a given topology includes a planar sub-topology which can be disconnected or not. In the first case, we start with the LA and integrate over one of the loop momenta. After that the obtained effective two-loop diagram is treated by the GA. In Fig. 3.21, propagators connected with a planar

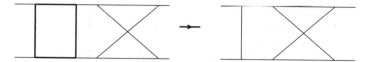

Fig. 3.21 Hybrid approach. LA is applied first to the one-loop subloop, H $(1 \to 2)$

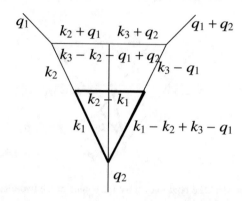

Fig. 3.22 Hybrid method, an example. The bold lines show the choice of the one-loop sub-diagram in the first step of H $(1 \to 2)$

subloop in the form of the one-loop box are transformed into an MB form in a first step, defining a new effective propagator. In this way an effective two-loop diagram is created.

Another example of this kind is a non-planar vertex shown in Fig. 3.22. A derivation of the corresponding representation is given in the first section of the file MB_AMBREnew_Springer.nb in the auxiliary material in [16]. Typically, the combination of three different loop momenta in one propagator usually does not take place; they appear in Fig. 3.22 due to the procedure of generation of diagrams using some automatic codes like FeynArts or qgraph. In this particular case, that plays no role because in the first iteration the momentum k_1 circulates only in the one-loop sub-diagram shown in bold, and for GA, in the second iteration, the momentum flow is not important. Nonetheless, the momentum flow is important in general each loop momentum should go through a minimal possible topological construction. For example, due to some specific construction procedure, k_1 could appear in more than three propagators. In this case a preliminary shift of loop momenta is needed for a successful/efficient application of HA.

In Fig. 3.23, a non-planar disconnected subgraph can be identified, and we go in the opposite direction; first, we apply the GA to a two-loop sub-diagram, basically a non-planar one, and the remaining one-loop integral is processed in a simple way by the LA.

Yet another example for HA $(2 \to 1)$ is shown in Fig. 3.24. In the first iteration, the two-loop non-planar vertex effectively has two off-shell external legs, so after transformation of Eq. (3.93), in contrast to the non-planar vertex from example

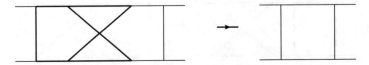

Fig. 3.23 Hybrid approach. The GA is applied first to the two-loop sub-diagram, H $(2 \to 1)$

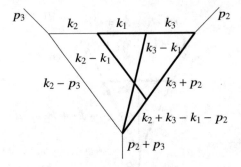

Fig. 3.24 Hybrid approach. The bold lines show the choice of the two-loop sub-diagram in the first step of the H $(2 \to 1)$ procedure

in Sect. 3.9.2, one gets a six-dimensional representation. The remaining three propagators form one-loop vertex and give no additional MB integration.

3.11   Minimal Dimensions of Symanzik Polynomials: Summary

For the two-loop skeleton diagram in Fig. 3.15, rescaling of U gives a two-dimensional MB integral (see Eq. (3.97))

$$U_{2\text{-loop}} = v_1 v_2 + v_2 v_3 + v_1 v_3.$$

For the three-loop Ladder skeleton diagram in Fig. 3.17, rescaling of U gives a 7-dim MB integral (see Eq. (3.105))

$$U_{3\text{-loop(I)}} = v_1 v_2 v_3 + v_1 v_2 v_4 + v_2 v_3 v_4 + v_1 v_2 v_5 + v_1 v_3 v_5 + v_2 v_3 v_5 + v_1 v_4 v_5 + v_3 v_4 v_5.$$

For the three-loop Mercedes skeleton diagram in Fig. 3.17, rescaling of U gives a 15-dim MB integral (see Eq. (3.106))

$$U_{3\text{-loop(II)}} = v_1 v_2 v_3 + v_1 v_2 v_4 + v_1 v_3 v_4 + v_1 v_2 v_5 + v_1 v_3 v_5 + v_2 v_3 v_5 + v_2 v_4 v_5 + v_3 v_4 v_5$$

$$+ v_1 v_2 v_6 + v_2 v_3 v_6 + v_1 v_4 v_6 + v_2 v_4 v_6 + v_3 v_4 v_6 + v_1 v_5 v_6 + v_3 v_5 v_6 + v_4 v_5 v_6.$$

Finally, using CW Theorem 3.1, we get further simplifications:

- For any two-loop diagram: $\delta(1 - v_1 - v_2)$, $U(\mathbf{v}) = v_3 + v_1 v_2$.
  There is no additional term in U, the situation is as simple as in the one-loop case, see the discussion around Eq. (3.102).
- For any three-loop diagram: $\delta(1 - v_1 - v_2 - v_3)$.
  - Ladder—U generates two additional MB integrations; see a discussion around Eq. (3.115).
  - Mercedes—U generates four additional MB integrations; see a discussion around Eq. (3.109).

For the F polynomial situation is more complicated and additional dimensions of MB integrals depend on kinematics and internal masses. As discussed, the expression $F = F_0 + U \sum x_i m_i^2$ must be treated individually for each FI, without or with suitable expansions of F_0 and U terms.

3.12 Beyond Three Loops

At four loops planar diagrams can still be treated with LA; the complexity of U and F leaves not much space for constructing good MB representations using the GA. Interestingly, the hybrid method opens a possibility to build MB representations beyond three loops. For one of the possible implementations, see [39]. However, these issues go beyond the content of this textbook.

3.13 MB and the Method of Brackets

In [40] a very interesting approach to the construction of MB representations was introduced. The approach is based on the 'Method of Brackets' defined in [41–43], which is a multi-fold construction of sums starting directly from a Schwinger-parameterized Feynman integral presented in Chap. 3. The technique of brackets transforms the parameter integral into a series-like structure called "brackets expansion" with four basic rules.

1. Exponential function expansion

$$\exp(-xA) = \sum_n \frac{(-1)^n}{\Gamma(n+1)} x^n A^n. \tag{3.116}$$

If the argument of exponential function is $\exp(xA)$,

$$\exp(xA) = \sum_n \frac{(-1)^n}{\Gamma(n+1)} x^n (-A)^n. \tag{3.117}$$

The reason for this is to associate to each expansion the factor $\phi_n = \frac{(-1)^n}{\Gamma(n+1)}$ as a simple convention.

2. Integration symbol and its equivalent bracket

This rule corresponds to the definition of the bracket symbol. The structure $\int x^{a_1+a_2+\ldots+a_n-1}\, dx$ is replaced by its respective bracket representation

$$\int x^{a_1+a_2+\ldots+a_n-1} dx = \langle a_1 + a_2 + \ldots + a_n \rangle. \tag{3.118}$$

3. Polynomials expansion

For polynomials the following representation in terms of series of brackets is used:

$$(A_1 + \ldots + A_r)^{\pm\mu}$$
$$= \sum_{n_1} \ldots \sum_{n_r} \phi_{n_1} \ldots \phi_{n_r} \, (A_1)^{n_1} \ldots (A_r)^{n_r} \times \frac{\langle \mp\mu + n_1 + \ldots + n_r \rangle}{\Gamma(\mp\mu)}. \tag{3.119}$$

This rule is derived using rules 1 and 2 after applying Schwinger's parametrization to this polynomial.

4. Finding the solution

For the case of a generic series of brackets J

$$J = \sum_{n_1} \ldots \sum_{n_r} \phi_{n_1} \ldots \phi_{n_r} \, F(n_1, \ldots, n_r) \tag{3.120}$$
$$\times \langle a_{11} n_1 + \ldots + a_{1r} n_r + c_1 \rangle \ldots \langle a_{r1} n_1 + \ldots + a_{rr} n_r + c_r \rangle,$$

the solution is obtained using the general formula

$$J = \frac{1}{|\det(A)|} \Gamma(-n_1^*) \ldots \Gamma(-n_r^*) F(n_1^*, \ldots, n_r^*) \tag{3.121}$$

Table 3.3 The number of MB integrations of the representation constructed by the method of brackets and AMBRE for cases (a)–(e) given in Fig. 3.25. P (NP) stands for planarity (non-planarity) of the diagram. The last column gives improvements discussed in the text when using the HA method

Diagram	Method of brackets	AMBRE	Planarity	AMBRE *
(a)	7	13	NP	**4**, H(1 → 2)
(b)	1	2	P	**1**
(c)	7	9	NP	**5**, H(1 → 2)
(d)	7	8	NP	**8**, H(1 → 2)
(e)	5	3	P	**3**

where $\det(\mathbf{A})$ is evaluated by the following expression

$$\det(\mathbf{A}) = \begin{vmatrix} a_{11} & \cdots & a_{1r} \\ \vdots & \ddots & \vdots \\ a_{r1} & \cdots & a_{rr} \end{vmatrix}, \tag{3.122}$$

and $\{n_i^*\}$ $(i = 1, \ldots, r)$ is the solution of the linear system obtained by the vanishing of the brackets

$$\begin{cases} a_{11}n_1 + \ldots + a_{1r}n_r = -c_1 \\ \qquad \vdots \qquad\qquad \vdots \\ a_{r1}n_1 + \ldots + a_{r}n_r = -c_r. \end{cases} \tag{3.123}$$

The value of J is not defined if the matrix $\mathbf{A}$ is not invertible.

The relation in Eq. (3.121) is the Generalized Ramanujan's Master Theorem (GRMT) derived in [42]. For simple application of the above formalism to a construction of MB integrals, based on RMTG, see Problem 3.15 and original discussion in [42], Problem 3.16.

Here we compare some basic results from this method of brackets with results by AMBRE packages, in which the above discussed three-loop non-planar cases were optimized.

By AMBRE* in Table 3.3, we denote AMBREv1 for planar cases and manual combination of AMBREv1 and AMBREv3 for the hybrid method. For the AMBRE branches v1 and v2 the latest versions of the packages from [13] are assumed. The AMBRE* results from Table 3.3 are given in the file MB_AMBREnew_Springer.nb in the auxiliary material in [16].

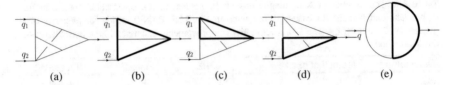

(a)   (b)   (c)   (d)   (e)

Fig. 3.25 Examples of two- and three-loop diagrams. Bold (thin) lines represent massive (massless) propagators. Taken from [40], results are discussed in Table 3.3

3.14   Phase Space MB Integrals

When computing predictions for physical observables such as cross sections or decay rates in perturbative quantum field theory, in addition to the Feynman integrals discussed above, we also encounter integrations over the momenta of outgoing external particles in a scattering or decay process. We will refer to such integrals as phase space integrals, since the associated integration measure is commonly referred to as the phase space measure. In particular, the measure for n outgoing particles of total momentum Q in d space-time dimensions reads

$$d\phi_n(p_1, \ldots, p_n; Q) = \prod_{k=1}^{n} \frac{d^d p_k}{(2\pi)^{d-1}} \delta_+(p_k^2 - m_k^2)(2\pi)^d \delta^{(d)}(p_1 + \ldots + p_n - Q), \quad (3.124)$$

where $\delta^{(d)}$ is the d-dimensional Dirac delta function while

$$\delta_+(p_k^2 - m_k^2) = \delta(p_k^2 - m_k^2)\Theta(p_k^0) \quad (3.125)$$

and the various factors of 2π are conventional. We see that in contrast to loop integrals, here the integrations over the momenta are constrained by both the mass-shell conditions (i.e., $p_k^2 = m_k^2$) and overall momentum conservation. Thus, it is less straightforward to develop general representations of phase space integrals than for loop integrals. Thus, in practical applications one often simply chooses some particular explicit parametrization of the components of the outgoing momenta, resolves the constraints, and attempts to evaluate the resulting parametric integral. We note in passing that the resolution of the constraints can lead to very elaborate expressions and so it can be crucial to correctly tailor the chosen parametrization to the problem at hand.

A Three-Particle Phase Space Integral

As an example, consider the three-particle phase space where all three outgoing momenta are massless, $p_1^2 = p_2^2 = p_3^2 = 0$. A typical situation is that one must compute the integral of some function of the invariants $s_{ij} = 2p_i \cdot p_j$, $i, j = 1, 2, 3$ over the full phase space. For example, consider the simple example

$$
\begin{aligned}
I &= \int d\phi_3(p_1, p_2, p_3; Q) \frac{1}{s_{13}s_{23}} \\
&= \int \frac{d^d p_1}{(2\pi)^{d-1}} \delta_+(p_1^2) \frac{d^d p_2}{(2\pi)^{d-1}} \delta_+(p_2^2) \frac{d^d p_3}{(2\pi)^{d-1}} \delta_+(p_3^2) \\
&\quad \times (2\pi)^d \delta^{(d)}(Q - p_1 - p_2 - p_3) \frac{1}{s_{13}s_{23}}.
\end{aligned}
\tag{3.126}
$$

In order to compute this integral, let us first use the Dirac delta function expressing overall momentum conservation to integrate out, e.g., the momentum p_3,

$$
I = (2\pi)^{3-2d} \int d^d p_1 \, \delta_+(p_1^2) d^d p_2 \, \delta_+(p_2)\delta_+(p_3^2)\delta_+[(Q - p_1 - p_2)^2] \frac{1}{s_{13}s_{23}}.
\tag{3.127}
$$

Now, since both the integration measure and the integrand are rotationally invariant, we can choose to work in the following convenient Lorentz frame:

$$
\begin{aligned}
Q^\mu &= \sqrt{s}(1, \mathbf{0}_{d-1}), \\
p_1^\mu &= E_1(1, \mathbf{0}_{d-2}, 1), \\
p_2^\mu &= E_2(1, \mathbf{0}_{d-3}, \sin\theta, \cos\theta), \\
p_3^\mu &= Q - p_1 - p_2.
\end{aligned}
\tag{3.128}
$$

Above $\mathbf{0}_j$ denotes a vector of j zeros. It is not difficult to show that in this frame we may write the various pieces of the integration measure as follows:

$$
\begin{aligned}
d^d p_1 \delta_+(p_1^2) &= \frac{E_1^{d-3}}{2} dE_1 \, d\Omega_{d-1} \Theta(E_1), \\
d^d p_2 \delta_+(p_2^2) &= \frac{E_2^{d-3}}{2} dE_2 \, d(\cos\theta) (\sin\theta)^{d-4} d\Omega_{d-2} \Theta(E_2),
\end{aligned}
\tag{3.129}
$$

where $d\Omega_j$ is the measure for the integration over the j angular variables on which the integrand does not depend (i.e., the angles of rotations that are necessary to bring a general configuration to the form of Eq. (3.128)). On the other hand, the invariants in this frame take the form

$$
\begin{aligned}
s_{12} &= 2E_1 E_2(1 - \cos\theta), \\
s_{13} &= 2E_1\sqrt{s} - 2E_1 E_2(1 - \cos\theta), \\
s_{23} &= 2E_2\sqrt{s} - 2E_1 E_2(1 - \cos\theta).
\end{aligned}
\tag{3.130}
$$

Then, our integral in Eq. (3.127) takes the form

$$I = (2\pi)^{3-2d}2^{-2}\Omega_{d-1}\Omega_{d-2} \int dE_1\, dE_2\, (E_1 E_2)^{d-3} \int_{-1}^{1} d(\cos\theta)(\sin\theta)^{d-4}$$
$$\times \delta[s - 2E_1\sqrt{s} - 2E_2\sqrt{s} + 2E_1 E_2(1 - \cos\theta)]\Theta(\sqrt{s} - E_1 - E_2)\Theta(E_1)\Theta(E_2)$$
$$\times \frac{1}{[2E_1\sqrt{s} - 2E_1 E_2(1 - \cos\theta)][2E_2\sqrt{s} - 2E_1 E_2(1 - \cos\theta)]}.$$

$$(3.131)$$

Here we have used that $\delta_+(p^2) = \delta(p^2)\Theta(p^0)$ and we have been careful to indicate all constraints on the energy integrals. It is possible to evaluate this integral by solving the delta function for one of the energies, which leads to an integration over the angle θ and the other energy, E_1. Being careful to determine the correct limits of integration on the remaining variables, we find that the result can be expressed in terms of just gamma functions; see Problem 3.17. Below we will have much more to say about how the MB method can be used to compute the angular parts of such phase space integrals.

However, in order to highlight the simplifications that can result from a well-chosen parametrization, let us not follow this path, but instead choose a new set of integration variables, namely, the invariants s_{12}, s_{13}, and s_{23}. Obviously we can use Eq. (3.130) to solve for E_1, E_2, and $\cos\theta$ in terms of these invariants. Performing the change of variables, we find simply that

$$I = (2\pi)^{3-2d}2^{-1-d}s^{\frac{2-d}{2}}\Omega_{d-1}\Omega_{d-2} \int ds_{12}\, ds_{13}\, ds_{23}\, (s_{12}s_{13}s_{23})^{\frac{d-4}{2}}$$
$$(3.132)$$
$$\times \delta(s - s_{12} - s_{13} - s_{23})\frac{1}{s_{13}s_{23}},$$

where all integrations run between 0 and s. This last integral is very easy to evaluate by resolving the delta function for s_{12} and we find simply

$$I = \frac{2^{-8+6\epsilon}\pi^{-\frac{5}{2}+2\epsilon}\Gamma^2(-\epsilon)}{\Gamma(1 - 3\epsilon)\Gamma\left(\frac{3}{2} - \epsilon\right)}s^{-1-2\epsilon}.$$

$$(3.133)$$

Here we have set $d = 4 - 2\epsilon$ and used that the total angular volume in p dimensions is $\Omega_p = \frac{2\pi^{p/2}}{\Gamma(p/2)}$ (see Eq. (3.138)).

As a final word of caution, we mention that beyond three-particle integrals, the parametrization in terms of invariants is oftentimes not as useful as our example might lead one to believe. The reason behind this is that although the measure of integration can be written in a very nice and compact form also for phase spaces of higher multiplicity (see, e.g., [44, 45]), the limits of integration are actually quite nontrivial to express in a way that is convenient for actual calculations.

In practical applications the parametrizations many times involve angles between three-momenta, defined in some suitable Lorentz frame, and so the evaluation of phase space integrals will involve the computation of integrals over these angles [46]. Such angular integrals are particularly well suited to treatment by the Mellin-Barnes method, as we will describe below.

To start, let us define what we will call an angular integral with n denominators:

$$\Omega_{j_1,\ldots,j_n} \equiv \int d\Omega_{d-1}(q) \frac{1}{(p_1 \cdot q)^{j_1} \ldots (p_n \cdot q)^{j_n}}. \tag{3.134}$$

Here $p_1^\mu, \ldots, p_n^\mu$ are fixed d-dimensional vectors, and we are integrating over angular variables of the massless d-dimensional vector q^μ. Thus $d\Omega_{d-1}(q)$ is the rotationally invariant angular measure in d dimensions, whose explicit expression will be given in Eq. (3.138). It turns out that the general integral in Eq. (3.134) admits a very nice representation in terms of Mellin-Barnes integrals which can be a suitable starting point for both analytic and numerical evaluation.

To derive this representation, we begin by noting that the overall normalization of the p_i^μ and q^μ clearly does not play an essential role, and so without loss of generality, we can simply choose to normalize these vectors in whatever way is most convenient. Thus, we choose a Lorentz frame where

$$p_1^\mu = (1, \mathbf{0}_{d-2}, \beta_1),$$

$$p_2^\mu = (1, \mathbf{0}_{d-3}, \beta_2 \sin \chi_2^{(1)}, \beta_2 \cos \chi_2^{(1)}),$$

$$p_3^\mu = (1, \mathbf{0}_{d-4}, \beta_3 \sin \chi_3^{(2)} \sin \chi_3^{(1)}, \beta_3 \cos \chi_3^{(2)} \sin \chi_3^{(1)}, \beta_3 \cos \chi_3^{(1)}),$$

$$\vdots \tag{3.135}$$

$$p_n^\mu = (1, \mathbf{0}_{d-1-n}, \beta_n \prod_{k=1}^{n-1} \sin \chi_n^{(k)},$$

$$\beta_n \cos \chi_n^{(n-1)} \prod_{k=1}^{n-2} \sin \chi_n^{(k)}, \ldots, \beta_n \cos \chi_n^{(2)} \sin \chi_n^{(1)}, \beta_n \cos \chi_n^{(1)}).$$

Again, $\mathbf{0}_j$ denotes a vector of j zeros, and thus in words, we have chosen a frame where the direction of p_1^μ fixes the d-th axis, then p_2^μ fixes the plane of the d-th and $(d-1)$-st axis, and so on. Notice that we have written all vectors in d-dimensional polar coordinates and used the freedom to fix the normalizations to fix each zeroth component to be one, $p_j^0 = 1$. In this frame, q^μ can be written in the following form, again using d-dimensional polar coordinates:

$$
q^\mu = \left(1, ..\text{"angles"}.., \cos\vartheta_n \prod_{k=1}^{n-1}\sin\vartheta_k, \cos\vartheta_{n-1}\prod_{k=1}^{n-2}\sin\vartheta_k, \ldots, \cos\vartheta_2\sin\vartheta_1, \cos\vartheta_1\right).
$$
$$(3.136)$$

Here ..."angles".. denotes those angular variables on which the integral does not depend and that can thus be integrated trivially. Also, as before, we have used the freedom to fix the normalization to set $q^0 = 1$. It can be shown that the angular measure appearing in Eq. (3.134) above can be written as follows:

$$
d\Omega_{d-1}(q) = \prod_{k=1}^{n} d(\cos\vartheta_k)(\sin\vartheta_k)^{-k+1-2\epsilon}\, d\Omega_{d-1-n}(q) \tag{3.137}
$$

where we have now set $d = 4 - 2\epsilon$. Notice that $d\Omega_{d-1-n}(q)$ corresponds to the angular measure of the variables denoted simply as ..."angles".. in Eq. (3.136). For later use, we note that the total angular volume in p dimensions is simply (Problem 3.18)

$$
\Omega_p = \int d\Omega_p(q) = \frac{2\pi^{\frac{p}{2}}}{\Gamma\left(\frac{p}{2}\right)}. \tag{3.138}
$$

For a sanity check, notice that the angular volume Ω_p is just the integral of the angular part of the total volume measure in polar coordinates in p dimensions, i.e., the surface of the p-dimensional unit sphere. Hence in $p = 2$ we expect $\Omega_2 = 2\pi$, which is clearly in agreement with Eq. (3.138). For $p = 3$ Eq. (3.138) gives $\Omega_3 = 2\pi^{\frac{3}{2}}/\Gamma(\frac{3}{2})$, but $\Gamma(\frac{3}{2}) = \sqrt{\pi}/2$, so in fact we obtain the expected result of $\Omega_3 = 4\pi$.

Then, inserting Eq. (3.137) into Eq. (3.134), we find the explicit integral representation

$$
\Omega_{j_1,\ldots,j_n} = \int d\Omega_{d-1-n}(q) \int_{-1}^{1} \prod_{k=1}^{n}\left[d(\cos\vartheta_k)(\sin\vartheta_k)^{-k+1-2\epsilon}\right]
$$
$$
\times \prod_{k=1}^{n}\left\{1 - \beta_k \sum_{l=1}^{k}\left[\left(\delta_{lk} + (1-\delta_{lk})\cos\chi_k^{(l)}\right)\cos\vartheta_l \prod_{m=1}^{l-1}\left(\sin\chi_k^{(m)}\sin\vartheta_m\right)\right]\right\}^{-j_k}.
$$
$$(3.139)$$

As usual, δ_{kl} is equal to one if k and l coincide; otherwise it is zero. We will spell out this formula for small values of n (in particular $n = 1$ and $n = 2$) explicitly when we discuss examples later; see Eqs. (3.162), (3.164), (3.167), and (3.170).

In general, the parametric integral in Eq. (3.139) is quite elaborate. For example, for $j_k = 1$ and $\beta_k = 1$, already for $n = 2$ the integrand has a line singularity inside the (two real dimensional) integration domain. Thus even in this simple case, already the resolution of singularities as $\epsilon \to 0$ is nontrivial. However, the Mellin-Barnes method provides a particularly nice way to approach the evaluation of Eq. (3.139). We proceed as follows: First, let us go back to the original definition (Eq. (3.134)) and use Feynman parametrization to combine all denominators:

$$
\Omega_{j_1,\dots,j_n} = \int d\Omega_{d-1}(q) \frac{\Gamma(j)}{\prod_{k=1}^{n} \Gamma(j_k)}
$$
$$
\times \int_0^1 \left[\prod_{k=1}^{n} dx_k (x_k)^{j_k - 1} \right] \delta \left(\sum_{k=1}^{n} x_k - 1 \right) \left[\left(\sum_{k=1}^{n} x_k p_k \right) \cdot q \right]^{-j} , \tag{3.140}
$$

where $j = j_1 + \dots + j_n$ is the sum of the exponents. Now we make an important observation: we can exploit the rotational invariance of the original expression and evaluate the integral in a frame where the direction of the weighted sum of momenta $x_1 p_1^\mu + \dots + x_n p_n^\mu$ points along the d-th direction. In this frame, we have simply

$$
\sum_{k=1}^{n} x_k p_k^\mu = (1, \mathbf{0}_{d-2}, \beta) \qquad \text{and} \qquad q^\mu = (1, ..\text{"angles"}.., \sin \vartheta, \cos \vartheta). \tag{3.141}
$$

A quick computation shows that the variable β introduced here can be expressed as the solution of the following equation:

$$
1 - \beta^2 = \left(\sum_{k=1}^{n} x_k p_k^\mu \right)^2 = \sum_{k=1}^{n} \sum_{l=k+1}^{n} 2 x_k x_l (p_k \cdot p_l) + \sum_{k=1}^{n} x_k^2 p_k^2 \tag{3.142}
$$

Thus, Eq. (3.140) takes the form

$$
\Omega_{j_1,\dots,j_n} = \frac{\Gamma(j)}{\prod_{k=1}^{n} \Gamma(j_k)} \int_0^1 \left[\prod_{k=1}^{n} dx_k (x_k)^{j_k - 1} \right] \delta \left(\sum_{k=1}^{n} x_k - 1 \right)
$$
$$
\times \int d\Omega_{d-2}(q) \int_{-1}^1 d(\cos \vartheta)(\sin \vartheta)^{-2\epsilon} [1 - \beta \cos \vartheta]^{-j} . \tag{3.143}
$$

The only nontrivial angular integration above can be performed in terms of a Gauss $_2F_1$ hypergeometric function via the substitution $\cos \vartheta \to 2s - 1$, while Ω_{d-2} is given in Eq. (3.138):

$$
\begin{aligned}
\int d\Omega_{d-2}(q) &\int_{-1}^{1} d(\cos\vartheta)(\sin\vartheta)^{-2\epsilon}\,[1-\beta\cos\vartheta]^{-j} \\
&= 2^{2-2\epsilon}\pi^{1-\epsilon}(1+\beta)^{-j}\frac{\Gamma(1-\epsilon)}{\Gamma(2-2\epsilon)}{}_2F_1\left(j,1-\epsilon,2-2\epsilon,\frac{2\beta}{1+\beta}\right).
\end{aligned}
\tag{3.144}
$$

This result can be put in a more convenient form by using the quadratic hypergeometric identity

$$
{}_2F_1(a,b,2b,z) = \left(1-\frac{z}{2}\right)^{-a}{}_2F_1\left[\frac{a}{2},\frac{a+1}{2},b+\frac{1}{2},\left(\frac{z}{2-z}\right)^2\right]
\tag{3.145}
$$

which leads to

$$
\begin{aligned}
\int d\Omega_{d-2}(q) &\int_{-1}^{1} d(\cos\vartheta)(\sin\vartheta)^{-2\epsilon}\,[1-\beta\cos\vartheta]^{-j} \\
&= 2^{2-2\epsilon}\pi^{1-\epsilon}\frac{\Gamma(1-\epsilon)}{\Gamma(2-2\epsilon)}{}_2F_1\left(\frac{j}{2},\frac{j+1}{2},\frac{3}{2}-\epsilon,\beta^2\right).
\end{aligned}
\tag{3.146}
$$

The utility of this later form lies in the fact that it is β^2 and not β itself that has a simple expression in terms of the dot-products of the p_i^μ as evidenced by Eq. (3.142). In fact, it turns out that it is convenient to write Eq. (3.142) in a more compact form by introducing the variables v_{kl} such that

$$
v_{kl} \equiv \begin{cases} \frac{p_k \cdot p_l}{2}, & k \neq l \\ \frac{p_k^2}{4}, & k = l \end{cases}.
\tag{3.147}
$$

Then Eq. (3.142) takes the simple form

$$
1-\beta^2 = 4\sum_{k=1}^{n}\sum_{l=k}^{n}x_k x_l v_{kl}.
\tag{3.148}
$$

The reason for this particular choice of normalization will become clear shortly.

To continue, we substitute Eq. (3.146) into Eq. (3.143) and obtain

$$
\begin{aligned}
\Omega_{j_1,\dots,j_n} = \frac{\Gamma(j)}{\prod_{k=1}^{n}\Gamma(j_k)}\int_0^1 &\left[\prod_{k=1}^{n} dx_k (x_k)^{j_k-1}\right]\delta\left(\sum_{k=1}^{n}x_k-1\right) \\
&\times 2^{2-2\epsilon}\pi^{1-\epsilon}\frac{\Gamma(1-\epsilon)}{\Gamma(2-2\epsilon)}{}_2F_1\left(\frac{j}{2},\frac{j+1}{2},\frac{3}{2}-\epsilon,1-4\sum_{k=1}^{n}\sum_{l=k}^{n}x_k x_l v_{kl}\right).
\end{aligned}
\tag{3.149}
$$

At this point, we have traded all angular integrations for integrals over Feynman parameters, x_j. However, the dependence of the integrand on the x_j is quite complicated. This is where the Mellin-Barnes representation comes to our help. To further manipulate Eq. (3.149), we proceed as follows: First, we will represent the hypergeometric function as a one-dimensional Mellin-Barnes integral. Then, we will transform the double sum in the argument of the hypergeometric function to a product of factors by using the basic Mellin-Barnes formula. This will then allow us to perform all integrations over the Feynman parameters, leading to our final result: a pure Mellin-Barnes representation of the general n denominator angular integral.

To implement the steps outlined above, first use the following well-known Mellin-Barnes representations of the $_2F_1$ hypergeometric function,

$$
_2F_1(a, b, c, x) = \frac{\Gamma(c)}{\Gamma(a)\Gamma(b)\Gamma(c-a)\Gamma(c-b)}
$$
$$
\times \int_{-i\infty}^{+i\infty} \frac{dz_0}{2\pi i} \Gamma(a+z_0)\Gamma(b+z_0)\Gamma(c-a-b-z_0)\Gamma(-z_0)(1-x)^{z_0},
$$

(3.150)

to write

$$
_2F_1\left(\frac{j}{2}, \frac{j+1}{2}, \frac{3}{2} - \epsilon, 1 - 4\sum_{k=1}^{n}\sum_{l=k}^{n} x_k x_l v_{kl}\right)
$$

$$
= \frac{\Gamma\left(\frac{3}{2} - \epsilon\right)}{\Gamma\left(\frac{j}{2}\right)\Gamma\left(\frac{j+1}{2}\right)\Gamma\left(\frac{3-j}{2} - \epsilon\right)\Gamma\left(\frac{2-j}{2} - \epsilon\right)} \int_{-i\infty}^{+i\infty} \frac{dz_0}{2\pi i}
$$

$$
\times \Gamma\left(\frac{j}{2} + z_0\right)\Gamma\left(\frac{j+1}{2} + z_0\right)\Gamma(1 - j - \epsilon - z_0)\Gamma(-z_0)\left(4\sum_{k=1}^{n}\sum_{l=k}^{n} x_k x_l v_{kl}\right)^{z_0}
$$

$$
= 2^{-j} \frac{\Gamma(2 - 2\epsilon)}{\Gamma(1 - \epsilon)\Gamma(2 - j - 2\epsilon)\Gamma(j)} \int_{-i\infty}^{+i\infty} \frac{dz_0}{2\pi i}
$$

$$
\times \Gamma(j + 2z_0)\Gamma(1 - j - \epsilon - z_0)\Gamma(-z_0)\left(\sum_{k=1}^{n}\sum_{l=k}^{n} x_k x_l v_{kl}\right)^{z_0}.
$$

(3.151)

To write the last equality, we used the doubling relation for the gamma function,

$$
\Gamma(2x) = \frac{2^{2x-1}}{\sqrt{\pi}} \Gamma(x)\Gamma\left(x + \frac{1}{2}\right),
$$

(3.152)

to simplify the arguments of the gamma functions. Notice, in particular, that the factor of 4^{z_0} in the second line is cancelled, which explains our normalization of the variables in Eq. (3.147). Then, substituting Eq. (3.151) into Eq. (3.149), we find

$$
\Omega_{j_1,\ldots,j_n} = \frac{2^{2-j-2\epsilon}\pi^{1-\epsilon}}{\prod_{k=1}^{n}\Gamma(j_k)\Gamma(2-j-2\epsilon)} \int_0^1 \left[\prod_{k=1}^{n} dx_k (x_k)^{j_k-1}\right] \delta\left(\sum_{k=1}^{n} x_k - 1\right)
$$

$$
\times \int_{-i\infty}^{+i\infty} \frac{dz_0}{2\pi i} \Gamma(j+2z_0)\Gamma(1-j-\epsilon-z_0)\Gamma(-z_0) \left(\sum_{k=1}^{n}\sum_{l=k}^{n} x_k x_l v_{kl}\right)^{z_0}.
$$

$$(3.153)$$

Next, we write the double sum to the z_0-th power in a factorized form. Using the basic Mellin-Barnes formula, it is not difficult to show that (see Problem 3.19)

$$
\left(\sum_{k=1}^{n}\sum_{l=k}^{n} x_k x_l v_{kl}\right)^{z_0} = \frac{1}{\Gamma(-z_0)} \int_{-i\infty}^{+\infty} \left[\prod_{k=1}^{n-1}\prod_{l=k}^{n} \frac{dz_{kl}}{2\pi i} \Gamma(-z_{kl})(x_k x_l v_{kl})^{z_{kl}}\right]
$$

$$
\times \Gamma\left(-z_0 + \sum_{k=1}^{n-1}\sum_{l=k}^{n} z_{kl}\right) (x_n^2 v_{nn})^{z_0 - \sum_{k=1}^{n-1}\sum_{l=k}^{n} z_{kl}}
$$

$$(3.154)$$

Substituting Eq. (3.154) into Eq. (3.153), we obtain

$$
\Omega_{j_1,\ldots,j_n} = \frac{2^{2-j-2\epsilon}\pi^{1-\epsilon}}{\prod_{k=1}^{n}\Gamma(j_k)\Gamma(2-j-2\epsilon)} \int_0^1 \left[\prod_{k=1}^{n} dx_k (x_k)^{j_k-1}\right] \delta\left(\sum_{k=1}^{n} x_k - 1\right)
$$

$$
\times \int_{-i\infty}^{+i\infty} \frac{dz_0}{2\pi i} \Gamma(j+2z_0)\Gamma(1-j-\epsilon-z_0)
$$

$$
\times \int_{-i\infty}^{+\infty} \left[\prod_{k=1}^{n-1}\prod_{l=k}^{n} \frac{dz_{kl}}{2\pi i} \Gamma(-z_{kl})(x_k x_l v_{kl})^{z_{kl}}\right]
$$

$$
\times \Gamma\left(-z_0 + \sum_{k=1}^{n-1}\sum_{l=k}^{n} z_{kl}\right) (x_n^2 v_{nn})^{z_0 - \sum_{k=1}^{n-1}\sum_{l=k}^{n} z_{kl}}.
$$

$$(3.155)$$

Now let us change the variable of integration from z_0 to $z_{nn} \equiv z_0 - \sum_{k=1}^{n-1}\sum_{l=k}^{n} z_{kl}$. Then $z_0 = \sum_{k=1}^{n-1}\sum_{l=k}^{n} z_{kl} + z_{nn} = \sum_{k=1}^{n}\sum_{l=k}^{n} z_{kl}$, i.e., z_0 is simply the sum of all $\frac{n(n+1)}{2}$ integration variables. To avoid any confusion, let us denote this sum simply as z in the following:

$$
z = \sum_{k=1}^{n}\sum_{l=k}^{n} z_{kl}. \tag{3.156}
$$

Then we find

$$\Omega_{j_1,\dots,j_n} = \frac{2^{2-j-2\epsilon}\pi^{1-\epsilon}}{\prod_{k=1}^n \Gamma(j_k)\Gamma(2-j-2\epsilon)} \int_0^1 \left[\prod_{k=1}^n dx_k (x_k)^{j_k-1}\right] \delta\left(\sum_{k=1}^n x_k - 1\right)$$

$$\times \left[\prod_{k=1}^n \prod_{l=k}^n \frac{dz_{kl}}{2\pi i}\Gamma(-z_{kl})(x_k x_l v_{kl})^{z_{kl}}\right]\Gamma(j+2z)\Gamma(1-j-\epsilon-z).$$

$$(3.157)$$

Collecting all factors of the xs, we find

$$\Omega_{j_1,\dots,j_n} = \frac{2^{2-j-2\epsilon}\pi^{1-\epsilon}}{\prod_{k=1}^n \Gamma(j_k)\Gamma(2-j-2\epsilon)} \int_0^1 \left[\prod_{k=1}^n dx_k (x_k)^{j_k-1+z_k}\right] \delta\left(\sum_{k=1}^n x_k - 1\right)$$

$$\times \left[\prod_{k=1}^n \prod_{l=k}^n \frac{dz_{kl}}{2\pi i}\Gamma(-z_{kl})(v_{kl})^{z_{kl}}\right]\Gamma(j+2z)\Gamma(1-j-\epsilon-z),$$

$$(3.158)$$

where we have introduced

$$z_k = \sum_{l_1}^k z_{lk} + \sum_{l=k}^n z_{kl}. \tag{3.159}$$

In words, z_k is the sum of all variables that involve k as one of their indices, such that z_{kk} itself is counted twice, i.e., $z_k = z_{1k} + \dots + z_{k-1k} + 2z_{kk} + z_{kk+1} + \dots + z_{kn}$. To finish the computation, realize that the integration over the Feynman parameters can now be performed using the formula

$$\int_0^1 \left[\prod_{k=1}^N dx_k (x_k)^{p_k-1}\right] \delta\left(\sum_{k=1}^N x_k - 1\right) = \frac{\prod_{k=1}^N \Gamma(p_k)}{\Gamma\left(\sum_{k=1}^N p_k\right)}. \tag{3.160}$$

Applying the above formula with $p_k = j_k + z_k$ and noting that $\sum_{k=1}^n (j_k + z_k) = j + 2z$ (recall j is the sum of all exponents j_k), we arrive at our final result:

$$\Omega_{j_1,\dots,j_n} = \frac{2^{2-j-2\epsilon}\pi^{1-\epsilon}}{\prod_{k=1}^n \Gamma(j_k)\Gamma(2-j-2\epsilon)} \left[\prod_{k=1}^n \prod_{l=k}^n \int_{-i\infty}^{+i\infty} \frac{dz_{kl}}{2\pi i}\Gamma(-z_{kl})(v_{kl})^{z_{kl}}\right]$$

$$\times \left[\prod_{k=1}^n \Gamma(j_k+z_k)\right]\Gamma(1-j-\epsilon-z).$$

$$(3.161)$$

We have thus derived an $\frac{n(n+1)}{2}$-fold Mellin-Barnes integral representation for the general angular integral with n denominators.

Before presenting some examples, let us make some comments. First, notice that the derivation of Eq. (3.161) implicitly assumes that the exponents j_k are not zero or negative integers, and indeed, Eq. (3.161) is clearly ill-defined if any exponent is a non-positive integer. The case of an exponent being zero is obviously uninteresting from a practical point of view: it simply signals that the given denominator is not actually present in the integrand. On the other hand, negative integer powers, i.e., dot-products $p_j \cdot q$ in the *numerator*, do sometimes appear in practical applications. In such cases, when say $-j_k \in \mathbb{N}^+$, we can attempt to analytically continue Eq. (3.161) to the required value of j_k, e.g., by setting $j_k \rightarrow j_k + \delta$ and performing the analytic continuation $\delta \rightarrow 0$. In practice, this analytic continuation can be performed using the same methods and tools that we employ to analytically continue Mellin-Barnes integrals in the parameter of dimensional regularization ϵ to zero and that will be explained in depth in Sect. 4.2.

Second, the derivation of Eq. (3.161) also assumes that all variables v_{kl} are non-zero (in fact, positive). However, it may well happen that some v_{kl} is zero, say when some momentum p_i in Eq. (3.135) is massless, which implies $v_{ii} = 0$. In such cases Eq. (3.161) clearly cannot be used as it stands. However, it is straightforward to adapt the derivation to such cases, since if some v_{ij} is identically zero, the only modification is that the corresponding term is missing from the sum in Eq. (3.148). Then, the integration over z_{ij} in Eq. (3.154) is absent, but the rest of the derivation goes through as before. The final result is that we must drop all integrations that correspond to variables which are identically zero. Practically this amounts to the following simple changes in Eq. (3.161): first, the double product over k and l in the first bracket in Eq. (3.154) is restricted to those values of k and l for which $v_{kl} \neq 0$, and second, the sums defining z and z_k in Eqs. (3.156) and (3.159) are similarly restricted to those z_{kl} for which $v_{kl} \neq 0$.

Finally, we mention that the general angular integral $\Omega_{j_1, \ldots, j_n}$ can be expressed in a compact way with the H-function of several variables introduced in Eq. (2.196). The details of this representation are given in [47].

Let us now turn to some examples. First, we consider the angular integral with a single massless denominator,

$$\Omega_j(0, \epsilon) = \int d\Omega_{d-2} \int_{-1}^{1} d\cos(\theta_1)(\sin\theta_1)^{-2\epsilon}(1 - \cos\theta_1)^{-j} . \tag{3.162}$$

The single momentum p_1 is massless and so $v_{11} = 0$; hence the discussion above regarding variables that are identically zero applies. Noting that $z_1 = z = 0$, we obtain the zero-dimensional Mellin-Barnes representation

$$\Omega_j(0, \epsilon) = 2^{2-j-2\epsilon} \pi^{1-\epsilon} \frac{\Gamma(1 - j - \epsilon)}{\Gamma(2 - j - 2\epsilon)} . \tag{3.163}$$

This result is simple to verify, since the integral can be performed in terms of gamma functions after setting $\cos(\theta_1) \rightarrow 2s - 1$. Then using Eq. (3.138) for the angular volume Ω_{d-2}, we obtain Eq. (3.163) immediately.

Although we have already derived an expression for the angular integral with one massive denominator,

$$\Omega_j(v_{11}, \epsilon) = \int d\Omega_{d-2} \int_{-1}^{1} d\cos(\theta_1)(\sin\theta_1)^{-2\epsilon}(1 - \beta_1\cos\theta_1)^{-j}, \qquad (3.164)$$

in Eq. (3.146), for the sake of completeness, let us nevertheless discuss how this result may be derived from our master formula Eq. (3.161). Since now $v_{11} \neq 0$ and furthermore $z_1 = 2z_{11}$, $z = z_{11}$ (see Eqs. (3.159) and (3.156)), we find the one-dimensional Mellin-Barnes integral representation

$$\Omega_j(v_{11}, \epsilon) = \frac{2^{2-j-2\epsilon}\pi^{1-\epsilon}}{\Gamma(j)\Gamma(2 - j - 2\epsilon)} \int_{-i\infty}^{+i\infty} \frac{dz_{11}}{2\pi i}$$
$$\times \Gamma(-z_{11})\Gamma(j + 2z_{11})\Gamma(1 - j - \epsilon - z_{11})(v_{11})^{z_{11}}. \qquad (3.165)$$

This Mellin-Barnes integral can be evaluated by repeating the steps that lead from Eq. (3.149)–(3.153): after writing $\Gamma(j + 2z_{11})$ as a product of the gamma functions $\Gamma(\frac{j}{2} + z_{11})$ and $\Gamma(\frac{j+1}{2} + z_{11})$ using the relation in Eq. (3.152), the resulting integral is of the form given in Eq. (3.150) and can be evaluated in terms of a $_2F_1$ hypergeometric function. In fact, we may read off the final result simply by comparing our integral in Eq. (3.165) to the right-hand side of Eq. (3.151),

$$\Omega_j(v_{11}, \epsilon) = 2^{2-2\epsilon}\pi^{1-\epsilon}\frac{\Gamma(1 - \epsilon)}{\Gamma(2 - 2\epsilon)} {}_2F_1\left(\frac{j}{2}, \frac{j+1}{2}, \frac{3}{2} - \epsilon, 1 - 4v_{11}\right). \qquad (3.166)$$

This result clearly agrees with our earlier one in Eq. (3.146) since $v_{11} = (1 - \beta_1^2)/4$ (see Eq. (3.147)).

Turning to a less trivial example, consider now the angular integral with two massless denominators,

$$\Omega_{j,k}(v_{12}, \epsilon) = \int d\Omega_{d-3} \int_{-1}^{1} d(\cos\theta_1)(\sin\theta_1)^{-2\epsilon} \int_{-1}^{1} d(\cos\theta_2)(\sin\theta_2)^{-1-2\epsilon}$$
$$\times (1 - \cos\theta_1)^{-j}(1 - \cos\chi_2^{(1)}\cos\theta_1 - \sin\chi_2^{(1)}\sin\theta_1\cos\theta_2)^{-k}. \qquad (3.167)$$

Since $p_1^2 = p_2^2 = 0$, only v_{12} is non-zero, and we must drop from Eq. (3.161) the integrations corresponding to v_{11} and v_{22} as discussed above. Hence, we obtain a one-dimensional Mellin-Barnes representation. Using Eqs. (3.159) and (3.156), we see that $z_1 = z_2 = z = z_{12}$, so

$$\Omega_{j,k}(v_{12}, \epsilon) = \frac{2^{2-j-k-2\epsilon}\pi^{1-\epsilon}}{\Gamma(j)\Gamma(k)\Gamma(2 - j - k - 2\epsilon)} \int_{-i\infty}^{+i\infty} \frac{dz_{12}}{2\pi i}$$

$$\times \Gamma(-z_{12})\Gamma(j + z_{12})\Gamma(k + z_{12})\Gamma(1 - j - k - \epsilon - z_{12})(v_{12})^{z_{12}}.$$

$$(3.168)$$

This Mellin-Barnes integral can be evaluated immediately in terms of a $_2F_1$ hypergeometric function using Eq. (3.150), and we find

$$\Omega_{j,k}(v_{12}, \epsilon) = 2^{2-j-k-2\epsilon}\pi^{1-\epsilon}\frac{\Gamma(1 - j - \epsilon)\Gamma(1 - k - \epsilon)}{\Gamma(1 - \epsilon)\Gamma(2 - j - k - 2\epsilon)}{_2F_1}(j, k, 1 - \epsilon, 1 - v_{12}).$$

$$(3.169)$$

Before moving on to the next example, let us make some comments. First, the argument of the hypergeometric function is simply related to the angle enclosed by the spatial parts of the vectors p_1 and p_2. Using Eq. (3.147), we see immediately that $v_{12} = (1 + \cos \chi_2^{(1)})/2$. Second, although Eq. (3.168) was derived under the assumption that both j and k are not zero or negative integers, nevertheless, the final result in Eq. (3.169) applies in such cases as well. Finally, we note that in practical applications we often require the expansion of this result in ϵ around zero for specific integers j and k. We may arrive at such an expansion by resolving the poles of the Mellin-Barnes integral in Eq. (3.168) using the methods of Sect. 4.1 and expanding the integrand. Then the expansion coefficients will generically be given by (sums of) one-dimensional finite Mellin-Barnes integrals that no longer depend on ϵ. In Chap. 5 we will present several tools that can be used to analytically compute these integrals.

As our last example, let us consider the generalization of the previous integral to the case when the momentum p_1 is massive,

$$\Omega_{j,k}(v_{11}, v_{12}, \epsilon) = \int d\Omega_{d-3} \int_{-1}^{1} d(\cos \theta_1)(\sin \theta_1)^{-2\epsilon} \int_{-1}^{1} d(\cos \theta_2)(\sin \theta_2)^{-1-2\epsilon}$$

$$\times (1 - \beta_1 \cos \theta_1)^{-j}(1 - \cos \chi_2^{(1)} \cos \theta_1 - \sin \chi_2^{(1)} \sin \theta_1 \cos \theta_2)^{-k}.$$

$$(3.170)$$

Since now only v_{22} is identically zero, we obtain a two-dimensional Mellin-Barnes representation. Using $z_1 = 2z_{11} + z_{12}$, $z_2 = z_{12}$ and $z = z_{11} + z_{12}$ (see Eqs. (3.159) and (3.156)), we find

$$
\begin{aligned}
\Omega_{j,k}(v_{11}, v_{12}, \epsilon) = & \frac{2^{2-j-k-2\epsilon}\pi^{1-\epsilon}}{\Gamma(j)\Gamma(k)\Gamma(2-j-k-2\epsilon)} \int_{-i\infty}^{+i\infty} \frac{dz_{11}}{2\pi i}\frac{dz_{12}}{2\pi i} \\
& \times \Gamma(-z_{11})\Gamma(-z_{12})\Gamma(j+2z_{11}+z_{12})\Gamma(k+z_{12}) \\
& \times \Gamma(1-j-k-\epsilon-z_{11}-z_{12})(v_{11})^{z_{11}}(v_{12})^{z_{12}}.
\end{aligned}
\tag{3.171}
$$

We will show in Sect. 5.3.3 that this integral can be evaluated in a closed form in terms of the Appell function of the first kind,

$$
\begin{aligned}
\Omega_{j,k}(v_{11}, v_{12}, \epsilon) = & 2^{2-j-k-2\epsilon}\pi^{1-\epsilon}\frac{\Gamma(1-k-\epsilon)}{\Gamma(2-k-2\epsilon)}v_{12}^{-j}F_1\bigg(j, 1-k-\epsilon, \\
& 1-k-\epsilon, 2-k-2\epsilon, \frac{2v_{12}-1-\sqrt{1-4v_{11}}}{2v_{12}}, \frac{2v_{12}-1+\sqrt{1-4v_{11}}}{2v_{12}}\bigg).
\end{aligned}
\tag{3.172}
$$

As in the case of the massless two-denominator angular integral, it is interesting to note that although Eq. (3.171) was derived under the assumptions that j and k are not zero or negative integers, nevertheless, the solution in Eq. (3.172) is valid even for these cases. For practical applications we are often interested in the ϵ expansion of the final result in Eq. (3.172). In Sect. 5.1.6, we will present methods that allow to perform this expansion for integer values of the parameters j and k, based on the double sum representation of the Appell F_1 function given in Eq. (2.189). Further details and some more examples of angular integrals can be found in [47].

Problems

Problem 3.1   Prove the identity in Eq. (3.4).
<u>Hint</u>: For proofs of relations including also Symanzik polynomials, see [1,48].

Problem 3.2   Prove the equivalence of the Lee-Pomeransky representation of Eq. (3.17) with the Feynman parametrization of Eq. (3.13).
<u>Hint</u>: Insert $1 = \int_0^\infty d\eta\, \delta(\eta - \sum_{j=1}^N z_j)$, substitute $z_j = \eta\, x_j$ for $j = 1, \ldots, N$, and identify the corresponding gamma functions.

Problem 3.3   It is hard by a naked eye to see how cutting lines in Fig. 3.3 we get appropriate spanning tree T and k−forest $\mathcal{T}_k$ for a given diagram (here $k = 2$). Rearrange cut diagrams in Fig. 3.3 to see it explicitly, in a similar way as shown in Fig. 3.1.
<u>Hint</u>: For a complete solution, see [30], Section 4.2.2.

Problem 3.4 Generate F and U polynomials for integral in Fig. 3.3. Do the same to find Symanzik polynomials as in Eqs. (3.85) and (3.86).

Hint: Use MB.m or the Mathematica module which can be found in [49].

Problem 3.5 Show that Eq. (3.29) gives Eq. (3.30).

Hint: Use Cauchy's residue theorem.

Problem 3.6 Following Sect. 3.4, show that for (3.25) the MB master formula in Eq. (3.31) gives

$$
I_{MB} \sim \int dz_1 dz_2 dz_3 \, (-s x_1 x_2)^{z_1} (-q_2^2 x_1 x_3)^{z_2} (-q_3^2 x_2 x_3)^{z_3}
$$
$$
\times \left(x_1 m_1^2 + x_2 m_2^2 + x_3 m_3^2 \right)^{-z_1 - z_2 - z_3 - N_\nu + d/2} .
\tag{3.173}
$$

Problem 3.7 Assign external and internal momenta to the diagrams in Fig. 3.6. Show that after integrating one internal momentum (removing corresponding lines where the integrated internal momenta are the only internal momenta), not all vertices in the non-planar case conserve momenta.

Problem 3.8 Built up the matrix M_Γ of Eq. (3.52) for the transformation of z-variables in the MB integrals given in Eqs. (3.45) and (3.90).

Hint: See the solution in the file MB_Zmatrix_Springer.nb in the auxiliary material in [16].

Problem 3.9 Analyze a proof of the CW theorem using a notion of sector decomposition [35] as given in the Appendix of [34].

Problem 3.10 For the diagrams in Fig. 3.18, find a suitable set of transformations of variables analogously as in Eq. (3.96), and derive Eq. (3.105).

Hint: See the solution in the file MB_3L_Springer.nb in the auxiliary material in [16].

Problem 3.11 Consider the two-loop sunset diagram in Fig. 3.16. Derive corresponding MB integral, and find relations among gamma functions in the numerator which cancel against $\Gamma[0]$ in the denominator.

Hint: Find suitable transformation of integration variables in the MB integral and use 1BL. See the solution in the file MB_Gamma0_Springer.nb in the auxiliary material in [16].

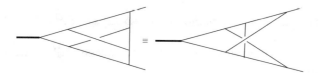

Fig. 3.26 Apparently two the same topologies

Problem 3.12 Consider the banana diagram given in Chap. 2, Fig. 2.4. The partial fractioning for the banana's F polynomial is discussed in Note 3.9.2 "Applying the CW Theorem" and Eq. (3.103). It eventually leads to square roots, which cannot be integrated in an iterative way using HPL and MPL formalism (so solutions go beyond MPLs). For a pedagogical example in $d = 2$, see [50] and Sect. 5.3.4.

Problem 3.13 Starting from the skeleton diagram on the right side of Fig. 3.17, generate diagrams in Fig. 4.22 in [30]. At this level it is already more complicated to recognize different topologies by eye; the two diagrams in Fig. 3.26 are actually the same; for topologies recognition, see, e.g., [17, 51].

Problem 3.14 Apply the LA method to get a four-dimensional MB representation for the three-loop integral I_{SE6l0m} of Eq. (3.110).
Hint: For a solution, see the file MB_SE6l0m_Springer.nb in the auxiliary material in [16].

Problem 3.15 The method of brackets consists of a small set of simple rules on how to rewrite a Schwinger-parametrized integral into a form to which this master formula can be applied, as discussed in Sect. 3.13. To see all the steps in practice, follow a non-planar two-loop didactic example discussed in Section E.5 in [38].

Problem 3.16 For applying the method of brackets and GRMT to solve some one-loop cases, follow examples discussed in [42].

Problem 3.17 Evaluate the integrals in Eq. (3.131) and verify the result in Eq. (3.133). You will have to pay careful attention to the limits of integration imposed by the positivity of energies as well as the range of $\cos\theta$.
Hint: You may find it useful to partial fraction the integrand. See also the file MB_PSint_Springer.nb in the auxiliary material in [16].

Problem 3.18 Using the basic d-dimensional Gaussian integral $\int d^d k e^{-\frac{k^2}{2}} = (2\pi)^{\frac{d}{2}}$, prove
Eq. (3.137) for integer p.

Hint: This is rather an elementary derivation (see, for instance, [52]) however often applied
(see, e.g., Eq. (3.12)).

Problem 3.19 Prove Eq. (3.154).

Hint: The result is actually just an application of the general formula in Eq. (3.31); the only
nontrivial part is keeping track of the indexing. In this regard, it may help to write out the
double sum explicitly, $\sum_{k=1}^{n} \sum_{l=k}^{n} x_k x_l v_{kl} = x_1 x_1 v_{11} + x_1 x_2 v_{12} + \ldots + x_{n-1} x_n v_{n-1n} + x_n x_n v_{nn}$.

References

1. N. Nakanishi, *Graph Theory and Feynman Integrals* (Gordon and Breach, New York, 1971)
2. K. Symanzik, Dispersion Relations and Vertex Properties in Perturbation Theory. Prog. Theor. Phys. **20**(5), 690–702 (1958). https://doi.org/10.1143/PTP.20.690
3. E. Panzer, Feynman integrals and hyperlogarithms, Ph.D. thesis, Humboldt U. (2015). arXiv: 1506.07243. https://doi.org/10.18452/17157.
4. G. Heinrich, Collider Physics at the Precision Frontier. Phys. Rept. **922**, 1–69 (2021). arXiv: 2009.00516. https://doi.org/10.1016/j.physrep.2021.03.006
5. R.N. Lee, A.A. Pomeransky, Critical points and number of master integrals. JHEP **11**, 165 (2013). arXiv:1308.6676. https://doi.org/10.1007/JHEP11(2013)165
6. P.A. Baikov, Explicit solutions of the multiloop integral recurrence relations and its application. Nucl. Instrum. Meth. A **389**, 347–349 (1997). arXiv:hep-ph/9611449. https://doi.org/10.1016/S0168-9002(97)00126-5
7. R.N. Lee, Calculating multiloop integrals using dimensional recurrence relation and D-analyticity. Nucl. Phys. B Proc. Suppl. **205–206**, 135–140 (2010). arXiv:1007.2256. https://doi.org/10.1016/j.nuclphysbps.2010.08.032
8. M. Czakon, Automatized analytic continuation of Mellin-Barnes integrals. Comput. Phys. Commun. **175**, 559–571 (2006). arXiv:hep-ph/0511200. https://doi.org/10.1016/j.cpc.2006.07.002
9. E. Whittaker, G. Watson, in *A Course of Modern Analysis*. Cambridge Mathematical Library, 1996 Edition (Cambridge University, Cambridge, 1927)
10. J. Gluza, K. Kajda, T. Riemann, AMBRE - a Mathematica package for the construction of Mellin-Barnes representations for Feynman integrals. Comput. Phys. Commun. **177**, 879–893 (2007). arXiv:0704.2423. https://doi.org/10.1016/j.cpc.2007.07.001

11. J. Gluza, K. Kajda, T. Riemann, V. Yundin, Numerical Evaluation of Tensor Feynman Integrals in Euclidean Kinematics. Eur. Phys. J. C **71**, 1516 (2011). arXiv:1010.1667. https://doi.org/10.1140/epjc/s10052-010-1516-y

12. I. Dubovyk, J. Gluza, T. Riemann, J. Usovitsch, Numerical integration of massive two-loop Mellin-Barnes integrals in Minkowskian regions, PoS LL2016 (2016) 034. arXiv:1607.07538

13. AMBRE webpage: http://jgluza.us.edu.pl/ambre, Backup: https://web.archive.org/web/20220119185211/http://prac.us.edu.pl/~gluza/ambre/

14. MB Tools webpage. http://projects.hepforge.org/mbtools/

15. M. Hidding, DiffExp, a Mathematica package for computing Feynman integrals in terms of one-dimensional series expansions. Comput. Phys. Commun. **269**, 108125 (2021). arXiv:2006.05510. https://doi.org/10.1016/j.cpc.2021.108125

16. https://github.com/idubovyk/mbspringer. http://jgluza.us.edu.pl/mbspringer

17. K. Bielas, I. Dubovyk, J. Gluza, T. Riemann, Some Remarks on Non-planar Feynman Diagrams. Acta Phys. Polon. **B44**(11), 2249–2255 (2013). arXiv:1312.5603. https://doi.org/10.5506/APhysPolB.44.2249

18. J. Blumlein, I. Dubovyk, J. Gluza, M. Ochman, C.G. Raab, T. Riemann, C. Schneider, Non-planar Feynman integrals, Mellin-Barnes representations, multiple sums. PoS LL2014 (2014) 052. arXiv:1407.7832

19. H. Cheng, T. Wu, *Expanding protons: Scattering at High Energies* (MIT Press, Cambridge, Massachusetts, 1987)

20. B. Tausk, Non-planar massless two-loop Feynman diagrams with four on- shell legs. Phys. Lett. **B469**, 225–234 (1999). arXiv:hep-ph/9909506

21. G. Heinrich, V. Smirnov, Analytical evaluation of dimensionally regularized massive on-shell double boxes. Phys. Lett. **B598**, 55–66 (2004). arXiv:hep-ph/0406053

22. K. Bielas, I. Dubovyk, PlanarityTest 1.2.1 (Aug 2017), a Mathematica package for testing the planarity of Feynman diagrams. http://us.edu.pl/~gluza/ambre/planarity/, [17]

23. K. Kajda, I. Dubovyk, AMBRE 2.1.1 and 1.3.1 (Aug 2017), Mathematica packages representing Feynman integrals by Mellin-Barnes integrals. http://jgluza.us.edu.pl/ambre/, [11]

24. I. Dubovyk, AMBRE 3.1.1 (Aug 2017), a Mathematica package representing Feynman integrals by Mellin-Barnes integrals. http://jgluza.us.edu.pl/ambre/, [11, 12]

25. A. Smirnov, Mathematica program MBresolve.m version 1.0 (Jan 2009), available at the MB Tools webpage. http://projects.hepforge.org/mbtools/ [29]

26. D. Kosower, Mathematica program barnesroutines.m version 1.1.1 (Jul 2009), available at the MB Tools webpage. http://projects.hepforge.org/mbtools/

27. I. Dubovyk, T. Riemann, J. Usovitsch, Numerical calculation of multiple MB-integral representations for Feynman integrals. J. Usovitsch. MBnumerics, a Mathematica/Fortran package at http://jgluza.us.edu.pl/ambre/

28. K. Bielas, I. Dubovyk, PlanarityTest 1.3, a Mathematica package for testing the planarity of Feynman diagrams. http://jgluza.us.edu.pl/ambre/planarity/

29. A.V. Smirnov, V.A. Smirnov, On the Resolution of Singularities of Multiple Mellin- Barnes Integrals. Eur. Phys. J. **C62**, 445 (2009). arXiv:0901.0386

30. I. Dubovyk, Mellin-Barnes representations for multiloop Feynman integrals with applications to 2-loop electroweak Z boson studies, Ph.D. thesis, Hamburg U. (2019). https://ediss.sub.uni-hamburg.de/handle/ediss/6052

31. I. Dubovyk, A. Freitas, J. Gluza, T. Riemann, J. Usovitsch, Complete electroweak two-loop corrections to Z boson production and decay. Phys. Lett. **B783**, 86–94 (2018). arXiv:1804.10236, https://doi.org/10.1016/j.physletb.2018.06.037

32. I. Dubovyk, A. Freitas, J. Gluza, T. Riemann, J. Usovitsch, The two-loop electroweak bosonic corrections to $sin^2\theta_{\text{eff}}^{bb}$. Phys. Lett. **B762**, 184–189 (2016). arXiv:1607.08375. https://doi.org/10.1016/j.physletb.2016.09.012

33. S. Jahn, SecDec: a toolbox for the numerical evaluation of multi-scale integrals. PoS RADCOR2017 (2018) 017. arXiv:1802.07946. https://doi.org/10.22323/1.290.0017

34. G. Heinrich, S. Jahn, S.P. Jones, M. Kerner, F. Langer, V. Magerya, A. Pöldaru, J. Schlenk, E. Villa, Expansion by regions with pySecDec. arXiv:2108.10807
35. T. Binoth, G. Heinrich, An automatized algorithm to compute infrared divergent multi-loop integrals. Nucl. Phys. **B585**, 741–759 (2000). arXiv:hep-ph/0004013v.2
36. B. Jantzen, A.V. Smirnov, V.A. Smirnov, Expansion by regions: revealing potential and Glauber regions automatically. Eur. Phys. J. C **72**, 2139 (2012). arXiv:1206.0546. https://doi.org/10.1140/epjc/s10052-012-2139-2
37. P. Cvitanovic, T. Kinoshita, Feynman-Dyson rules in parametric space. Phys. Rev. **D10**, 3978–3991 (1974). https://doi.org/10.1103/PhysRevD.10.3978
38. A. Blondel, et al., Standard model theory for the FCC-ee Tera-Z stage, in *Mini Workshop on Precision EW and QCD Calculations for the FCC Studies : Methods and Techniques.* CERN Yellow Reports: Monographs, vol. 3/2019 (CERN, Geneva, 2018). arXiv:1809.01830. https://doi.org/10.23731/CYRM-2019-003
39. R. Boels, B.A. Kniehl, G. Yang, Master integrals for the four-loop Sudakov form factor. Nucl. Phys. **B902**, 387–414 (2016). arXiv:1508.03717. https://doi.org/10.1016/j.nuclphysb.2015.11.016
40. M. Prausa, Mellin-Barnes meets Method of Brackets: a novel approach to Mellin-Barnes representations of Feynman integrals. Eur. Phys. J. **C77**(9), 594 (2017). arXiv:1706.09852. https://doi.org/10.1140/epjc/s10052-017-5150-9
41. I. Gonzalez, I. Schmidt, Optimized negative dimensional integration method (NDIM) and multiloop Feynman diagram calculation. Nucl. Phys. **B769**, 124–173 (2007). arXiv:hep-th/0702218. https://doi.org/10.1016/j.nuclphysb.2007.01.031
42. I. Gonzalez, V.H. Moll, *Definite integrals by the method of brackets. Part 1.* arXiv:0812.3356
43. I. Gonzalez, Method of Brackets and Feynman diagrams evaluation. Nucl. Phys. Proc. Suppl. **205–206**, 141–146 (2010). arXiv:1008.2148. https://doi.org/10.1016/j.nuclphysbps.2010.08.033
44. A. Gehrmann-De Ridder, T. Gehrmann, G. Heinrich, Four particle phase space integrals in massless QCD. Nucl. Phys. B **682**, 265–288 (2004). arXiv:hep-ph/0311276. https://doi.org/10.1016/j.nuclphysb.2004.01.023
45. G. Heinrich, Towards $e^+e^- \to 3$ jets at NNLO by sector decomposition. Eur. Phys. J. C **48**, 25–33 (2006). arXiv:hep-ph/0601062. https://doi.org/10.1140/epjc/s2006-02612-9
46. C. Anastasiou, C. Duhr, F. Dulat, B. Mistlberger, Soft triple-real radiation for Higgs production at N3LO. JHEP **07**, 003 (2013). arXiv:1302.4379. https://doi.org/10.1007/JHEP07(2013)003
47. G. Somogyi, Angular integrals in d dimensions. J. Math. Phys. **52**, 083501 (2011). arXiv:1101.3557. https://doi.org/10.1063/1.3615515
48. N. Nakanishi, Parametric integral formulas and analytic properties in perturbation theory. Prog. Theor. Phys. Supplement **18**, 1 (1961). http://ptps.oxfordjournals.org/content/18/1.full.pdf
49. A.V. Smirnov, Tools-UF. https://www.ttp.kit.edu/~asmirnov/Tools-UF.htm
50. J.L. Bourjaily, et al., *Functions Beyond Multiple Polylogarithms for Precision Collider Physics* (2022). arXiv:2203.07088
51. M. Gerlach, F. Herren, M. Lang, tapir: A tool for topologies, amplitudes, partial fraction decomposition and input for reductions. arXiv:2201.05618
52. E. Zeidler, *Quantum Field Theory II: Quantum Electrodynamics: A Bridge between Mathematicians and Physicists.* Springer, Berlin, 2008. https://doi.org/10.1007/978-3-540-85377-0

Resolution of Singularities

4

Abstract

We discuss strategies for extracting singularities of Feynman integrals connected with dimensional regularization (ϵ poles) using Mellin-Barnes representations. Based on a concrete example, we outline an algorithmic procedure for the Laurent expansion of MB integrals in the parameter ϵ. Some peculiarities in the analytic continuation for $\epsilon \to 0$ are discussed.

4.1 Where Do the Poles Come From?

In Chap. 1 we discussed the nature of singularities which appear in FI. In the context of MB integrals, we are especially interested in singularities connected with the dimensional parameter $d = n - 2\epsilon$ (n must not necessarily be four [1]; see Problem 4.1). The ϵ regulator is transformed from Feynman parametrization of integrals to the arguments of gamma functions, as was discussed on the occasion of construction of MB integrals in Chap. 3. We will discuss now how these singularities can be resolved. MB representations allow in a systematic way to treat this class of IR and UV singularities.

4.2 Resolving Poles: Straight Line and Deformed Contours

After defining the Mellin-Barnes representation for the Feynman integral, we are interested in further processing, aiming at analytical and numerical solutions. Let us then evaluate further Eq. (3.50). This integral depends on the dimensional parameter ϵ, which in the four-dimensional physical limit tends to be zero. Naturally, we need the Laurent expansion of the constructed MB integral representations. Then we

© The Author(s), under exclusive license to Springer Nature Switzerland AG 2022
I. Dubovyk et al., *Mellin-Barnes Integrals*, Lecture Notes in Physics 1008,
https://doi.org/10.1007/978-3-031-14272-7_4

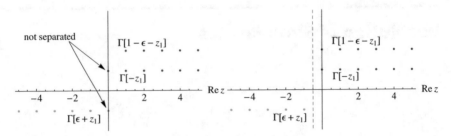

Fig. 4.1 The poles of the MB representation in Eq. (3.50) for negative (positive) values of $\Re[z]$ denoted in red (blue). Left plot shows situation when $\epsilon \to 0$ which leads to non-separated left and right poles. The right plot presents a proper shift which separates the poles ($\epsilon \to 1$, $\Re(z) \to -1/2$)

can use the Cauchy theorem to get a representation in terms of a sum of contour integrals, valid at $\epsilon = 0$. As discussed in Sect. 3.3, it is important that the Mellin-Barnes representation is only well defined, if the integration contour separates the left poles $\Gamma(\ldots+z)$ from the right poles $\Gamma(\ldots-z)$ and in general for the combination of gamma functions with $\epsilon = 0$ that will not be the case (if the contour is chosen to be a straight line parallel to the imaginary axis). In practice two solutions to this problem have been proposed:

- The "Tausk method"—which fixes the contours parallel to the imaginary axis and accounts for the poles crossing in the analytic continuation [2]. The idea of it is presented in Fig. 4.1. This method was implemented in the MB program [3] written in `Mathematica` and implemented for numerical calculations in [4] (unpublic) (see Appendix A.1).
- The "Smirnov method"—which identifies gamma functions responsible for the generation of poles in ϵ. The general observation is that $\Gamma[a+z_1]\Gamma[b-z_1]$ with a and b which can depend on other variables z_i generates the pole $\Gamma[a+b]$. So, for instance, $\Gamma[1+z_1]\Gamma[-1-z_1-\epsilon]$ generates directly a pole $\Gamma[-\epsilon]$. The algorithm for identifying all poles in ϵ for a given set of gamma functions is implemented in the MBresolve program [5] (see Appendices A.1 and C (Smirnov lecture)).

Graphically we can see what happens if we put $\epsilon \to 0$ in the derived MB representation (the left plot of Fig. 4.1). We note that left and right poles are not separated from each other. To avoid that, we take $\epsilon \neq 0$, $\epsilon \to 1$ is a good option (see the right plot of Fig. 4.1). Using this scheme according to the Tausk method, the final result ($\epsilon = 0$) will be obtained from the case where the poles are separated ($\epsilon \neq 0$). Depending on whether the poles crossed the contours from left or right (when $\epsilon \to 0$), one should add or subtract the residue of the integrand on that pole.

4.2.1 Bromwich Contours and Separation of Poles

To discuss a pole structure of Eq. (3.50), let us write this integral in a form which helps identify individual gamma functions (what matters are gamma functions in numerators, which we assign as G_1, G_2, G_3; and possible poles of gamma functions in the denominator makes the integral to vanish):

$$aux = \int_{c-i\infty}^{c+i\infty} \frac{dz_1}{2\pi i} (ms)^{z_1} (-s)^{-\epsilon} \frac{G_1(1 - \epsilon - z_1)^2 G_2(-z_1) G_3(\epsilon + z_1)}{\Gamma(2 - 2\epsilon - 2z_1)}, \quad ms = -\frac{m^2}{s}.$$

(4.1)

According to Fig. 4.1, the following gamma functions contribute to the left and right poles in Eq. (3.50):

- Left poles: $G_3(\epsilon + z_1)$
- Right poles: $G_1(1 - \epsilon - z_1), G_2(-z_1)$

As discussed in Chap. 3, positive arguments of gamma function make the function regular (see Fig. 2.5). However, it appears that, for $\epsilon = 0$, the set of inequalities built from arguments of G_1, G_2, G_3

$$1 - \epsilon - z_1 > 0, \tag{4.2}$$

$$-z_1 > 0, \tag{4.3}$$

$$\epsilon + z_1 > 0, \tag{4.4}$$

results in no solution in z_1. This can be immediately seen comparing Eqs. (4.3) and (4.4).

At this stage, to define contours of integration for MB integrals, we follow the Tausk approach. We fix a contour of integration for MB integrals parallel to the imaginary axis (Bromwich contours; see Chap. 2).

> To separate poles of gamma functions we start with $\epsilon \neq 0$ and with real parts of integration variables z_i for which MB integrands are well-defined (positive arguments of gamma functions in a numerator of the integrand). Then we consider poles which appear when a limit $\epsilon \to 0$ is taken.

Taking non-zero ϵ, e.g., $\epsilon = 1/4$, we can already find z_1 which satisfies Eqs. (4.2)–(4.4), e.g., $z_1 = -1/4$.

So, in Step 1, taking $\epsilon = 1/4$, $\Re(z_1) = -1/4$, we get

$$subst0 = G_1(3/4), G_2(1/4), G_3(1/4). \tag{4.5}$$

We can see that with such a choice of ϵ, z_1, the arguments of all gamma functions in the numerator of the integrand are positive (the integral is well defined). The very next step in the calculation of Eq. (3.50) is the analytic continuation, i.e., we have to finally go down to the case where $\epsilon \to 0$.

So in Step 2, taking $\epsilon = 0$, $\mathcal{R}(z_1) = -1/4$, we get

$$\text{subst1} = G_1(5/4), G_2(1/4), G_3(-1/4). \tag{4.6}$$

As we can see, the analytic continuation in $\epsilon \to 0$ hits the pole at $\mathcal{R}(z_1) = -1/4$ for gamma $G_3[\epsilon + z_1]$. So, the original integral in Eq. (4.1) can be decomposed in the following way:

$$aux(\epsilon = -1/4, \mathcal{R}(z_1) = -1/4) = aux(\epsilon = 0, \mathcal{R}(z_1) = -1/4)$$
$$+ \text{Res}_{z \to -\epsilon} aux(\mathcal{R}(z_1) = -1/4). \tag{4.7}$$

As aux is a one-dimensional function in z_1, the residue of the aux function at $z = \epsilon$ returns a function:

$$\text{Res}_{z \to -\epsilon} aux(\mathcal{R}(z_1) = -1/4) = ms^{-\epsilon} \Gamma[\epsilon]. \tag{4.8}$$

We will solve Eq. (4.7) analytically using Cauchy's residue theorem in Chap. 5 and purely numerically in Chap. 6, taking the integral $aux(\epsilon = 0, \mathcal{R}(z_1) = -1/4)$ in the complex plane.

4.2.2   Analytic Continuation in ϵ

Let us discuss more complicated, two-loop example, with some massive propagators (see Fig. 4.2), which has been generated in Mathematica using two following commands. To get it, in addition kinematic variables must be defined for the 4PF, which is a subject of Problem 2 in Chap. 3:

```
In[1]:= int = PR[k1, 0, n1]*PR[k2 - p2, 0, n2]*PR[k1 + k2, m,
    ⤷ n3]*PR[k2 + p3, 0, n4]*PR[k2 - p1 - p2, m, n5];
In[2]:= PlanarityTest[{int}, {k1, k2}, DrawGraph -> True];
```

$$\tag{4.9}$$

MB representation for this case is (see Chap. 3 for details of its construction)

$$G(1)_{\text{B5l2m}} = -(-m^2)^{-\epsilon - z_1 + z_2} (-s)^{z_3} (-t)^{-1 - \epsilon + z_1 - z_2 - z_3} \tag{4.10}$$

$$\int_{c_1 - i\infty}^{c_1 + i\infty} \frac{dz_1}{2\pi i} \int_{c_2 - i\infty}^{c_2 + i\infty} \frac{dz_2}{2\pi i} \int_{c_3 - i\infty}^{c_3 + i\infty} \frac{dz_3}{2\pi i}$$

Fig. 4.2 Two-loop topology for Feynman integral with five internal lines (propagators, two massive internal lines). The figure generated by `PlanarityTest` [6, 7] (see Appendix A.1). We will call this topology B512m, as this is a 4PF (Box) with five internal lines, of which two are massive.

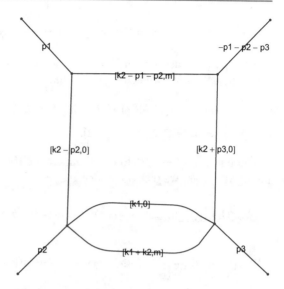

$$\left(G_1[1 - 2\epsilon - z_1]G_2[1 + z_1]G_3[\epsilon + z_1]G_4[-z_2]G_5[-\epsilon + z_1 - z_2 - z_3]G_6[-z_3] \right.$$

$$\left. \times \frac{G_7[1 + z_3]G_8[-z_1 + z_3]G_9[1 + \epsilon - z_1 + z_2 + z_3]G_{10}[1 - z_1 + 2z_2 + 2z_3]}{\Gamma[2 - 2\epsilon]\Gamma[1 - 2\epsilon + z_1]\Gamma[1 - z_1 + 2z_3]} \right).$$

Notation $G(1)_{B512m}$ means that MB representation is for a two-loop scalar integral (see Eq. (1.22)). As in Eq. (4.1), to track a discussion of identified poles in ϵ, we numbered gamma functions in the numerator. This integral is three-dimensional. Similarly to the previous case, we are looking for singular resolution in ϵ, i.e., we want to regularize the integral in the limit $\epsilon \to 0$. Taking $\epsilon = 0$ in arguments of all gamma functions in the numerator of the Eq. (4.10), we can seek a solution for a set of inequalities analogous to those in Eqs. (4.2)–(4.4). There are cases where we get no solution, see Problem 4.2. Incidentally, to find that some system of inequalities has no solutions can be quite tricky, fortunately systems like `Mathematica` are able to solve such problems with eye blink speed using command `FindInstance`.[1]

It is useful to see how it works in details for $G(1)_{B512m}$ as in a case of *aux* function of Eq. (4.1) we had one-dimensional integral which terminated in a number in Eq. (4.8) after taking a single residue. Learning by doing, we will be able to understand the way how a general algorithm for analytic continuation in ϵ works.

Taking $\epsilon = 1/10$, we find that the arguments of the G_1–G_{10} functions in Eq. (4.10) are positive:

---

[1] The shortest `Mathematica` description of the function: FindInstance[expr,vars] finds an instance of vars that makes the statement expr be True.

$$\text{rule0} = \{z_1 \rightarrow -\frac{1}{20}, z_2 \rightarrow -\frac{5}{16}, z_3 \rightarrow -\frac{1}{40}\} \tag{4.11}$$

For z_1, z_2, z_3 values as in Eq. (4.11), and $\epsilon = 0$ (which we aim at), we get the following values of arguments for G_1–G_{10}:

$$G_1[7/6]G_2[5/6]G_3[-(1/6)]G_4[11/24]G_5[3/8]^2G_6[1/12]G_7[11/12]$$

$$G_8[1/12]G_9[5/8]G_{10}[1/12]. \tag{4.12}$$

Coming from $\epsilon = 1/10$ to $\epsilon = 0$, function $G_3[\epsilon + z_1]$ changes sign, and it goes through the pole. We pick up residue at $z_1 = -\epsilon$:

$$Res[G(1)_{B5l2m}]|_{z_1=-\epsilon} = -(m^2)^{z_2}(-s)^{z_3}(-t)^{-2\epsilon-z_2-z_3}\int_{c_2-i\infty}^{c_2+i\infty}dz_2\int_{c_3-i\infty}^{c_3+i\infty}dz_3$$

$$G_1[1-\epsilon]^2G_2[-z_2]G_3[-2\epsilon-z_2-z_3]^2G_4[-z_3]G_5[1+z_3]G_6[\epsilon+z_3]$$

$$\frac{G_7[1+2\epsilon+z_2+z_3]G_8[1+\epsilon+2z_2+2z_3]}{G[1-3\epsilon]\Gamma[2-2\epsilon]\Gamma[1+\epsilon+2z_3]}. \tag{4.13}$$

Now, we are left with two variables z_2, z_3 and a new value of ϵ, which is, according to Eqs. (4.11) and (4.13), equal to $\epsilon = -z_1 = 1/20$. With this new value of ϵ, we continue analytic continuation with $\epsilon \rightarrow 0$.

Taking z_2, z_3 as in Eq. (4.11), and $\epsilon = 0$, we get for arguments of G_1–G_8 functions in Eq. (4.13):

$$G_1[1]G_2[5/16]G_3[27/80]G_4[1/40]G_5[39/40]G_6[-(1/40)]$$

$$G_7[53/80]G_8[13/40]. \tag{4.14}$$

This time, coming from $\epsilon = 1/20$ to $\epsilon = 0$, function $G_6[\epsilon + z_3]$ changes sign, and it goes through the pole. We pick up residue at $z_3 = -\epsilon$:

$$Res[G(1)_{B5l2m}]|_{z_1=-\epsilon,z_3=-\epsilon} = -(m^2)^{z_2}(-s)^{-\epsilon}(-t)^{-\epsilon-z_2}\int_{c_2-i\infty}^{c_2+i\infty}dz_2$$

$$\frac{G_1[1-\epsilon]^2G_2[\epsilon]G_3[-\epsilon-z_2]^2G_4[-z_2]G_5[1+\epsilon+z_2]G_6[1-\epsilon+2z_2]}{\Gamma[1-3\epsilon]\Gamma[2-2\epsilon]}. \tag{4.15}$$

Please note that if we took instead of 1/10 in Eq. (4.11) a larger value, 1/3 say, in the last step also G_8 would result in a negative value of the argument, and we would have to take additional residue at $1 + \epsilon + 2z_2 + 2z_3 = 0$, adding one more integral of lower dimension.

Now, we are left with one variable z_2, and a new value of ϵ, which is according to Eq. (4.11) $\epsilon = -z_3 = 1/40$. With this new value of ϵ, we continue analytic continuation with $\epsilon \rightarrow 0$.

Taking z_2 as in Eq. (4.11), and $\epsilon = 0$, we get for arguments of G_1–G_6 functions in Eq. (4.15):

$$G_1[1]G_2[0]G_3[5/16]^2 G_4[5/16]G_5[11/16]G_6[(3/8)]. \qquad (4.16)$$

We can see that there is no negative arguments anymore, and we ended the procedure.[2] The final result is

$$
\begin{aligned}
G(1)_{\text{B5l2m}\{\epsilon \to \frac{1}{10}, z_1 \to -\frac{1}{20}, z_2 \to -\frac{5}{16}, z_3 \to -\frac{1}{12}\}} \\
= G_2(1)_{\text{B5l2m}\{\epsilon \to 0, z_1 \to -\frac{1}{20}, z_2 \to -\frac{5}{16}, z_3 \to -\frac{1}{12}\}} \\
+ Res[G(1)_{\text{B5l2m}\{\epsilon \to 0, z_1 \to -\frac{1}{20}, z_2 \to -\frac{5}{16}, z_3 \to -\frac{1}{12}\}}]\big|_{z_1 = -\epsilon} \\
+ Res[G(1)_{\text{B5l2m}\{\epsilon \to 0, z_1 \to -\frac{1}{20}, z_2 \to -\frac{5}{16}, z_3 \to -\frac{1}{12}\}}]\big|_{z_1 = -\epsilon, z_3 = -\epsilon}
\end{aligned}
\qquad (4.17)
$$

The whole procedure of analytic continuation in ϵ is automatized in the MB.m package [3] (see MBoptimezedRules command there). Using this command, we can easily find a solution for a set of inequalities (Problem 4.1), which written in terms of substitutions can take a form:

```
In[3]:= rules = MBoptimizedRules[fin, eps -> 0, {}, {eps}]]
Out[3]:= {{eps -> 11/64, {z1 -> -3/64, z2 -> -7/32, z3 ->
↪    -1/32}}
```

The notebook file with a complete solution can be found in the file MB_B5l2m2_Springer.nb in the auxiliary material in [8]. Algorithm 1 summarizes the main steps we discussed above for analytic continuation in ϵ. More elaborated algorithms can be found in [3] and [4].

4.2.3   Analytic Continuation in Auxiliary Parameters

In order to find proper integration paths for the MB integrations, i.e., the condition that all arguments of gamma functions are positive, it may happen that starting with $\epsilon \neq 0$ no solutions for the inequalities like those in Eqs. (4.2)–(4.4) can be found. Fortunately, we have the freedom to increase the number of manipulated variables and add to the set of ϵ and z_i parameters the powers of propagators n_k (see, e.g., Eq. (4.9)). We do not fix power of propagators; instead we take $n_k \to n_k + \eta_k$ ($\eta \in \mathbb{R}$). Then, of course, after finding a solution for ϵ, z_i, η_k, analytic continuation must be done $\epsilon \to 0$ *and* $\eta \to 0$. This can be automatized by a modification of Algorithm 1. Namely, instead of starting with $F(\{z_i\}, \epsilon)$, we take $F(\{z_i\}, \epsilon, \eta_1)$

---

[2] Function $G_2[0]$ is actually $\Gamma[\epsilon]$, which can be Taylor expanded, adding to the singularity at ϵ expansion of the final functions in Eq. (4.17).

Algorithm 1: Pseudocode for analytic continuation in ϵ, basic steps

---

1 Let $F(\{z_i\}, \epsilon)$ be a MB integral over z_i variables and ϵ is the dimensional regulator to be
   analytically continued to 0 ;
2 **for** *a set of* $\{\Gamma_j\}$ *in the numerator of F* **do**
3   Solve[$Arg(\Gamma_j) > 0$, $\{\{z_i\}, \epsilon\} \subseteq X$] ;
4 **end**
5 **for** $\{z_i\}$ *Looking for change of sign in arguments of gammas* **do**
6   Find $\{\Gamma_k\} \in \Omega_0$; $k < j$: Solve[$Arg(\Gamma_j) < 0$, $\{\{z_i\}, \epsilon\} \subseteq X, \epsilon = 0$] ;
7 **end**
8 **if** $\Omega_0 = \emptyset$ **then**
9        $F(\{z_i\}, \epsilon)$ is a solution ;
10   **else**
11   Take residue of F for each $\Gamma_i \in \Omega_0$, Res[F]$|_{Arg[\Gamma_i]=0} \equiv R_i$ For each $\{\Gamma_k\} \in \Omega_0$
                 and $\{z_i\} \in X$, Solve[$Arg(\Gamma_k) == 0$, $\{\epsilon_k\}, \{X, \epsilon_k\} \equiv X_k$] ;
12   For $\epsilon = \epsilon_k$ and $\{z_i, \epsilon\} \subseteq X$ and $F(\{z_i\}, \epsilon) = R_i$;
13   **go to** 2
14   **end**
15 **end**

---

where η_1 is the power of the first propagator. It is better to start with one-by-one propagator as analytic continuation of each η_i may proliferate additional integrals. If we cannot find a valid rule, ν_2 is treated analogously, and the procedure can be continued until we find a set of $\{z_i, \epsilon, \eta_j\}$ which is successful. If it happens that an analytic continuation in only one additional parameter η is not sufficient, the program will stop with a proper remark. We introduced the auxiliary file MBnum.m [7] to MB.m which realizes this procedure and the subsequent automatized analytic continuation, ϵ-expansion, and numerics. The complete solution is in the file MB_B5nf_Springer.nb in the auxiliary material in [8].

The integral under consideration is depicted in Fig. 4.3.

This is an interesting integral as MBRules or MBoptimezedRules gives solutions only for n_5 (and n_6). There are the following additional steps (see the frame below) connected with n_5 in analytic continuation ($n_5 \to n_5 + \eta$). Please note that in the last line, MBexpand is used with factor 1, and for ϵ expansion at two loops, the factor is Exp[2*eps*EulerGamma]:

```
In[4]:= rules = MBoptimizedRules[fin,eta->0,{},{eps,eta}]}
Out[4]:= {{eps->-9/8, eta->25/16}, {z1->-3/8, z2->-1/2,
 ↪   z3->-31/32}}
In[5]:= Step1cont = MBcontinue[fin,eta->0,rules]};
Out[5]:= 10 integrals found ...
In[6]:= after = MBexpand[Step1cont,1,{eta,0,0}]}
```

$$(4.18)$$

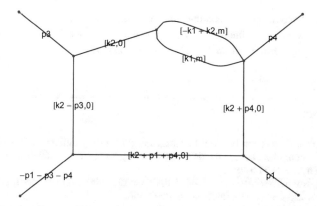

Fig. 4.3 Two-loop topology for Feynman integral with five internal lines (propagators, two massive internal lines). The figure generated by `PlanarityTest` [6, 7] (see Appendix A.1). We will call this topology `B5nf`, with four internal lines and one SE insertion with two massive propagators

In the examples for resolution of singularities, we explored the "Tausk method," used in `MB.m`. It is an intuitively easy approach to follow and apply. However, we should acknowledge that the "Smirnov method" is a good alternative. In fact, it has been used in [9] or [10] for a simple reason that it does not need additional regulators as in Sect. 4.2.3 (so no proliferation of additional `MB` integrals during analytic continuation). The file `MB_B5nf_Springer.nb` in the auxiliary material in [8] shows a difference between applying `MB.m` and `MBresolve.m`.

Problems

Problem 4.1 MB integrals must not necessarily be considered for $n = 4$ where in general the dimensional parameter is $d = n - 2\epsilon$. See, for instance, [1] or [11] for application in recent studies. Make analytic continuation for some MB representations for general n.
<u>Hint</u>: For example, see [7].

Problem 4.2 Show that for $\epsilon \to 0$, the set of solutions for inequalities constructed from arguments of gamma functions being positive in the numerator of the MB representation represented by Fig. 4.3 is empty. Find a solution, this time with $\epsilon \neq 0$.
<u>Hint</u>: The solution can be found using semidefinite programming [12], e.g., in Python. Alternatively, it can be coded in `Mathematica`, using the aforementioned function `FindInstance`. See the file `MB_B5nf_Springer.nb` in the auxiliary material in [8].

Problem 4.3 Analyze analytic continuation in ϵ and η parameters and identify poles of gamma functions, finding corresponding residues in the file MB_B512m2_Springer.nb in the auxiliary material in [8].

References

1. O.V. Tarasov, Connection between Feynman integrals having different values of the space-time dimension. Phys. Rev. D **54**, 6479–6490 (1996). arXiv:hep-th/9606018, https://doi.org/10.1103/PhysRevD.54.6479
2. B. Tausk, Non-planar massless two-loop Feynman diagrams with four on- shell legs. Phys. Lett. **B469**, 225–234 (1999). arXiv:hep-ph/9909506
3. M. Czakon, Automatized analytic continuation of Mellin-Barnes integrals. Comput. Phys. Commun. **175**, 559–571 (2006). arXiv:hep-ph/0511200, https://doi.org/10.1016/j.cpc.2006.07.002
4. C. Anastasiou, A. Daleo, Numerical evaluation of loop integrals. JHEP **10**, 031 (2006). arXiv:hep-ph/0511176, https://doi.org/10.1088/1126-6708/2006/10/031
5. A.V. Smirnov, V.A. Smirnov, On the resolution of singularities of multiple Mellin-Barnes integrals. Eur. Phys. J. **C62**, 445 (2009). arXiv:0901.0386
6. K. Bielas, I. Dubovyk, J. Gluza, T. Riemann, Some remarks on non-planar Feynman diagrams. Acta Phys. Polon. **B44**(11), 2249–2255 (2013). arXiv:1312.5603, https://doi.org/10.5506/APhysPolB.44.2249
7. AMBRE, webpage: http://jgluza.us.edu.pl/ambre, Backup: https://web.archive.org/web/20220119185211/http://prac.us.edu.pl/~gluza/ambre/
8. https://github.com/idubovyk/mbspringer, http://jgluza.us.edu.pl/mbspringer
9. J. Gluza, K. Kajda, D. A. Kosower, Towards a basis for planar two-loop integrals. arXiv:1009.0472
10. I. Dubovyk, A. Freitas, J. Gluza, T. Riemann, J. Usovitsch, Electroweak pseudo-observables and Z-boson form factors at two-loop accuracy. JHEP **08**, 113 (2019). arXiv:1906.08815, https://doi.org/10.1007/JHEP08(2019)113
11. I. Dubovyk, A. Freitas, J. Gluza, K. Grzanka, M. Hidding, J. Usovitsch, Evaluation of multi-loop multi-scale Feynman integrals for precision physics. arXiv:2201.02576
12. J. Matouek, B. Gärtner, *Understanding and Using Linear Programming* (Springer-Verlag New York, Secaucus, NJ, USA, 2006). https://www.springer.com/gp/book/9783540306979

Analytic Solutions

5

Abstract

We discuss analytic solutions of MB representations for virtual loop and real phase space integrals. We present examples of how MB loop integrals can be transformed to nested sums over residues and derive analytic solutions for some one- and two-dimensional sums. The evaluation of real phase space angular integrals using MB representations and corresponding sums is also examined. We show how higher-dimensional MB integrals can be decoupled by changing MB variables and how MB integrals can be expressed via real Euler-type integrals. We also present a brief discussion on the symbolic evaluation of Euler-type integrals as a way of obtaining analytic solutions. Finally, we consider approximations of MB integrals in ratios of both kinematic parameters and masses present in propagators of FI, as well as by the method of expansion by regions.

5.1 Residues and Symbolic Summations

Summing over residues and performing the limit $\epsilon \to 0$ may be exchanged. In any case: When crossing a pole by letting $\epsilon \to 0$, take the residue and add it to the expression. Generate lower-dimensional sums. Then look at the resulting Mellin-Barnes integrals and for the series of remaining poles. Take the sum over residues when closing the contour to the right or to the left.

This is the only recipe, doesn't it look simple? Well, at first sight, yes. Indeed, some sums can be done immediately (see, e.g., Sect. 1.7 and evaluation of a binomial sum in the file MB_miscellaneous_Springer.nb in the auxiliary material in [1]).

© The Author(s), under exclusive license to Springer Nature Switzerland AG 2022
I. Dubovyk et al., *Mellin-Barnes Integrals*, Lecture Notes in Physics 1008,
https://doi.org/10.1007/978-3-031-14272-7_5

Mathematically, Mellin-Barnes integrals are simply complex contour integrals whose integrands have a special structure: they involve products of Euler gamma functions whose arguments depend on polynomials of the integration variables, as well as fixed quantities raised to powers that again can depend on the variables of integration. Thus, it is quite natural to try to apply Cauchy's residue theorem to perform the integrations over the Mellin-Barnes variables. In order to do so, we must first find a suitable closed contour to which we can apply the theorem. As the Mellin-Barnes integration path itself runs from $-i\infty$ to $+i\infty$, the obvious (and it turns out correct) guess is to add to this path a semi-circle "at infinity" and obtain a closed contour in this way. We are allowed to close the contour as long as the integrand vanishes at complex infinity.

5.1.1   Choosing the Contour

In practical applications, we can usually fulfill this requirement if we close the contour in the correct direction. In order to get a feeling for why this is true and what complications might arise, let us consider the following simple one-dimensional Mellin-Barnes integral:

$$\int_{-i\infty}^{+i\infty} \frac{dz}{2\pi i} \Gamma(a+z)\Gamma(b-z)x^z \,. \tag{5.1}$$

Let us consider how the integrand behaves at complex infinity. First, using the asymptotic formula for $\Gamma(z)$ as $|z| \to \infty$, Eq. (2.108), we have

$$\Gamma(a+z) \overset{|z|\to\infty}{\sim} \sqrt{\frac{2\pi}{z}} z^a e^{z(\ln z - 1)} \,. \tag{5.2}$$

Furthermore, using

$$\Gamma(1-b+z)\Gamma(b-z) = \frac{\pi}{\sin[\pi(b-z)]} \tag{5.3}$$

we obtain

$$\Gamma(b-z) \overset{|z|\to\infty}{\sim} \frac{1}{\sqrt{2\pi z}} z^b e^{-z(\ln z - 1)} \frac{\pi}{\sin[\pi(b-z)]} \,. \tag{5.4}$$

Then, the full integrand behaves as

$$\Gamma(a+z)\Gamma(b-z)x^z \overset{|z|\to\infty}{\sim} \sqrt{\frac{2\pi}{z}} z^a e^{z(\ln z - 1)} \frac{1}{\sqrt{2\pi z}} z^b e^{-z(\ln z - 1)} \frac{\pi}{\sin[\pi(b-z)]} e^{z\ln x}$$

$$= z^{a+b-1} \frac{\pi}{\sin[\pi(b-z)]} e^{z\ln x} \,. \tag{5.5}$$

This clearly goes to zero for $|z| \to \infty$, provided we choose $\Re(z) > 0$ (i.e., we close the contour to the right) if $0 < x < 1$ and $\Re(z) < 0$ (i.e., we close the contour to the left) if $1 < x$ (see Problem 5.1). (Obviously we must avoid the poles at $z = b + n$, $n \in \mathbb{N}$.) Hence, as promised, we can arrange to have the integrand go to zero along the semi-circle at infinity if we choose the correct contour.

Specific Contours

This example is not entirely generic, e.g., consider

$$\int_{-i\infty}^{+i\infty} \frac{dz}{2\pi i} \Gamma(z) A^z \tag{5.6}$$

where for $z \to -\infty$ the $\Gamma(z)$ goes to zero faster than A^z goes to infinity for any A, so the contour must always be closed to the left. Notice, however, that this particular integral does not come from applying the basic MB identity Eq. (1.44) to some expression, since the gamma function of the form $\Gamma(\ldots + z)$ is not accompanied by another gamma function with z-dependence $\Gamma(\ldots - z)$.

For multi-dimensional integrals, a similar analysis can be performed for each integration variable separately, taking into account all gamma functions whose argument involves the particular integration variable. In order to make this precise, let us consider the generic multi-dimensional Mellin-Barnes integral in Eq. (1.51):

$$\int_{-i\infty}^{+i\infty} \cdots \int_{-i\infty}^{+i\infty} \prod_{j=1}^{n} \frac{dz_j}{2\pi i} f(z_1, \ldots, z_n, x_1, \ldots, x_p, a_1, \ldots, a_q) \frac{\prod_k \Gamma(A_k + V_k)}{\prod_l \Gamma(B_l + W_l)}. \tag{5.7}$$

Recall that in this expression the x_j are fixed parameters, while A_k and B_l are linear combinations of the parameters a_i. The latter take the form $a_i = n_i + b_i\epsilon$, with $n_i \in \mathbb{N}$ and $b_i \in \mathbb{R}$. Last, V_k and W_l are linear combinations of the integration variables z_i, and f in practice is a product of powers of the x_i with exponents that are linear combinations of a_i and z_i (more generally it can be an analytic function). Let us analyze the behavior of the integrand at complex infinity in the variable z_n. Clearly this is no loss of generality, since we may simply relabel the integration variables such that the variable of interest becomes z_n. Then our integral takes the form

$$\int_{-i\infty}^{+i\infty} \cdots \int_{-i\infty}^{+i\infty} \prod_{j=1}^{n-1} \frac{dz_j}{2\pi i} g(z_2, \ldots, z_n, x_1, \ldots, x_p, a_1, \ldots, a_q) \frac{\prod_k' \Gamma(A_k + V_k)}{\prod_l' \Gamma(B_l + W_l)}$$

$$\times \int_{-i\infty}^{+i\infty} \frac{dz_n}{2\pi i} \prod_p \Gamma(C_p + z_n)^{\alpha_p} \prod_q \Gamma(D_q - z_n)^{\beta_q} X^{z_n}. \tag{5.8}$$

Above, the primes on the products $\prod_k'$ and $\prod_l'$ mean that only those factors for which V_k and W_l do not involve z_n are included, while gamma functions whose arguments depend on z_n are made explicit in the second line. Notice that we have separated the gamma functions that depend on z_n as $\Gamma(\ldots + z_n)$ and those whose argument involves $-z_n$ as $\Gamma(\ldots - z_n)$ and C_p, D_q are the corresponding linear combinations of parameters *and* other integration variables. We have moreover made explicit the (integer) powers $\alpha_p, \beta_q \in \mathbb{Z}$ for the gamma functions involving z_n as well as the piece of the f function which involves z_n. Thus X is a product of the parameters x_j, perhaps raised to powers whose exponents do not depend on any other integration variable z_j. We note that technically this is not the most general expression possible, as linear combinations of the integration variables in the arguments of gamma functions can also appear where the integration variables appear with coefficients different from ± 1 (e.g., $\Gamma(1 + 2z_n)$, etc.). However, the analysis of the behavior of the integrand at complex infinity that follows is easy to repeat for such a more general integrand which is considered, for instance, in [2].

Turning finally to examining the behavior of the integrand in Eq. (5.8) at complex infinity in z_n, we can use the following asymptotic formulae for gamma functions, discussed in Sect. 2.5:

$$\Gamma(a + z) \xrightarrow{|z| \to \infty} \sqrt{\frac{2\pi}{z}} z^a e^{z(\ln z - 1)}, \tag{5.9}$$

and

$$\Gamma(b - z) \xrightarrow{|z| \to \infty} \frac{1}{\sqrt{2\pi z}} z^b e^{-z(\ln z - 1)} \frac{\pi}{\sin[\pi(b - z)]}, \tag{5.10}$$

to obtain

$$\prod_p \Gamma(C_p + z_n)^{\alpha_p} \prod_q \Gamma(D_q - z_n)^{\beta_q} X^{z_n} \xrightarrow{z_n \to \infty} (2\pi)^{\frac{1}{2}\sum_p \alpha_p - \frac{1}{2}\sum_q \beta_q}$$

$$\times z_n^{z_n(\sum_p \alpha_p - \sum_q \beta_q) + \sum_p (C_p - \frac{1}{2})\alpha_p + \sum_q (D_q - \frac{1}{2})\beta_q} e^{-(\sum_p \alpha_p - \sum_q \beta_q)z_n} \tag{5.11}$$

$$\times \prod_q \left(\frac{\pi}{\sin[\pi(D_q - z_n)]} \right)^{\beta_q} X^{z_n}.$$

The behavior of this expression at complex infinity is controlled by the expression on the second line. In particular, we see that if the difference $\sum_p \alpha_p - \sum_q \beta_q$ is negative (recall α_p and β_q are integers), then for large and positive $\Re(z)$, the expression goes to zero because of the presence of the factor $z_n^{-c \cdot z_n}$ with $c > 0$. So in this case, we must close the contour "to the right," i.e., such that $\Re(z) \to +\infty$. On the other hand, if $\sum_p \alpha_p - \sum_q \beta_q$ is positive, the contour must be closed "to the left" such that $\Re(z) \to -\infty$. However, it may happen that $\sum_p \alpha_p - \sum_q \beta_q = 0$, and in fact, this is the typical situation in practical applications.

Balanced Mellin-Barnes Integrals

The Mellin-Barnes integral

$$\int_{-i\infty}^{+i\infty} \frac{dz_j}{2\pi i} \prod_{k=1}^{n_+} \Gamma(a_k + z_j)^{\alpha_k} \prod_{l=1}^{n_-} \Gamma(a_l - z_j)^{\beta_l} , \qquad \alpha_k, \beta_l \in \mathbb{Z} \qquad (5.12)$$

is *balanced* in the variable z_j if $\sum_{k=1}^{n_+} \alpha_k = \sum_{l=1}^{n_-} \beta_l$.

> Informally, an integral is balanced if for each function of the form $\Gamma(\ldots + z_j)$ the integrand contains also a function of the form $\Gamma(\ldots - z_j)$, with gamma functions in the denominator being counted as appearing a negative number of times.

A multi-dimensional Mellin-Barnes integral is balanced if it is balanced in each integration variable. The Mellin-Barnes representations we encounter in actual applications to Feynman integrals are typically balanced by construction. This is because the basic identity expressing the sum $(A + B)^{-\lambda}$ as a Mellin-Barnes integral

$$\frac{1}{(A + B)^\lambda} = \frac{1}{\Gamma(\lambda)} \int_{-i\infty}^{+i\infty} \frac{dz}{2\pi i} \Gamma(\lambda + z)\Gamma(-z) A^{-\lambda-z} B^z \qquad (5.13)$$

is obviously balanced, and it is not hard to see that repeated applications of this formula will also lead to balanced integrands, e.g.,

$$\frac{1}{(A + B + C)^\lambda} = \frac{1}{\Gamma(\lambda)} \int_{-i\infty}^{+i\infty} \frac{dz_1}{2\pi i} \Gamma(\lambda + z_1)\Gamma(-z_1) A^{-\lambda-z_1} (B + C)^{z_1}$$

$$= \frac{1}{\Gamma(\lambda)} \int_{-i\infty}^{+i\infty} \frac{dz_1}{2\pi i} \frac{dz_2}{2\pi i} \Gamma(\lambda + z_1)\Gamma(-z_1 + z_2)\Gamma(-z_2) A^{-\lambda-z_1} B^{z_1-z_2} C^{z_2} . \qquad (5.14)$$

The integrand above is seen to be balanced both in z_1 and z_2, and we call attention to the fact that one specific gamma function may play a role for determining if the integrand is balanced for more than one variable, as is the case with $\Gamma(-z_1 + z_2)$ above. This particular gamma function must be included when counting both $\Gamma(\ldots - z_1)$ functions and $\Gamma(\ldots + z_2)$ functions. Hence, the Mellin-Barnes integrals we encounter when evaluating Feynman integrals are essentially balanced by construction and so typically $\sum_p \alpha_p - \sum_q \beta_q = 0$ in Eq. (5.8) in practical applications. Finally, we note that polygamma functions

$$\psi(z) = \frac{\Gamma'(z)}{\Gamma(z)}, \qquad \psi^{(n)}(z) = \frac{d^n \psi(z)}{dz^n} , \qquad (5.15)$$

do not enter the balance counting, so an integral like

$$\int_{-i\infty}^{+i\infty} \frac{dz}{2\pi i} \Gamma(1+z)\Gamma(-z)\psi(2+z) \tag{5.16}$$

is considered balanced.

For balanced integrals, i.e., when $\sum_p \alpha_p - \sum_q \beta_q = 0$ in Eq. (5.8), the asymptotic expression simplifies, and we find

$$\prod_p \Gamma(C_p + z_n)^{\alpha_p} \prod_q \Gamma(D_q - z_n)^{\beta_q} X^{z_n} \xrightarrow{|z_n| \to \infty} z_n^{\sum_p (C_p - \frac{1}{2})\alpha_p + \sum_q (D_q - \frac{1}{2})\beta_q}$$

$$\times \prod_q \left(\frac{\pi}{\sin[\pi(D_q - z_n)]} \right)^{\beta_q} X^{z_n} . \tag{5.17}$$

Away from the poles at $z = b + n$, $n \in \mathbb{N}$, this expression essentially has the form $\sim z_n^c X^{z_n}$ with some constant c. Then we find that it is the magnitude of X which determines if the integrand vanishes at complex infinity in a particular direction. Assuming that X is real and positive, the integrand clearly goes zero for $|z_n| \to \infty$, provided we choose $\Re(z) > 0$ and close the contour to the right if $0 < X < 1$, while for $1 < X$ we must demand $\Re(z) < 0$ and close the contour to the left. (Obviously the poles at $z = b + n$, $n \in \mathbb{N}$ need to be avoided.) Hence, as promised, we can arrange to have the integrand go to zero along the semi-circle at infinity if we choose the correct contour.

Before moving on, let us note one further complication which can arise. If both $\sum_p \alpha_p - \sum_q \beta_q = 0$ *and* $X = 1$, then the integrand simply behaves as a power of z_n at infinity. Evidently the exponent is given by $\sum_p (C_p - \frac{1}{2})\alpha_p + \sum_q (D_q - \frac{1}{2})\beta_q$ which may be negative, positive, or zero (e.g., at $\epsilon = 0$), and the integrand is not guaranteed to vanish at complex infinity. In this case one option is to introduce an auxiliary variable $0 < X < 1$, obtain the solution with this $X \neq 1$, and finally take the $X \to 1$ limit. We will come back briefly to this point in the next section.

5.1.2   From MB Integrals to Sums

Let us then assume that we have identified the proper closed contour for some Mellin-Barnes integration variable z. It is now in principle a simple matter to apply Cauchy's residue theorem to perform the integration in z and obtain the result in the form of a sum over residues. As an example, consider the simple one-dimensional Mellin-Barnes integral:

$$I(x) = \int_{\bar{z}-i\infty}^{\bar{z}+i\infty} \frac{dz}{2\pi i} \Gamma(1+z)\Gamma(-z)x^z . \tag{5.18}$$

For definiteness, let us assume that the contour runs parallel to the imaginary axis with constant real part that is between zero and one, $0 < \bar{z} < 1$, and that x is real and positive. By our previous analysis, we must close the contour to the right or the left depending on the magnitude of x. For $x < 1$, we must have $\mathcal{R}(z) \to +\infty$, and so our closed contour encircles the poles of $\Gamma(-z)$ at $z = 1, 2, \ldots$. Then, using Eq. (2.160),

$$\text{Res}_{z \to n} \Gamma(-z) = -\frac{(-1)^n}{n!}, \qquad n \in \mathbb{N} \tag{5.19}$$

we immediately find (note that our contour runs clockwise)

$$I(x) = -2\pi i \sum_{n=1}^{\infty} \frac{1}{2\pi i} \left[-\Gamma(1+n) \frac{(-1)^n}{n!} x^n \right] = \sum_{n=1}^{\infty} (-x)^n . \tag{5.20}$$

The sum is convergent for $0 < x < 1$, and it is of course elementary. We obtain

$$I(x) = -\frac{x}{1+x}, \qquad 0 < x < 1. \tag{5.21}$$

Elementary Sums

We recall some elementary sums that we will make use of in the following. We begin with the simple geometric series and discuss some simple generalizations. To start recall that

$$\sum_{n=0}^{\infty} x^n = \frac{1}{1-x}, \tag{5.22}$$

where the sum is convergent for $|x| < 1$. One simple derivation of this result proceeds to show that the Nth partial sum s_N is simply $S_N = \frac{1+x^{N+1}}{1-x}$. Then for $|x| < 1$, we find Eq. (5.22) by taking the $N \to \infty$ limit. However, Eq. (5.22) can be established also by considering the Taylor expansion of $f(x) = (1-x)^{-1}$ around $x = 0$. Clearly $f(0) = 1$, $f'(0) = 1 \cdot (1-x)^{-2}|_{x=0} = 1$, $f''(0) =$

$1 \cdot 2 \cdot (1-x)^{-2}|_{x=0} = 2!$, and so on; hence $f^{(n)}(0) = 1 \cdot 2 \cdot \ldots \cdot n(1-x)^{-n-1}|_{x=0} = n!$. Thus

$$(1-x)^{-1} = 1 + x + \frac{2!}{2!}x^2 + \ldots = \sum_{n=0}^{\infty} x^n. \tag{5.23}$$

The advantage of this second method is that it allows to consider some immediate generalizations. In particular, consider first $f(x) = (1-x)^{-m}$ for some positive integer m. In this case $f(0) = 1$, $f'(0) = m(1-x)^{-m-1}|_{x=0} = m$, $f''(0) = m(m+1)(1-x)^{-m-2}|_{x=0} = m(m+1)$, and in general $f^{(n)}(0) = m(m+1)\cdots(m+n-1)(1-x)^{-m-n}|_{x=0} = m(m+1)\cdots(m+n-1)$. Using the ratio test, one can show that the Taylor series converges for $|x| < 1$ and thus

$$(1-x)^{-m} = 1 + mx + \frac{m(m+1)}{2!}x^2 + \ldots = \sum_{n=0}^{\infty} \frac{m(m+1)\cdots(m+n-1)}{n!}x^n \tag{5.24}$$

so we find the summation formula

$$\sum_{n=0}^{\infty} \frac{(m+n-1)!}{n!(m-1)!}x^n = \frac{1}{(1-x)^m}. \tag{5.25}$$

We may also apply the same arguments for fractional powers. So let us set $\alpha = \frac{p}{q}$ where p and q are positive integers and examine the Taylor expansion of the function $f(x) = (1+x)^{\alpha}$ around $x = 0$. We see that $f(0) = 1$, while $f'(0) = \alpha(1+x)^{\alpha-1}|_{x=0} = \alpha$, $f''(0) = \alpha(\alpha-1)(1+x)^{\alpha-2}|_{x=0} = \alpha(\alpha-1)$, and in general $f^{(n)}(0) = \alpha(\alpha-1)\cdots(\alpha-n+1)(1+x)^{\alpha-n}|_{x=0} = \alpha(\alpha-1)\cdots(\alpha-n+1)$. The convergence of the Taylor series for $|x| < 1$ can be established with the ratio test. Thus

$$(1+x)^{\alpha} = 1 + \alpha x + \frac{\alpha(\alpha-1)}{2!}x^2 + \cdots = \sum_{n=0}^{\infty} \frac{\alpha \cdot (\alpha-1) \cdot \ldots \cdot (\alpha-n+1)}{n!}x^n. \tag{5.26}$$

Introducing the *generalized binomial coefficient* $\binom{\alpha}{n}$,

$$\binom{\alpha}{n} = \frac{\alpha \cdot (\alpha-1) \cdot \ldots \cdot (\alpha-n+1)}{n!}, \tag{5.27}$$

we can write

$$\sum_{n=0}^{\infty} \binom{\alpha}{n} x^n = (1+x)^{\alpha}. \tag{5.28}$$

This result is known as the *binomial theorem for fractional exponents*.

We note finally that all of the sums above are special cases of the Gauss hypergeometric sum introduced in Sect. 2.171. In particular, they are all just realizations of the general formula

$$2F_1(-\alpha, \beta, \beta, -z) = (1 + z)^\alpha \qquad \beta \text{ arbitrary.} \tag{5.29}$$

Finally, let us recall one more well-known but useful trick for evaluating sums where some positive integer power of the index of summation appears in the summand. Consider, e.g.,

$$\sum_{n=0}^{\infty} n^k x^n, \qquad k \in \mathbb{N}^+. \tag{5.30}$$

One approach to dealing with this sum is to notice that formally

$$x \frac{d}{dx} \frac{1}{1-x} = x \frac{d}{dx} \sum_{n=0}^{\infty} x^n = x \sum_{n=0}^{\infty} n x^{n-1}, \tag{5.31}$$

and so

$$\sum_{n=0}^{\infty} n x^n = \frac{x}{(1-x)^2}. \tag{5.32}$$

Obviously this operation can be iterated k times once the specific value of k is known to compute the sum in Eq. (5.30). The same trick can be useful also for the generalized sums discussed above.

Continuing with our example, for $1 < x$ we must close the contour to the left. Then the contour encircles the poles of $\Gamma(1 + z)$ at $z = -1, -2, \ldots$ *and* the pole of $\Gamma(-z)$ at $z = 0$. Using

$$\text{Res}_{z \to -n} \Gamma(1 + z) = \frac{(-1)^{n-1}}{(n-1)!}, \qquad n \in \mathbb{N}^+, \tag{5.33}$$

we obtain (notice that the contour now runs counter-clockwise)

$$I(x) = 2\pi i \left\{ \sum_{n=1}^{\infty} \frac{1}{2\pi i} \left[\frac{(-1)^{n-1}}{(n-1)!} \Gamma(n) x^{-n} \right] - \Gamma(1+0) x^0 \right\} = -\sum_{n=1}^{\infty} (-x)^{-n} - 1. \tag{5.34}$$

The sum is convergent now for $1 < x$ and is again elementary. We find

$$I(x) = -\frac{x}{1+x}, \qquad 1 < x. \tag{5.35}$$

So in this case, we indeed obtain the same functional form for both $0 < x < 1$ and $1 < x$. On the one hand, this should not come as a surprise. Indeed, our starting Mellin-Barnes integral only differed from the basic Mellin-Barnes formula of Eq. (1.44) by the position of the contour: it did not separate all poles of $\Gamma(1 + z)$ and $\Gamma(-z)$ as the pole of $\Gamma(-z)$ at $z = 0$ was on the wrong side of the contour. So we could have evaluated the integral simply by shifting the contour to some position between -1 and 0 and accounting for the residue of the pole at $z = 0$. So letting $\bar{z}' \in (-1, 0)$, we compute for all $0 < x$

$$\begin{aligned} I(x) &= \int_{\bar{z}-i\infty}^{\bar{z}+i\infty} \frac{dz}{2\pi i} \Gamma(1+z)\Gamma(-z)x^z \\ &= \int_{\bar{z}'-i\infty}^{\bar{z}'+i\infty} \frac{dz}{2\pi i} \Gamma(1+z)\Gamma(-z)x^z - 1 \\ &= \Gamma(1)(1+x)^{-1} - 1 = -\frac{x}{1+x}, \qquad 0 < x. \end{aligned} \tag{5.36}$$

On the other hand, this phenomena of obtaining the same functional form of the solution for both $x < 1$ and $x > 1$ is not generic as the next example demonstrates.

Regions of Validity of the Solution

To illustrate this, we consider the Mellin-Barnes integral

$$I(x) = \int_{\bar{z}-i\infty}^{\bar{z}+i\infty} \frac{dz}{2\pi i} \frac{\Gamma(-z)}{\Gamma(3-z)} x^z, \tag{5.37}$$

where again let us assume that the contour is parallel to the imaginary axis and $\bar{z}$ is between zero and one. Applying our analysis regarding the choice of contours, we see that once more we must close the contour to the right if $0 < x < 1$ and to the left if $1 < x$. In the former case, the closed contour, running clockwise, encircles the poles of the integrand at $z = 1$ and $z = 2$. Notice that in this case there are in fact no other poles inside the contour, since

$$\frac{\Gamma(-z)}{\Gamma(3-z)} = \frac{1}{(2-z)(1-z)(-z)} = -\frac{1}{2(z-2)} + \frac{1}{z-1} - \frac{1}{2z}. \tag{5.38}$$

It is also trivial to read off the residues at the poles:

$$\operatorname{Res}_{z\to 1} \frac{\Gamma(-z)}{\Gamma(3-z)} = 1, \qquad \operatorname{Res}_{z\to 2} \frac{\Gamma(-z)}{\Gamma(3-z)} = -\frac{1}{2}; \tag{5.39}$$

hence, the sum over poles is finite, and we simply have

$$I(x) = -\sum_{n=1}^{2} 1 \cdot x^n = -x + \frac{x^2}{2}, \qquad 0 < x < 1. \tag{5.40}$$

On the other hand, when $1 < x$, the contour runs counter-clockwise and encircles the single pole at $z = 0$ with residue

$$\text{Res}_{z \to 0} \frac{\Gamma(-z)}{\Gamma(2-z)} = -\frac{1}{2}. \tag{5.41}$$

Then the sum over poles involves just a single term, and we find

$$I(x) = \sum_{n=0}^{0} \left(-\frac{1}{2}\right) \cdot x^n = -\frac{1}{2}, \qquad 1 < x. \tag{5.42}$$

So finally we have

$$I(x) = \begin{cases} -x + \frac{x^2}{2} & \text{if } 0 < x < 1 \\[2mm] -\frac{1}{2} & \text{if } 1 \leq x \end{cases}. \tag{5.43}$$

The lesson to take away from these simple examples is that one must be mindful of both the direction in which the contour is closed and the range over which the summation must be taken.

In the case of multi-dimensional Mellin-Barnes integrals, closing the contours for each integration variable in different directions can lead to different multiple sum representations of the result, each convergent in different ranges of the variables (see [3] for further details).

Finally, let us briefly comment on the case of $x = 1$. First, notice that if $x = 1$ in Eq. (5.18), then we are precisely in the situation discussed in Eq. (5.17): $\sum_p (C_p - \frac{1}{2})\alpha_p + \sum_q (D_q - \frac{1}{2})\beta_q = (1 - \frac{1}{2})^1 + (0 - \frac{1}{2})^1 = 0$ and $X = 1$. So it is not evident how the contour should be closed. In fact, we can also see that this case requires special care since if we proceed naively and close the contour, say, to the right, we obtain (just set $x = 1$ in Eq. (5.20))

$$I(1) = \sum_{n=1}^{\infty} (-1)^n. \tag{5.44}$$

Clearly this sum does not converge! Had we closed the contour to the left, the same divergent sum would have appeared. In such situations, one option is in fact to compute $I(x)$ instead of $I(1)$ directly and take the $x \to 1$ limit. In our case, we find simply

$$I(1) = \lim_{x \to 1} I(x) = \lim_{x \to 1} -\frac{x}{1+x} = -\frac{1}{2}. \tag{5.45}$$

The correctness of this result can be checked simply by evaluating the integral numerically. This is easily done in Mathematica:

```
In[1]:= NIntegrate[I*1/(2 Pi I)*Gamma[1 + z] Gamma[-z] /. {z ->
 ↪  1/2 + I t}, {t, -Infinity, Infinity}]
Out[1]:= -0.5 + 0. I
```

The extra factor of i is the Jacobian associated with the change of integration variable $z \to 1/2 + it$ which we have used to integrate along the correct contour.

In order to highlight how these considerations apply in the multi-dimensional case, let us look at the following example. Consider

$$I(x, y) = \int_{\bar{z}_1 - i\infty}^{\bar{z}_1 + i\infty} \int_{\bar{z}_2 - i\infty}^{\bar{z}_2 + i\infty} \frac{dz_1}{2\pi i} \frac{dz_2}{2\pi i} \Gamma(1 + z_1)\Gamma(1 + z_2)\Gamma(-z_1 - z_2)x^{z_1} y^{z_2}. \tag{5.46}$$

Let us assume that both $\bar{z}_1$ and $\bar{z}_2$ are between zero and one and that $\bar{z}_1 + \bar{z}_2 < 1$. Consider the situation when both x and y are positive but smaller than one, $0 < x, y < 1$. Then, starting with the z_1 integration, we have that the contour in this variable must be closed to the right and will thus encircle the poles of $\Gamma(-z_1 - z_2)$, which lie at $z_1 = n - z_2$, $n \in \mathbb{N}$. However, notice that not all of these poles are inside the contour! In fact, the pole at $z_1 = -z_2$, corresponding to $n = 0$, has a negative real part (since $\bar{z}_2 > 0$) and hence is not inside the closed contour. The next pole at $z_1 = 1 - z_2$ (corresponding to $n = 1$) is inside the contour only if $\Re(z_1) < 1 - \Re(z_2)$, and we see now the significance of the condition $\bar{z}_1 + \bar{z}_2 < 1$ above. The rest of the poles at $z_1 = n - z_2$ for $n \geq 2$ are all inside the contour since $\Re(z_1) < n - \Re(z_2)$ is clearly satisfied for $0 < \bar{z}_1, \bar{z}_2 < 1$ and $n \geq 2$. Then, using

$$\mathrm{Res}_{z_1 \to n - z_2} \left[\Gamma(1 + z_1)\Gamma(1 + z_2)\Gamma(-z_1 - z_2)x^{z_1} y^{z_2} \right] =$$
$$= \Gamma(1 + n - z_2)\Gamma(1 + z_2)\frac{-(-1)^n}{n!}x^{n - z_2} y^{z_2}, \tag{5.47}$$

we can perform the integration for z_1 and obtain

$$I(x, y) = \int_{\bar{z}_2 - i\infty}^{\bar{z}_2 + i\infty} \frac{dz_2}{2\pi i}(-2\pi i)\frac{1}{2\pi i} \sum_{n_1 = 1}^{\infty} \Gamma(1 + n_1 - z_2)\Gamma(1 + z_2)\frac{-(-1)^{n_1}}{n_1!}x^{n_1 - z_2} y^{z_2}$$

$$= \int_{\bar{z}_2 - i\infty}^{\bar{z}_2 + i\infty} \frac{dz_2}{2\pi i} \sum_{n_1 = 1}^{\infty} \Gamma(1 + n_1 - z_2)\Gamma(1 + z_2)\frac{(-x)^{n_1}}{n_1!}\left(\frac{y}{x}\right)^{z_2}. \tag{5.48}$$

Notice how the summation over n_1 runs from one to infinity, as explained above. Next, we would like to perform the integration over z_2 by using the residue theorem. Examining the behavior of the integrand at infinity, we see that we must close the integration contour to the right if $0 < y/x < 1$ or to the left if $1 < y/x$. Let us assume the former for the sake of example, so that the closed contour encircles the residues of $\Gamma(1 + n_1 - z_2)$ at $z_2 = 1 + n_1 + n_2$ with $n_2 \in \mathbb{N}$. In order to determine the range of summation in n_2, we note that $1 + n_1 \geq 2$, since $n_1 \geq 1$, and so each pole of $\Gamma(1 + n_1 - z_2)$ lies inside the closed contour since $Re(z_2) < 1$. The residues at $z_2 = 1 + n_1 + n_2$ are easy to compute, and we find

$$\text{Res}_{z_2 \to 1+n_1+n_2} \left[\Gamma(1 + n_1 - z_2)\Gamma(1 + z_2)\frac{(-x)^{n_1}}{n_1!} \left(\frac{y}{x}\right)^{z_2} \right]$$
$$= \frac{-(-1)^{n_2}}{n_2!}\Gamma(2 + n_1 + n_2)\frac{(-x)^{n_1}}{n_1!} \left(\frac{y}{x}\right)^{1+n_1+n_2} . \tag{5.49}$$

Then the integration over z_2 gives the result

$$I(x, y) = -(2\pi i)\frac{1}{2\pi i} \sum_{n_2=0}^{\infty} \sum_{n_1=1}^{\infty} \frac{-(-1)^{n_2}}{n_2!}\Gamma(2 + n_1 + n_2)\frac{(-x)^{n_1}}{n_1!} \left(\frac{y}{x}\right)^{1+n_1+n_2}$$
$$= \frac{y}{x} \sum_{n_1=1}^{\infty} \sum_{n_2=0}^{\infty} \frac{\Gamma(n_1 + n_2 + 2)}{\Gamma(n_1 + 1)\Gamma(n_2 + 1)}(-y)^{n_1} \left(-\frac{y}{x}\right)^{n_2}, \qquad 0 < y < x < 1. \tag{5.50}$$

In the last step, we have exchanged the order of the summations which we are allowed to do if the sums are absolutely convergent. We will not pursue this example any further, as hopefully it is clear how one would derive similar multiple sum representations for the cases we have not considered explicitly, e.g., for $0 < x < y < 1$ and so on.

Evaluating Sums with `Mathematica`

The double sum in Eq. (5.50) is actually elementary and can be computed in `Mathematica` immediately (see the file `MB_basicsums_Springer.nb` in the auxiliary material in [1]):

```
In[2] := summand = y/x Gamma[n1+n2+2]/Gamma[n1+1]/Gamma[n2+1]
↪    (-y)^n1 (-y/x)^n2;
In[3] := sum = Sum[summand, {n2, 0, Infinity}]
Out[3] := (x (-y)^n1 y ((x+y)/x)^-n1 Gamma[2+n1])/((x+y)^2
↪    Gamma[1+n1])
In[4] := sum = Sum[sum, {n1, 1, Infinity}]
Out[4] := -((x^2 y^2 (2 x+2 y+x y))/((x+y)^2 (x+y+x y)^2))
```

It is simple to understand both steps of the computation. First, consider the inner summation over n_2 for some fixed $n_1 \in \mathbb{N}^+$. This sum can easily be brought to the form of the elementary sum in Eq. (5.25):

$$\sum_{n_2=0}^{\infty} \frac{\Gamma(n_1+n_2+2)}{\Gamma(n_1+1)\Gamma(n_2+1)} \left(-\frac{y}{x}\right)^{n_2} = \sum_{n_2=0}^{\infty} \frac{(n_1+n_2+1)!}{n_1!n_2!} \left(-\frac{y}{x}\right)^{n_2}$$

$$= (n_1+1) \sum_{n_2=0}^{\infty} \frac{(n_1+2+n_2-1)!}{(n_1+2-1)!n_2!} \left(-\frac{y}{x}\right)^{n_2}$$

$$= (n_1+1) \left(1+\frac{y}{x}\right)^{-n_1-2},$$

$$(5.51)$$

where in the last line we used Eq. (5.25) with $m = n_1 + 2$ and $n = n_2$. The final summation in n_1 is then essentially a geometric sum, with the generalization that n_1 also appears in the numerator of the summand, as in Eq. (5.30). We have seen that such sums are easy to evaluate, and we can use Eqs. (5.22) and (5.32) to find the final result (notice though that the n_1 summation runs from $n_1 = 1$, not $n_1 = 0$).

The previous examples make it clear that computing Mellin-Barnes integrals by the use of the residue theorem generally leads to multiple sums involving gamma functions as well as powers of parameters. Thus, our next task is to develop tools for the evaluation of such sums. However, before we turn to this, let us examine one further example which will serve also to motivate the class of sums we will examine. Thus, consider the following Mellin-Barnes integral:

$$I(x) = \int_{\bar{z}-i\infty}^{\bar{z}+i\infty} \frac{dz}{2\pi i} \Gamma(z)^2 \Gamma(-z)^2 x^z, \qquad (5.52)$$

with $0 < \bar{z} < 1$ and $0 < x < 1$. We can perform the integration using Cauchy's theorem by closing the contour to the right, which thus encircles the poles of the integrand at $z = n$ for $n = 1, 2, \ldots$. Since $0 < \bar{z} < 1$, the pole at $z = 0$ is not inside the contour. We can compute the residue, and we find

$$\mathrm{Res}_{z \to n} \left[\Gamma(z)^2 \Gamma(-z)^2 x^z \right] = \frac{x^n}{n^2} \left[\ln x + 2\psi^{(0)}(n) - 2\psi^{(0)}(n+1) \right]. \qquad (5.53)$$

We will explain one approach to computing such symbolic residues in detail in Sect. 5.1.4; however in many situations, they are simple to obtain also with Mathematica:

```
In[5]:= Assuming[n > 0 && Element[n, Integers],
  Residue[Gamma[z]^2 Gamma[-z]^2 x^z, {z, n}]]
Out[5]:= ((-1)^(-2 n) x^n Gamma[n]^2 (Log[x]+2 PolyGamma[0,n] -
  ↪    2 PolyGamma[0,1+n]))/(n!)^2
```

Since the contour runs clockwise, we find

$$I(x) = -(2\pi i)\frac{1}{2\pi i}\sum_{n=1}^{\infty}\frac{x^n}{n^2}\left[\ln x + 2\psi^{(0)}(n) - 2\psi^{(0)}(n+1)\right]. \tag{5.54}$$

From Eq. (2.142) we see that the polygamma function can be written as

$$\psi^{(0)}(n) = -\gamma + \sum_{k=1}^{n-1}\frac{1}{k}, \tag{5.55}$$

and so we obtain the following double sum representation of the integral

$$I(x) = -(2\pi i)\frac{1}{2\pi i}\sum_{n=1}^{\infty}\frac{x^n}{n^2}\left[\ln x + 2\sum_{k=1}^{n}\frac{1}{k} - 2\sum_{k=1}^{n-1}\frac{1}{k}\right]. \tag{5.56}$$

Notice that this representation involves *nested sums*: the upper limit of the inner summation over k depends on the index of the outer summation over n. For completeness, we note that in this particular case, the nesting of the sums is illusory, since using $\psi^{(0)}(n+1) = \psi^{(0)}(n) + \frac{1}{n}$, which follows immediately from Eq. (5.55), and we obtain simply

$$I(x) = -\sum_{n=1}^{\infty}\frac{x^n}{n^2}\left(\ln x - \frac{2}{n}\right). \tag{5.57}$$

The sums that appear here are no longer elementary in the sense that the results can no longer be expressed with elementary functions such as rational functions and logarithms. Hence, we must develop tools and techniques to evaluate such sums. Furthermore, we emphasize that in more elaborate applications, evaluating Mellin-Barnes integrals via the residue theorem produces genuine multiple nested sums. Thus, next we turn our attention to a class of such sums.

5.1.3   Z- and S-Nested Sums

Motivated by the previous examples, we define a class of functions called *Z-sums* by the following recursive formula [4]

$$Z(n) = \begin{cases} 1 & \text{if } n \geq 0, \\ 0 & \text{if } n < 0, \end{cases}$$

$$Z(n; m_1, \ldots, m_k; x_1, \ldots, x_k) = \sum_{i=1}^{n}\frac{x_1^i}{i^{m_1}}Z(i-1; m_2, \ldots, m_k; x_2, \ldots, x_k). \tag{5.58}$$

Here k is called the *depth* of the sum, while $w = m_1 + \ldots + m_k$ is the *weight* of the sum. Unfolding the recursion, one obtains the equivalent definition

$$Z(n; m_1, \ldots, m_k; x_1, \ldots, x_k) = \sum_{n \geq i_1 > i_2 > \ldots > i_k > 0} \frac{x_1^{i_1}}{i_1^{m_1}} \cdots \frac{x_k^{i_k}}{i_k^{m_k}} . \tag{5.59}$$

We note that the upper limit of summation, n, can be infinity. In much the same way, we also define *S-sums* by

$$S(n) = \begin{cases} 1 & \text{if } n > 0, \\ 0 & \text{if } n \leq 0, \end{cases} \tag{5.60}$$

$$S(n; m_1, \ldots, m_k; x_1, \ldots, x_k) = \sum_{i=1}^{n} \frac{x_1^i}{i^{m_1}} S(i; m_2, \ldots, m_k; x_2, \ldots, x_k) .$$

Once again, we can unfold the recursion to arrive at the equivalent definition

$$S(n; m_1, \ldots, m_k; x_1, \ldots, x_k) = \sum_{n \geq i_1 \geq i_2 \geq \ldots \geq i_k > 0} \frac{x_1^{i_1}}{i_1^{m_1}} \cdots \frac{x_k^{i_k}}{i_k^{m_k}} . \tag{5.61}$$

Again, $n = \infty$ is allowed. The difference between Eqs. (5.59) and (5.61) is only in the upper summation boundary (see the discussion of the property **S-Z1** further in this section).

Z- and S-sums define a rather general class of functions which includes as special cases classical polylogarithms $\text{Li}_m(x)$, Nielsen's polylogarithms $S_{n,p}(x)$, the harmonic polylogarithms of Remiddi and Vermaseren $H_{m_1,\ldots,m_k}(x)$, as well as Goncharov's multiple polylogarithms $G(a_1, \ldots, a_w; z)$ discussed in Chap. 2 (see in particular Sects. 2.3 and 2.4). Objects such as multiple ζ-values $\zeta(m_1, \ldots, m_k)$, Euler-Zagier sums $Z_{m_1,\ldots,m_k}(n)$ and harmonic sums $S_{m_1,\ldots,m_k}(n)$, which are perhaps less than well-known are also special cases of Z- and S-sums [4].

For $n = \infty$ Z-sums are by definition the multiple polylogarithms of Goncharov, introduced in Eq. (2.88):

$$Z(\infty; m_1, \ldots, m_k; x_1, \ldots, x_k) = \text{Li}_{m_k,\ldots,m_1}(x_k, \ldots, x_1) . \tag{5.62}$$

Notice that the indices and arguments on the right hand side are reversed with respect to the left hand side. In particular the classical polylogarithms of order n are simply the depth-one Z-sums to infinity:

$$Z(\infty; m, x) = \sum_{i=1}^{\infty} \frac{x^i}{i^m} = \text{Li}_m(x) . \tag{5.63}$$

We next turn to developing the basic tools that are useful for performing the summation of residues when evaluating Mellin-Barnes integrals, but first let us remark that already at this stage we can write a symbolic solution for the example in Eq. (5.57). Indeed, we have simply

$$I(x) = -\sum_{n=1}^{\infty} \frac{x^n}{n^2} \left(\ln x - \frac{2}{n} \right) = -\ln x\, Z(\infty; 2; x) + 2Z(\infty; 3, x)\,. \qquad (5.64)$$

Since depth-one sums to infinity simply correspond to the classical polylogarithms, we have

$$I(x) = -\sum_{n=1}^{\infty} \frac{x^n}{n^2} \left(\ln x - \frac{2}{n} \right) = -\ln x\, \mathrm{Li}_2(x) + 2\mathrm{Li}_3(x)\,. \qquad (5.65)$$

Of course, in most real applications, we will not simply obtain sums that can be instantly recognized to be a Z-sum or an S-sum. Hence, we must study the properties of these functions and work out algorithms for expressing sums over residues in terms of them, when possible. To do so, we begin by describing some basic operations on Z- and S-sums.

The four essential properties we will need are the following:

S-Z1 Z- and S-sums define the same class of functions and can be converted into each other. Therefore, we will mostly work with Z-sums in the following.

S-Z2 Z- and S-sums form an algebra. This means that (finite) products of Z- and S-sums can be written as linear combinations of Z- and S-sums of increased depth.

S-Z3 Sums with (positive) offsets can be algebraically reduced to regular Z- and S-sums. Offsets here refer either to sums with an upper limit of summation of the form $n + c$, where $c > 0$ is a fixed integer, or to denominators of the type $(i + c)^m$ in sums.

S-Z4 Z- and S-sums appear as expansion coefficients when expanding gamma functions around integer values.

First, the fact that Z- and S-sums can be converted into one another is quite straightforward, since the two definitions differ only in the upper summation boundary of the nested sums: $(i - 1)$ for Z-sums and i for S-sums. Hence sums of depth one ($k = 1$) simply coincide:

$$S(n; m_1; x_1) = Z(n; m_1; x_1), \tag{5.66}$$

both being equal to $\sum_{i_1} \frac{x_1^{i_1}}{i_1^{m_1}}$. For depths greater than one, the conversion is in principle a straightforward, if somewhat tedious computation, e.g.,

$$
\begin{aligned}
S(n; m_1, m_2; x_1, x_2) &= \sum_{i_1=1}^{n} \frac{x_1^{i_1}}{i_1^{m_1}} S(i_1; m_2; x_2) = \sum_{i_1=1}^{n} \frac{x_1^{i_1}}{i_1^{m_1}} \sum_{i_2=1}^{i_1} \frac{x_2^{i_2}}{i_2^{m_2}} \\
&= \sum_{i_1=1}^{n} \frac{x_1^{i_1}}{i_1^{m_1}} \left[\sum_{i_2=1}^{i_1-1} \frac{x_2^{i_2}}{i_2^{m_2}} + \frac{x_2^{i_1}}{i_1^{m_2}} \right] \\
&= \sum_{i_1=1}^{n} \frac{x_1^{i_1}}{i_1^{m_1}} Z(i_1 - 1; m_2; x_2) + \sum_{i_1=1}^{n} \frac{(x_1 x_2)^{i_1}}{i_1^{m_1+m_2}} \\
&= Z(n; m_1, m_2; x_1, x_2) + Z(n; m_1 + m_2; x_1 x_2).
\end{aligned}
\tag{5.67}
$$

In general, the conversion between Z- and S-sums can be performed in a recursive fashion using an easy generalization of the computation above. The details of this algorithm are discussed in [4]. Thus one can easily convert Z-sums to S-sums and vice versa. *However, some properties are more naturally expressed in terms of Z-sums, while others in terms of S-sums. Thus, we prefer to keep both definitions and convert between them when and if convenient.*

Second, Z- and S-sums form an algebra. This simply means that the product of two Z-sums (or S-sums) with the same upper limit of summation n can be expressed as a linear combination of single Z-sums (or S-sums) of the same upper limit of summation. In order to get a feeling for why this is true, let us try and express the product of $Z(n; m_1; x_1)$ and $Z(n; m_2; x_2)$ as a single Z-sum. We begin by using the definition of Z-sums to write

$$Z(n; m_1; x_1) Z(n; m_2; x_2) = \sum_{i=1}^{n} \frac{x_1^{i}}{i^{m_1}} Z(n) \sum_{j=1}^{n} \frac{x_2^{j}}{j^{m_2}} Z(n). \tag{5.68}$$

Now, let us rewrite the double summation over i and j as three terms using the relation

$$\sum_{i=1}^{n} \sum_{j=1}^{n} a_{ij} = \sum_{i=1}^{n} \sum_{j=1}^{i-1} a_{ij} + \sum_{j=1}^{n} \sum_{i=1}^{j-1} a_{ij} + \sum_{i=1}^{n} a_{ii}, \tag{5.69}$$

which leads to

$$Z(n; m_1; x_1)Z(n; m_2; x_2) = \sum_{i=1}^{n}\sum_{j=1}^{i-1}\frac{x_1^i}{i^{m_1}}\frac{x_2^j}{j^{m_2}}[Z(n)]^2 + \sum_{j=1}^{n}\sum_{i=1}^{j-1}\frac{x_1^i}{i^{m_1}}\frac{x_2^j}{j^{m_2}}[Z(n)]^2$$

$$+ \sum_{i=1}^{n}\frac{(x_1x_2)^i}{i^{m_1+m_2}}[Z(n)]^2 \,.$$

(5.70)

Next, we use $[Z(n)]^2 = Z(n)$, since $Z(n) = 1$ for $n \geq 0$, and the definition of Z-sums in Eq. (5.58) to obtain

$$Z(n; m_1; x_1)Z(n; m_2; x_2) = \sum_{i=1}^{n}\frac{x_1^i}{i^{m_1}}\sum_{j=1}^{i-1}\frac{x_2^j}{j^{m_2}}Z(n) + \sum_{j=1}^{n}\frac{x_2^j}{j^{m_2}}\sum_{i=1}^{j-1}\frac{x_1^i}{i^{m_1}}Z(n)$$

$$+ \sum_{i=1}^{n}\frac{(x_1x_2)^i}{i^{m_1+m_2}}Z(n)$$

$$= Z(n; m_1, m_2; x_1, x_2) + Z(n; m_2, m_1; x_2, x_1)$$

$$+ Z(n; m_1 + m_2; x_1x_2) \,.$$

(5.71)

As promised, we have expressed the product of two Z-sums as a linear combination of single Z-sums. The decomposition of a product of two general Z-sums into single Z-sums can be performed using a recursion on the depth of the sums. This is discussed further in [4].

Third, for any fixed positive integer c, the Z-sum $Z(n+c-1, \ldots)$ and the S-sum $S(n+c, \ldots)$ can be related to $Z(n-1, \ldots)$ and $S(n, \ldots)$. To see this, let us consider, e.g., $Z(n + 2 - 1; m_1, m_2; x_1, x_2)$. Start by resolving the outermost summation and splitting it as follows:

$$Z(n + 2 - 1; m_1, m_2; x_1, x_2) = \sum_{i=1}^{n-1}\frac{x_1^i}{i^{m_1}}Z(i - 1; m_2; x_2) + \frac{x_1^n}{n^{m_1}}Z(n - 1; m_2; x_2)$$

$$+ \frac{x_1^{n+1}}{(n + 1)^{m_1}}Z(n + 1 - 1; m_2; x_2) \,.$$

(5.72)

The first term is simply $Z(n-1; m_1, m_2; x_1, x_2)$ by definition, while the second term is already expressed in terms of a Z-sum to $n-1$. Moreover, in the third term, the depth of the remaining Z-sum is reduced. So repeating the above idea, we find

$$Z(n+1-1; m_2; x_2) = \sum_{j=1}^{n-1} \frac{x_2^j}{j^{m_2}} + \frac{x_2^n}{n^{m_2}} = Z(n-1; m_2; x_2) + \frac{x_2^n}{n^{m_2}}. \qquad (5.73)$$

Substituting Eq. (5.73) into Eq. (5.72), we obtain

$$Z(n+2-1; m_1, m_2; x_1, x_2) = Z(n-1; m_1, m_2; x_1, x_2) + \left(\frac{x_1^n}{n^{m_1}} + \frac{x_1^{n+1}}{(n+1)^{m_1}} \right)$$

$$\times Z(n-1; m_2; x_2) + \frac{x_2^n}{n^{m_2}} \frac{x_1^{n+1}}{(n+1)^{m_1}}. \qquad (5.74)$$

Hence, we have rewritten the original expression in terms of only Z-sums to $n-1$. Generally, offsets in Z-sums of higher depth, as well as S-sums, can be reduced in a recursive way (see [4] for a discussion). This procedure is called *the synchronization of sums.*

In practical calculations, we are also often faced with sums of the form

$$\sum_{i=1}^{n} \frac{x^i}{(i+c)^m} Z(i-1, \ldots), \qquad (5.75)$$

with c as a positive integer. We would like to relate this sum to sums where the offset c is zero. The basic idea is simple: if the depth of the subsum $Z(i-1, \ldots)$ is zero, we can shift c down by one using the identity

$$\sum_{i=1}^{n} \frac{x^i}{(i+c)^m} = \frac{1}{x} \sum_{i=1}^{n} \frac{x^i}{(i+c-1)^m} - \frac{1}{c^m} + \frac{x^n}{(n+c)^m}. \qquad (5.76)$$

This relation is easily derived by replacing the index of summation as $i \to i-1$ and accounting for the changed limits of summation. The derivation also makes it obvious that the last term on the right-hand side only contributes if n is not equal to infinity. Repeated applications of Eq. (5.76) then reduce the offset to zero. In order to see what happens if the depth of the subsum is not zero, let us consider, e.g.,

$$\sum_{i=1}^{n} \frac{x_1^i}{(i+1)^{m_1}} Z(i-1; m_2; x_2). \qquad (5.77)$$

We start by resolving the first (and in this case only) summation in the definition of the subsum and shifting the index of summation $i \to i - 1$, as before

$$\sum_{i=1}^{n} \frac{x_1^i}{(i+1)^{m_1}} Z(i-1; m_2; x_2) = \sum_{i=1}^{n} \frac{x_1^i}{(i+1)^{m_1}} \sum_{j=1}^{i-1} \frac{x_2^j}{j^{m_2}} Z(j-1)$$

$$= \sum_{i=2}^{n+1} \frac{x_1^{i-1}}{(i-1+1)^{m_1}} \sum_{j=1}^{i-1-1} \frac{x_2^j}{j^{m_2}} Z(j-1).$$

$$(5.78)$$

Next, separate the term with $i = n + 1$ to obtain a sum in i whose upper limit is n

$$\sum_{i=1}^{n} \frac{x_1^i}{(i+1)^{m_1}} Z(i-1; m_2; x_2) = \sum_{i=2}^{n} \frac{x_1^{i-1}}{(i-1+1)^{m_1}} \sum_{j=1}^{i-1-1} \frac{x_2^j}{j^{m_2}} Z(j-1)$$

$$+ \frac{x_1^n}{(n+1)^{m_1}} \sum_{j=1}^{n-1} \frac{x_2^j}{j^{m_2}} Z(j-1).$$

$$(5.79)$$

Obviously, this step is only necessary if n is finite; thus the last term is not present if the limit of summation is infinity. Now, rewrite the sum over j on the first line such that its upper limit of summation is $i - 1$

$$\sum_{i=1}^{n} \frac{x_1^i}{(i+1)^{m_1}} Z(i-1; m_2; x_2) =$$

$$= \sum_{i=2}^{n} \frac{x_1^{i-1}}{(i-1+1)^{m_1}} \left[\sum_{j=1}^{i-1} \frac{x_2^j}{j^{m_2}} Z(j-1) - \frac{x_2^{i-1}}{(i-1)^{m_2}} Z(i-1-1) \right]$$

$$+ \frac{x_1^n}{(n+1)^{m_1}} \sum_{j=1}^{n-1} \frac{x_2^j}{j^{m_2}} Z(j-1).$$

$$(5.80)$$

The two sums over j on the second and third lines can be evaluated as Z-sums to $i - 1$ and $n - 1$, while the contribution coming from the second term in the square brackets does not depend on j. In this contribution change the index of summation as $i \rightarrow i + 1$

$$\sum_{i=1}^{n} \frac{x_1^i}{(i+1)^{m_1}} Z(i-1; m_2; x_2) =$$

$$= \frac{1}{x_1} \sum_{i=2}^{n} \frac{x_1^i}{(i-1+1)^{m_1}} Z(i-1; m_2; x_2) - \sum_{i=1}^{n-1} \frac{x_1^i}{(i+1)^{m_1}} \frac{x_2^i}{(i)^{m_2}} Z(i-1)$$

$$+ \frac{x_1^n}{(n+1)^{m_1}} Z(n-1; m_2; x_2) .$$

(5.81)

To finish, we note the following: in the first term, we have lowered the offset by one, which was our goal, and in this particular case, the offset is already zero at this stage. Moreover, the lower limit of summation in this term can formally be put to $i = 1$, since the term with $i = 1$ would be zero because $Z(n, \ldots) = 0$ if $n < 0$ (the outermost sum is empty). In the second term, we have reduced the depth of the inner sum by one, and in this particular case, the inner sum is now depth zero. Finally, the last term is only present if n is not infinity. The only contribution that is not yet in the form of a Z-sum is the second term on the right-hand side, as products of $(i + 1)$ and i appear in the denominator. However, for specific integers m_1 and m_2, partial fractioning can be used to reduce the product in the second term to a sum of terms that contain either $(i + 1)$ or i in their denominator, but not both. In general, iterated applications of the formula

$$\frac{x_1^i}{(i+c_1)^{m_1}} \frac{x_2^i}{(i+c_2)^{m_2}} = \frac{1}{c_1 - c_2} \left[\frac{x_1^i}{(i+c_1)^{m_1}} \frac{x_2^i}{(i+c_2)^{m_2-1}} - \frac{x_1^i}{(i+c_1)^{m_1-1}} \frac{x_2^i}{(i+c_2)^{m_2}} \right]$$

(5.82)

will achieve the desired form. At this stage, all remaining sums will be of the form

$$\sum_{i=1}^{n} \frac{x^i}{i^m} Z(i-1, \ldots) ,$$

(5.83)

which are simply Z-sums. Finally, for offsets greater than one and subsums of general depth, we must apply the procedure outlined above iteratively, until the offset has been completely reduced to zero. The general case is discussed in [4].

Finally, the fourth property **S-Z4** that we will discuss is that Z-sums and S-sums appear in expansions of the gamma function around integer values. This property makes them particularly useful for computing ϵ-expansions of sums obtained from

Mellin-Barnes integrals using the residue theorem. Indeed, one can show that for positive integers n we have

$$\Gamma(n + \epsilon) = \Gamma(1 + \epsilon)\Gamma(n)$$
$$\times \left[1 + \epsilon Z_1(n-1) + \epsilon^2 Z_{11}(n-1) + \epsilon^3 Z_{111}(n-1) + \ldots + \epsilon^{n-1} Z_{11\ldots1}(n-1)\right],$$

(5.84)

where

$$Z_{\underbrace{11\ldots1}_{k}}(n-1) = Z(n-1; \underbrace{1, 1, \ldots, 1}_{k}; \underbrace{1, 1, \ldots, 1}_{k}).$$

(5.85)

Notice that in Eq. (5.84), since we are keeping the factor of $\Gamma(1+\epsilon)$ unexpanded on the right-hand side, the sum in the square brackets is finite. The expansion around negative integers takes the form (again $n > 0$)

$$\Gamma(-n + 1 + \epsilon) = \frac{\Gamma(1+\epsilon)}{\epsilon} \frac{(-1)^{n-1}}{\Gamma(n)}$$
$$\times \left[1 + \epsilon S_1(n-1) + \epsilon^2 S_{11}(n-1) + \epsilon^3 S_{111}(n-1) + \ldots\right],$$

(5.86)

with

$$S_{\underbrace{11\ldots1}_{k}}(n-1) = S(n-1; \underbrace{1, 1, \ldots, 1}_{k}; \underbrace{1, 1, \ldots, 1}_{k}).$$

(5.87)

In contrast to the expansion around positive integers, in this case the sum in the square brackets is infinite. Last, we note the inverse formula

$$\left[1 + \epsilon Z_1(n-1) + \epsilon^2 Z_{11}(n-1) + \epsilon^3 Z_{111}(n-1) + \ldots + \epsilon^{n-1} Z_{11\ldots1}(n-1)\right]^{-1} =$$
$$= 1 - \epsilon S_1(n-1) + \epsilon^2 S_{11}(n-1) - \epsilon^3 S_{111}(n-1) + \ldots,$$

(5.88)

which can be useful in practical applications.

5.1.4 Solving One-Dimensional Mellin-Barnes Integrals with Nested Sums

The tools developed in the previous section allow us to derive analytic solutions for a large and important class of one-dimensional Mellin-Barnes integrals. Thus, let us consider the integral

$$
\int_{z_0-i\infty}^{z_0+i\infty} \frac{dz}{2\pi i} \frac{\prod_{i=1}^{m} \Gamma(a_i + z) \prod_{j=1}^{n} \Gamma(b_j - z)}{\prod_{k=1}^{\bar{m}} \Gamma(c_k + z) \prod_{l=1}^{\bar{n}} \Gamma(d_l - z)} \prod_{p=1}^{r} \psi^{(h_p)}(e_p + z) \prod_{q=1}^{s} \psi^{(g_q)}(f_q - z) A^z ,
$$
(5.89)

where a_i, b_j, c_k, d_l, e_p, and f_q are assumed to be integers, not necessarily distinct. (Hence, e.g., $a_{i_1} = a_{i_2}$, etc. is allowed.) *Such expressions arise from dimensionally regularized Mellin-Barnes integrals after one has resolved the poles and expanded in ϵ (see Chap. 4).* It may also be possible to construct first the sum representation of the Mellin-Barnes integral and perform the ϵ-expansion on this sum, and we will come back briefly to this approach below. However, let us first concentrate on Eq. (5.89). To start, let us observe that because $a_i, \ldots, f_q \in \mathbb{Z}$, the integrand only has poles at integer values of z.

> Our aim will essentially be to show that one can write the residue of the integrand in Eq. (5.89) at a generic integer n explicitly and that the summation over the residues can be performed in terms of S- or Z-sums.

In order to be specific, let us assume that $0 < \Re(A) < 1$, so that we must close the contour of integration to the right, but the procedure can be adapted also to the case of $1 < \Re(A)$ as well.

To show the above, we first perform a change of integration variables, whose utility will become clear in a moment. Thus, set

$$
\bar{a} \equiv \min\{0, a_i\}, \quad \bar{d} \equiv \max\{0, d_l - 1\}, \quad \bar{e} \equiv \min\{0, e_p\}, \quad \bar{f} \equiv \max\{0, f_q - 1\}.
$$
(5.90)

Then define

$$
r \equiv \max\{|\bar{a}|, \bar{d}, |\bar{e}|, \bar{f}\}
$$
(5.91)

and let $z \to z + r$ in Eq. (5.89). Introducing the notation

$$
\bar{a}_i \equiv a_i + r, \quad \bar{b}_j \equiv b_j - r, \quad \bar{c}_k \equiv c_k + r, \quad \bar{d}_l \equiv d_l - r, \quad \bar{e}_p \equiv e_p + r, \quad \bar{f}_q \equiv f_q - r,
$$
(5.92)

we find

$$
A^r \int_{z_0-r-i\infty}^{z_0-r+i\infty} \frac{dz}{2\pi i} \frac{\prod_{i=1}^m \Gamma(\bar{a}_i + z) \prod_{j=1}^n \Gamma(\bar{b}_j - z)}{\prod_{k=1}^p \Gamma(\bar{c}_k + z) \prod_{l=1}^q \Gamma(\bar{d}_l - z)} \prod_{p=1}^r \psi^{(h_p)}(\bar{e}_p + z) \prod_{q=1}^s \psi^{(g_q)}(\bar{f}_q - z) A^z .
$$

$$(5.93)$$

The point of this transformation is that by construction $\bar{a}_i \geq 0$, $\bar{d}_l \leq 1$, $\bar{e}_p \geq 0$, and $\bar{f}_q \leq 1$, which will be convenient in what follows. Second, we shift the contour of integration to some standard contour. Let us choose, e.g., the straight-line contour parallel to the imaginary axis running from $\bar{z}_0 - i\infty$ to $\bar{z}_0 + i\infty$ with $\bar{z}_0 \in (0, 1)$. As $\bar{a}_i, \ldots, \bar{f}_q$ are all integers, such a contour avoids all poles of the integrand. Clearly, when shifting the contour from $z_0 - r$ to $\bar{z}_0$, we must account for the residues of the poles that are crossed, so our integral becomes

$$
A^r \int_{\bar{z}_0-i\infty}^{\bar{z}_0+i\infty} \frac{dz}{2\pi i} \frac{\prod_{i=1}^m \Gamma(\bar{a}_i + z) \prod_{j=1}^n \Gamma(\bar{b}_j - z)}{\prod_{k=1}^p \Gamma(\bar{c}_k + z) \prod_{l=1}^q \Gamma(\bar{d}_l - z)}
$$

$$
\times \prod_{p=1}^r \psi^{(h_p)}(\bar{e}_p + z) \prod_{q=1}^s \psi^{(g_q)}(\bar{f}_q - z) A^z + \sum \text{Residues} . \quad (5.94)
$$

Notice that as we shift the contour, we cross only finitely many poles at specific known integers; thus the sum of residues can be computed explicitly once the values of $a_i, \ldots, f_p$ as well as h_p, g_q and z_0, $\bar{z}_0$ are fixed.

Thus, we concentrate on the left-over integral. To proceed, we perform the synchronization of the gamma and ψ functions. This simply means that we apply the identities $\Gamma(1 + x) = x\Gamma(x)$ and $\psi^{(n)}(1 + x) = \psi^{(n)}(x) + (-1)^n n! z^{-n-1}$ repeatedly, until all gamma functions have been reduced to just $\Gamma(z)$ or $\Gamma(1 - z)$ and all ψ functions to $\psi^{(n)}(z)$ or $\psi^{(m)}(1 - z)$. Since $\bar{a}_i, \ldots, \bar{f}_q$ are all integers, we can achieve this in a finite number of steps. In fact, clearly

$$
\Gamma(\bar{a}_i + z) = (\bar{a}_i - 1 + z)(\bar{a}_i - 2 + z)\ldots z\Gamma(z), \qquad \bar{a}_i > 0,
$$

$$(5.95)$$

$$
\Gamma(\bar{b}_j - z) = \begin{cases} (\bar{b}_j - 1 - z)(\bar{b}_j - 2 - z)\ldots(1 - z)\Gamma(1 - z) & \text{if } \bar{b}_j > 1, \\ \frac{\Gamma(1-z)}{(-|\bar{b}_j|-z)(-|\bar{b}_j|+1-z)\ldots(-z)} & \text{if } \bar{b}_j < 1, \end{cases}
$$

$$(5.96)$$

$$
\Gamma(\bar{c}_k + z) = \begin{cases} (\bar{c}_k - 1 + z)(\bar{c}_k - 2 + z)\ldots z\Gamma(z) & \text{if } \bar{c}_k > 0, \\ \frac{\Gamma(z)}{(-|\bar{c}_k|+z)(-|\bar{c}_k|+1+z)\ldots(-1+z)} & \text{if } \bar{c}_k < 0, \end{cases}
$$

$$(5.97)$$

$$\Gamma(\bar{d}_l - z) = \frac{\Gamma(1-z)}{(-|\bar{d}_l| - z)(-|\bar{d}_l| + 1 - z)\dots(-z)}, \qquad \bar{d}_l < 1,$$

(5.98)

and furthermore

$$\psi^{(n_p)}(\bar{e}_p + z) = \psi^{(n_p)}(z) + (-1)^n n! \left(\frac{1}{(\bar{e}_p - 1 + z)^{n_p+1}} + \dots + \frac{1}{z^{n_p+1}} \right),$$

(5.99)

$$\psi^{(m_q)}(\bar{f}_q - z) = \psi^{(m_q)}(1 - z) - (-1)^n n! \left(\frac{1}{(-|\bar{f}_q| - z)^{m_q+1}} + \dots + \frac{1}{(-z)^{m_q+1}} \right).$$

(5.100)

Equation (5.99) holds for $\bar{e}_p > 0$, while Eq. (5.100) is true for $\bar{f}_q < 1$.

Then, we use the relation

$$\psi^{(m_q)}(1 - z) = (-1)^{m_q} \left[\psi^{(m_q)}(z) + \pi \frac{\partial^{m_q}}{\partial z^{m_q}} \cot(\pi z) \right]$$

(5.101)

to convert all $\psi^{(m_q)}(1 - z)$ functions to $\psi^{(m_q)}(z)$ functions.

After synchronizing the gamma and ψ functions in this way, the remaining one-dimensional integral in Eq. (5.94) takes the form

$$\int_{\bar{z}_0-i\infty}^{\bar{z}_0+i\infty} \frac{dz}{2\pi i} R(\bar{a}_i, \bar{b}_j, \bar{c}_k, \bar{d}_l, z) \frac{\Gamma^m(z)\Gamma^n(1-z)}{\Gamma^p(z)\Gamma^q(1-z)} \prod_{p=1}^r \left[\psi^{(n_p)}(z) + R_{n_p}(\bar{e}_p, z) \right]$$

$$\times \prod_{q=1}^s \left[(-1)^{m_q} \psi^{(m_q)}(z) + (-1)^{m_q} \pi \frac{\partial^{m_q}}{\partial z^{m_q}} \cot(\pi z) + R_{m_q}(\bar{f}_q, z) \right] A^z,$$

(5.102)

where $R(\bar{a}_i, \bar{b}_j, \bar{c}_k, \bar{d}_l, z)$, $R_{n_p}(\bar{e}_p, z)$, and $R_{m_q}(\bar{f}_q, z)$ are rational functions in z, such that their denominators only contain factors of $(n_i + z)$ with $n_i \in \mathbb{N}$. This last property is quite convenient, since it ensures that the poles of the integrand come from the gamma functions only and not from the rational functions R. However, it is easy to see that this property only holds if $\bar{a}_i \geq 0$, $\bar{d}_l \leq 1$, as well as $\bar{e}_p \geq 0$ and $\bar{f}_1 \leq 1$. This explains our initial transformation of the integration variable. Thus, after expanding the products, we find that our integral can be decomposed into a sum of integrals with the following general structure

$$\sum \int_{\bar{z}_0-i\infty}^{\bar{z}_0+i\infty} \frac{dz}{2\pi i} \Gamma^N(z)\Gamma^M(1-z) R'(\bar{a}_i, \dots, \bar{f}_q, z) A^z \prod_{p=1}^{P'} \psi^{n'_p}(z) \prod_q^{Q'} \pi \frac{\partial^{m'_q}}{\partial z^{m'_q}} \cot(\pi z).$$

(5.103)

Here $R'(\bar{a}_i, \ldots, \bar{f}_q, z)$ is again a rational function, generally a product of $R(\bar{a}_i, \bar{b}_j, \bar{c}_k, \bar{d}_l, z)$ and factors of $R_{n_p}(\bar{e}_p, z)$ and $R_{m_q}(\bar{f}_q, z)$. Then it is clear that just as for R, R_{n_p} and R_{m_q}, the denominator of R' also factorizes into factors of $(n_i + z)$ with $n_i \in \mathbb{N}$. Thus, after partial fractioning in z, we have simply

$$R'(\bar{a}_i, \ldots, \bar{f}_q, z) = \sum_i \frac{\alpha_i}{(n_i + z)^{p_i}} + \sum_j \beta_j z^j , \qquad (5.104)$$

where n_i and p_i are non-negative integers. The fact that all n_i above are non-negative is significant, because then all poles of the integrand to the right of the imaginary axis only come from $\Gamma(1 - z)$ or the derivatives of $\cot(\pi z)$, but not from the rational function R'. However, since we have chosen $\bar{z}_0 \in (0, 1)$ (and we assumed that $0 < Re A < 1$, so that we must close the contour to the right), it is precisely these poles, i.e., the ones at $z = 1, 2, \ldots$, whose residues contribute to the integral.

After this preliminary work, we now want to show that the residue of the pole at some generic $z \to n, n = 1, 2, \ldots$ can be expressed in terms of objects such as factors of $(c + n)^k$, $Z(n - 1, \ldots)$ and so forth, such that we will be able to perform the summation over n from one to infinity in terms of Z-sums. To do this, we will assume that our integrals are *balanced* (see Sect. 5.1.1).

Evidently, the integrals in Eq. (5.103) are balanced if $N = M$, and we will also require that $N = M > 0$. Now, computing the residue of the integrand at some generic positive integer n simply amounts to expanding the integrand in z around n, with the coefficient of $(z-n)^{-1}$ essentially giving the residue (up to a factor of $2\pi i$). Thus, the basic question becomes: *Can we perform this expansion symbolically around a generic n and what kind of objects appear in this expansion?* Remember that N, R' as well as the n'_p and m'_q are all assumed to be explicitly known; however the integer n must be kept symbolic. In order to perform the expansion and the computation of the residue, let us set $z = n + \delta$ so that the expansion is around $\delta = 0$, and recall the following facts:

1. The gamma functions appearing in the integrand in Eq. (5.103), $\Gamma(n + \delta)$ and $\Gamma(1 - n - \delta)$, may be expanded in δ using Eqs. (5.84) and (5.86). These expressions will generally yield S- and Z-sums as well as their products, which can be converted to an expression involving only the linear combination of, say, Z-sums as explained above. However, for the balanced integrals we are considering here, this procedure can be circumvented altogether by noting that

$$\Gamma(z)\Gamma(1 - z) = \frac{\pi}{\sin(\pi z)} . \qquad (5.105)$$

Computing the expansion of this expression around integers is made quite simple by observing that for $z = n + \delta$, $n \in \mathbb{N}$ we have

$$\frac{\pi}{\sin[\pi(n+\delta)]} = \frac{(-1)^n \pi}{\sin(\pi\delta)} = (-1)^n \left[\frac{1}{\delta} + \frac{\pi^2}{6}\delta + \frac{7\pi^4}{360}\delta^3 + O(\delta^5) \right]. \quad (5.106)$$

Thus the expansion of the product of gamma functions will simply yield a power of $[(-1)^n]^N$ times a Laurent series in δ starting at order $1/\delta^N$ with explicitly known numeric coefficients independent of n.

2. The expansion of the rational function R' in Eq. (5.104) is trivial. Recalling that by construction R' has no poles at positive integers, we may simply compute the Taylor series of R' around $z = n$:

$$R'(\bar{a}_i, \ldots, \bar{f}_q, n + \delta) = R'(\bar{a}_i, \ldots, \bar{f}_q, n) + \left. \frac{dR'(\bar{a}_i, \ldots, \bar{f}_q, z)}{dz} \right|_{z=n} \delta$$

$$+ \left. \frac{d^2 R'(\bar{a}_i, \ldots, \bar{f}_q, z)}{dz^2} \right|_{z=n} \frac{\delta^2}{2} + O(\delta^3). \quad (5.107)$$

Obviously the coefficients in this expansion can be computed explicitly for general n given any specific R', and they are simply rational functions of n. It is also clear that the denominator of each expansion coefficient continues to factorize into linear factors of $(n_i + n)$ and so can be partial fractioned to yield a form

$$\sum_i \frac{\rho_i}{(n_i + n)^{k_i}} + \sum_j \sigma_j n^j. \quad (5.108)$$

3. Clearly the expansion of the factor of A^z is trivial, and we have simply

$$A^{n+\delta} = A^n \left[1 + \ln A \delta + \ln^2 A \frac{\delta^2}{2} + O(\delta^3) \right]. \quad (5.109)$$

Thus the result involves a factor of A^n as well as powers of logarithms of A.

4. Turning to the product of ψ functions, we note that again these do not have poles for positive integers, so we may simply Taylor-expand around n. Using the definition $\psi^{(m+1)}(z) = d\psi^{(m)}(z)/dz$, we have simply

$$\psi^{(m)}(n + \delta) = \psi^{(m)}(n) + \psi^{(m+1)}(n)\delta + \psi^{(m+2)}(n)\frac{\delta^2}{2} + O(\delta^3), \quad (5.110)$$

and thus the expansion of a product of ψ functions will involve products of ψ functions at the integer value n. We now make the important observation that polygamma functions at positive integer values can be written in terms of Z-sums as

$$\psi^{(0)}(n) = -\gamma + Z(n-1;1;1), \tag{5.111}$$

$$\psi^{(m)}(n) = (-1)^{m+1}m!\left[\zeta_{m+1} - Z(n-1;m+1;1)\right], \qquad m > 0. \tag{5.112}$$

Thus, the expansion coefficients will in general involve products of Z-sums which can be reduced to linear combinations of such sums using the algebra of Z-sums as explained above.

5. Last, let us consider the product of derivatives of $\cot(\pi z)$ that appears in Eq. (5.103). First, it is trivial to show that

$$\frac{\partial \cot(\pi z)}{\partial z} = -\pi\left[1 + \cot^2(\pi z)\right], \tag{5.113}$$

which implies that the mth derivative of $\cot(\pi z)$ can be expressed as an $m + 1$-st degree polynomial in $\cot(\pi z)$. Hence, the product of derivatives will evaluate to a polynomial in $\cot(\pi z)$. Computing the expansion of this polynomial around an integer n is simplified by noticing that

$$\cot[\pi(n+\delta)] = \cot(\pi\delta) = \frac{1}{\pi}\left[\frac{1}{\delta} - \frac{\pi^2}{3}\delta - \frac{\pi^4}{45}\delta^3 + O(\delta^5)\right]. \tag{5.114}$$

Thus, the expansion of the product of derivatives of $\cot(\pi z)$ yields a Laurent series in δ whose coefficients are explicitly known numbers independent of n.

From the above, it is clear that the residue of the integrand in Eq. (5.103) can indeed be computed for a generic positive integer n and will involve at most the product of factors of $(n_i + n)^{k_i}$, A^n, Z-sums $Z(n-1,\ldots)$ (these are only present if the integrand involved polygamma functions) as well as numeric factors independent of n and logarithms of A raised to n-independent powers. Moreover, the summation over residues runs from $n = 1$ to $n = \infty$ by construction (recall our choice of $0 < \bar{z}_0 < 1$), so at most we encounter sums of the form

$$\sum_{n=1}^{\infty} \frac{A^n}{(n_i + n)^{k_i}} Z(n-1,\ldots). \tag{5.115}$$

These sums are then straightforward to evaluate using the summation techniques presented above. The result will involve Z-sums to infinity which we recall are nothing but multiple polylogarithms (see Eq. (5.62) and Chap. 2).

Solving One-Dimensional Mellin-Barnes Integrals with Nested Sums

The above approach to solving one-dimensional Mellin-Barnes integrals with nested sums involves the following steps:

1. Transform the variable of integration $z \to z + r$, where the constant r is given in Eq. (5.91).
2. Shift the contour of integration to the standard contour that runs parallel to the imaginary axis from $\bar{z}_0 - i\infty$ to $\bar{z}_0 + i\infty$ with $\bar{z}_0 \in (0, 1)$. If any poles of the integrand are crossed, compute their contribution using Cauchy's residue theorem.
3. Transform all gamma functions in the integrand to the form $\Gamma(z)$ and $\Gamma(1 - z)$ using Eqs. (5.95)–(5.98).
4. Bring all polygamma functions to the form $\psi^{(n)}(z)$ using Eqs. (5.99)–(5.101).
5. The first two steps generally introduce a rational function $R(z)$ of the integration variable z into the integrand. Apply partial fractioning in z to this rational function.
6. Compute the residue of the integrand symbolically at $z = n$, $n \in \mathbb{N}^+$. We have shown how to do this in terms of at most Z-sums $Z(n - 1, \ldots)$, as well as factors of the form $(c + n)^k$, with $k \in \mathbb{Z}$ and A^n.
7. Perform the summation using the algorithms for Z-sums discussed above. Note that many of these algorithms, along with several others, have been implemented in publicly available packages such as XSummer and Sigma (see Appendix A.2).
8. The result will involve Z-sums to infinity which can be expressed in terms of multiple polylogarithms if desired.

5.1.5   Application of Sums to Angular Integrations

In order to illustrate the procedure outlined above, we consider the angular integral with two massless denominators, discussed in Sect. 3.14. There we showed that this integral has a one-dimensional Mellin-Barnes representation:

$$
\Omega_{j,k} = \frac{2^{1-2\epsilon}\pi^{-\epsilon}\Gamma(1 - \epsilon)}{\Gamma(1 - 2\epsilon)} \int_{-1}^{1} d(\cos\theta)(\sin\theta)^{-2\epsilon} \int_{-1}^{1} d(\cos\phi)(\sin\phi)^{-1-2\epsilon}
$$

$$
\times (1 - \cos\theta)^{-j}(1 - \cos\chi\cos\theta - \sin\chi\sin\theta\cos\phi)^{-k}
$$

$$
= 2^{2-j-k-2\epsilon}\pi^{1-\epsilon}\frac{1}{\Gamma(j)\Gamma(k)\Gamma(2 - j - k - 2\epsilon)}
$$

$$
\times \int_{z_0-i\infty}^{z_0+i\infty} \frac{dz}{2\pi i}\Gamma(-z)\Gamma(j + z)\Gamma(k + z)\Gamma(1 - j - k - \epsilon - z)v^z,
$$

$$
\tag{5.116}
$$

where $v = (1 - \cos \chi)/2$. Let us now consider the specific case of $j = k = 1$ and extract the overall factor in the first line by defining

$$I_{1,1}(v, \epsilon) \equiv \frac{2^{-1+2\epsilon} \pi^\epsilon \Gamma(1 - 2\epsilon)}{\Gamma(1 - \epsilon)} \Omega_{j,k}$$

$$= \frac{\pi}{\Gamma(-\epsilon)} \int_{z_0-i\infty}^{z_0+i\infty} \frac{dz}{2\pi i} \Gamma(-z) \Gamma^2(1 + z) \Gamma(-1 - \epsilon - z) v^z. \tag{5.117}$$

This choice of normalization keeps the expanded expressions simpler and avoids the appearance of constants such as γ and $\ln 2$ in the final result. For $\epsilon = 0$ it is clearly impossible to find a straight-line contour parallel to the imaginary axis that separates the poles of $\Gamma(1 + z)$ and $\Gamma(-1 - z)$. Put differently, there is no z_0 such that the real parts of the arguments of all gamma functions are positive. However for a proper choice of $\epsilon \neq 0$, such z_0 does exist. Hence we choose, e.g., $z_0 = -\frac{1}{2}$ and $\epsilon = -1$, and analytically continue the integral to $\epsilon \to 0$, picking up the residues of the poles which are crossed as explained in Chap. 4. We find

$$I_{1,1}(v, \epsilon) = \pi \Gamma(-\epsilon) \Gamma(1 + \epsilon) v^{-1-\epsilon}$$

$$+ \frac{\pi}{\Gamma(-\epsilon)} \int_{-\frac{1}{2}-i\infty}^{-\frac{1}{2}+i\infty} \frac{dz}{2\pi i} \Gamma(-z) \Gamma^2(1 + z) \Gamma(-1 - \epsilon - z) v^z. \tag{5.118}$$

The ϵ expansion can now be performed under the integral sign. Keeping terms up to order ϵ^2, we obtain

$$I_{1,1}(v, \epsilon) = \frac{\pi}{v} \left[-\frac{1}{\epsilon} + \ln v - \left(\frac{\ln^2 v}{2} + \frac{\pi^2}{6} \right) \epsilon + \left(\frac{\ln^3 v}{6} + \frac{\pi^2 \ln v}{6} \right) \epsilon^2 \right]$$

$$+ \pi (-\epsilon + \gamma \epsilon^2) \int_{-\frac{1}{2}-i\infty}^{-\frac{1}{2}+i\infty} \frac{dz}{2\pi i} \Gamma(-z) \Gamma^2(1 + z) \Gamma(-1 - z) v^z$$

$$+ \pi \epsilon^2 \int_{-\frac{1}{2}-i\infty}^{-\frac{1}{2}+i\infty} \frac{dz}{2\pi i} \Gamma(-z) \Gamma^2(1 + z) \Gamma(-1 - z) \psi^{(0)}(-1 - z) v^z + O(\epsilon^3). \tag{5.119}$$

Let us now evaluate the two remaining Mellin-Barnes integrals using our summation tools. Starting with

$$I_1 = \int_{-\frac{1}{2}-i\infty}^{-\frac{1}{2}+i\infty} \frac{dz}{2\pi i} \Gamma(-z) \Gamma^2(1 + z) \Gamma(-1 - z) v^z, \tag{5.120}$$

we first recognize that in this particular case r, defined in Eq. (5.91), is simply zero, $r = 0$. So we do not need to shift the integration variable. Next, we choose a straight-line contour running between zero and one, e.g., $\bar{z}_0 = +\frac{1}{2}$. As our original contour was at $z_0 = -\frac{1}{2}$, we must pick up the residue of the integrand at $z = 0$ as we move to the new contour. This residue is just $1 - \ln v$; however we are crossing the pole from left to right which means we encircle the pole clockwise, hence

$$I_1 = -1 + \ln v + \int_{+\frac{1}{2}-i\infty}^{+\frac{1}{2}+i\infty} \frac{dz}{2\pi i} \Gamma(-z)\Gamma^2(1+z)\Gamma(-1-z)v^z . \qquad (5.121)$$

Concentrating on the left-over integral, our next step is the synchronization of gamma functions. Using Eqs. (5.95)–(5.98), we find

$$\begin{aligned}
I_1' &= \int_{+\frac{1}{2}-i\infty}^{+\frac{1}{2}+i\infty} \frac{dz}{2\pi i}\Gamma(-z)\Gamma^2(1+z)\Gamma(-1-z)v^z \\
&= \int_{+\frac{1}{2}-i\infty}^{+\frac{1}{2}+i\infty} \frac{dz}{2\pi i}\Gamma^2(1-z)\Gamma^2(z)\frac{1}{-1-z}v^z .
\end{aligned} \qquad (5.122)$$

As expected, the rational function in the integrand on the second line does not have poles for positive z. Now we are ready to compute the residue of the integrand at a generic positive integer n. Using Eqs. (5.105) and (5.106), we find

$$\mathrm{Res}_{z\to n}\Gamma(1-z)^2\Gamma^2(z)\frac{1}{-1-z}v^z = \frac{v^n}{(1+n)^2} - \frac{a^n}{(1+n)}\ln v , \quad n \in \mathbb{N}^+ , \qquad (5.123)$$

where we have used that $(-1)^{2n} = 1$ for all positive integers n. Recalling that $v = (1 - \cos \chi)/2$, it is clear that $0 \le v \le 1$; thus we must close the contour of integration to the right (clockwise) and so

$$I_1' = -\sum_{n=1}^{\infty}\left[\frac{v^n}{(1+n)^2} - \frac{a^n}{(1+n)}\ln v\right]. \qquad (5.124)$$

The sums that appear are of the form discussed in Eq. (5.76) with $c = 1$. It is then trivial to reduce the shift to zero, and we are left with Z-sums of depth one to infinity:

$$I_1' = -\frac{1}{v}\sum_{n=1}^{\infty}\frac{v^n}{(n)^2}+1+\frac{1}{v}\sum_{n=1}^{\infty}\frac{a^n}{(n)}\ln v - \ln v = -\frac{Z(\infty, 2, v)}{v}+1+\frac{Z(\infty, 1, v)}{v}\ln v - \ln v . \qquad (5.125)$$

Recalling that depth-one Z-sums to infinity simply correspond to classical polylogarithms (see Eq. (5.63)), thus we have

$$I_1' = -\frac{\text{Li}_2(v)}{v} - \frac{\ln(1-v)\ln(v)}{v} + 1 - \ln v \,, \tag{5.126}$$

where we have used that $\text{Li}_1(v) = -\ln(1-v)$. Thus the complete integral I_1 in Eq. (5.121) evaluates to

$$I_1 = -\frac{\text{Li}_2(v)}{v} - \frac{\ln(1-v)\ln(v)}{v} \,. \tag{5.127}$$

Turning to the last integral,

$$I_2 = \int_{-\frac{1}{2}-i\infty}^{-\frac{1}{2}+i\infty} \frac{dz}{2\pi i} \Gamma(-z)\Gamma^2(1+z)\Gamma(-1-z)\psi^{(0)}(-1-z)v^z \,, \tag{5.128}$$

once more we find that r (defined in Eq. (5.91)) is zero, so we can move on straight away to shifting the contour to one running between zero and one and once more we choose $\bar{z}_0 = +\frac{1}{2}$. As before, we must account for the residue of the pole at $z = 0$ which we cross from left to right, and we find

$$I_2 = -1 + \gamma + \frac{\pi^2}{6} - \gamma \ln v + \frac{\ln^2 v}{2}$$
$$+ \int_{+\frac{1}{2}-i\infty}^{+\frac{1}{2}+i\infty} \frac{dz}{2\pi i} \Gamma(-z)\Gamma^2(1+z)\Gamma(-1-z)\psi^{(0)}(-1-z)v^z \,. \tag{5.129}$$

In order to evaluate the remaining Mellin-Barnes integral, we first synchronize the gamma and ψ functions using Eqs. (5.95)–(5.98) as well as Eqs. (5.100) and (5.101), which leads to

$$I_2' = \int_{+\frac{1}{2}-i\infty}^{+\frac{1}{2}+i\infty} \frac{dz}{2\pi i} \Gamma(-z)\Gamma^2(1+z)\Gamma(-1-z)\psi^{(0)}(-1-z)v^z$$
$$= \int_{+\frac{1}{2}-i\infty}^{+\frac{1}{2}+i\infty} \frac{dz}{2\pi i} \Gamma^2(1-z)\Gamma^2(z)\frac{1}{-1-z}\left[-\frac{1}{-1-z} - \frac{1}{-z} + \psi^{(0)}(z) + \pi \cot(\pi z)\right]v^z \,. \tag{5.130}$$

Notice that as before, the rational functions which appear do not have poles for positive z. Computing the residues of the integrand at a generic positive integer n is straightforward using the methods outlined above, and the result reads

$$
\operatorname{Res}_{z \to n} \Gamma^2(1-z)\Gamma^2(z) \frac{1}{-1-z} \left[-\frac{1}{-1-z} - \frac{1}{-z} + \psi^{(0)}(z) + \pi \cot(\pi z) \right] v^z
$$

$$
= \frac{v^n}{(n+1)^2} Z(n-1,1,1) + \frac{v^n}{n+1} Z(n-1,2,1) - \ln v \frac{v^n}{n+1} Z(n-1,1,1)
$$

$$
+ \frac{v^n}{(n+1)^3} - (1+\gamma)\frac{v^n}{(n+1)^2} - \left(\frac{\ln^2 v}{2} - \gamma \ln v - \ln v + \frac{\pi^2}{6} \right) \frac{v^n}{n+1}
$$

$$
+ \frac{v^n}{n^2} - \ln v \frac{v^n}{n}, \quad n \in \mathbb{N}^+.
$$

$$(5.131)$$

Once more we close the contour to the right, picking up the residues at $n = 1, 2, \ldots$. Thus, we must sum the right-hand side of Eq. (5.131) from $n = 1$ to $n = \infty$. Clearly the sums can be evaluated in terms of Z-sums, after reducing the offset to zero where needed. By way of illustration, let us consider the summation of the first term on the right-hand side. Using Eq. (5.81), we obtain

$$
\sum_{n=1}^{\infty} \frac{v^n}{(n+1)^2} Z(n-1,1,1) = \frac{1}{v} \sum_{n=1}^{\infty} \frac{v^n}{n^2} Z(n-1,1,1) - \sum_{n=1}^{\infty} \frac{v^n}{n(1+n)^2} Z(n-1)
$$

$$
= \frac{1}{v} \sum_{n=1}^{\infty} \frac{v^n}{n^2} Z(n-1,1,1) + \sum_{n=1}^{\infty} \frac{v^n}{(1+n)^2} + \sum_{n=1}^{\infty} \frac{v^n}{1+n} - \sum_{n=1}^{\infty} \frac{v^n}{n}
$$

$$(5.132)$$

where we used $Z(n-1) = 1$ for $n \in \mathbb{N}^+$ and performed partial fractioning of the expression with respect to n. The second and third sums on the second line still have a non-zero offset, so we apply Eq. (5.76) and obtain

$$
\sum_{n=1}^{\infty} \frac{v^n}{(n+1)^2} Z(n-1,1,1)
$$

$$
= \frac{1}{v} \sum_{n=1}^{\infty} \frac{v^n}{n^2} Z(n-1,1,1) + \frac{1}{v} \sum_{n=1}^{\infty} \frac{v^n}{n^2} - 1 + \frac{1}{v} \sum_{n=1}^{\infty} \frac{v^n}{n} - 1 - \sum_{n=1}^{\infty} \frac{v^n}{n}
$$

$$
= \frac{1}{v} Z(\infty,2,1,v,1) + \frac{1}{v} Z(\infty,2,v) + \frac{1}{v} Z(\infty,1,v) - Z(\infty,1,v) - 2,
$$

$$(5.133)$$

where in the last line we used the definition of Z-sums to evaluate the summations. We can evaluate the rest of the sums coming from Eq. (5.131) in a similar fashion to arrive at the final result (recall that the contour is closed in the clockwise direction)

$$I_2' = -\frac{Z(\infty, 1, 2, v, 1)}{v} - \frac{Z(\infty, 2, 1, v, 1)}{v} + \ln v \frac{Z(\infty, 1, 1, v, 1)}{v} - \frac{Z(\infty, 3, v)}{v}$$
$$+ \gamma \frac{Z(\infty, 2, v)}{v} + \left(\frac{\ln^2 v}{2} - \gamma \ln v + \frac{\pi^2}{6}\right) \frac{Z(\infty, 1, v)}{v} + 1 - \gamma - \frac{\pi^2}{6} + \gamma \ln v$$
$$- \frac{\ln^2 v}{2}.$$

$$(5.134)$$

Then, the complete I_2 integral of Eq. (5.129) becomes

$$I_2 = -\frac{Z(\infty, 1, 2, v, 1)}{v} - \frac{Z(\infty, 2, 1, v, 1)}{v} + \ln v \frac{Z(\infty, 1, 1, v, 1)}{v} - \frac{Z(\infty, 3, v)}{v}$$
$$+ \gamma \frac{Z(\infty, 2, v)}{v} + \left(\frac{\ln^2 v}{2} - \gamma \ln v + \frac{\pi^2}{6}\right) \frac{Z(\infty, 1, v)}{v}.$$

$$(5.135)$$

This result can be written in terms of more familiar functions, since the depth-one sums are simply classical polylogarithms (see Eq. (5.63)), while the depth-two sums can be expressed in terms of harmonic polylogarithms using Eqs. (5.62) and (2.100):

$$Z(\infty, 1, 2, v, 1) = \text{Li}_{2,1}(1, v) = H_{1,2}(v) = H_{1,0,1}(v), \tag{5.136}$$

$$Z(\infty, 2, 1, v, 1) = \text{Li}_{1,2}(1, v) = H_{2,1}(v) = H_{0,1,1}(v), \tag{5.137}$$

$$Z(\infty, 1, 1, v, 1) = \text{Li}_{1,1}(1, v) = H_{1,1}(v), \tag{5.138}$$

where on the right-hand sides of the first two equations above, we have given the HPLs both in "m"- and "a"-notation ($H_{1,1}(v)$ are clearly the same in both notations). Finally, MPLs (and thus also HPLs) of weight 3 or less can always be written in terms of classical polylogarithms [5–7]. For the specific harmonic polylogarithms appearing above, we find

$$H_{1,2}(v) = 2\text{Li}_3(1 - v) + \text{Li}_2(v) \ln(1 - v) + \ln v \ln^2(1 - v) - \frac{\pi^2}{3} \ln(1 - v) - 2\zeta_3,$$

$$(5.139)$$

$$H_{2,1}(v) = -\text{Li}_3(1 - v) - \text{Li}_2(v) \ln(1 - v) - \frac{1}{2} \ln v \ln^2(1 - v) + \frac{\pi^2}{6} \ln(1 - v) + \zeta_3$$

$$(5.140)$$

$$H_{1,1}(v) = \frac{1}{2} \ln^2(1 - v). \tag{5.141}$$

Thus, I_2 can indeed be expressed with just classical polylogarithms:

$$I_2 = -\frac{\text{Li}_3(v)}{v} - \frac{\text{Li}_3(1-v)}{v} + \gamma\left(\frac{\text{Li}_2(v)}{v} + \frac{\ln(1-v)\ln v}{v}\right) - \frac{\ln(1-v)\ln^2 v}{2v} + \frac{\zeta_3}{v}.$$

(5.142)

Combining our previous results, the complete solution for the angular integral $I_{1,1}(v, \epsilon)$ up to and including terms at order ϵ^2 may be written as (see also the file MB_I11_massless_Springer.nb in the auxiliary material in [1])

$$I_{1,1}(v, \epsilon) = \frac{\pi}{v}\left\{-\frac{1}{\epsilon} + \ln v + \left[\text{Li}_2(v) - \frac{1}{2}\ln^2 v + \ln(1-v)\ln v - \frac{\pi^2}{6}\right]\epsilon\right.$$

$$- \left[\text{Li}_3(v) + \text{Li}_3(1-v) - \frac{1}{6}\ln^3(v) + \frac{1}{2}\ln(1-v)\ln^2 v\right.$$

$$\left.\left. - \frac{\pi^2}{6}\ln v - \zeta_3\right]\epsilon^2 + O(\epsilon^3)\right\}.$$

(5.143)

5.1.6 Expansions of Special Functions

In Sect. 2.7, we have shown that the generalized Gauss hypergeometric function $_AF_B$ is represented as a one-dimensional sum. There exist several multi-variable generalizations of these hypergeometric functions that can be defined through sums of depths greater than one. Some of the better known generalizations to the two-variable case are the Appell functions F_1, F_2, F_3, and F_4, which also have simple representations as two-dimensional Mellin-Barnes integrals. In this section, we briefly comment on the fact that the summation technology presented above can be employed to expand such functions around integer values of their parameters.

We start by considering the Gauss hypergeometric function $_2F_1$, given by the following sum:

$$_2F_1(\alpha_1, \alpha_2; \beta_1; z) = \frac{\Gamma(\beta_1)}{\Gamma(\alpha_1)\Gamma(\alpha_2)} \sum_{n=0}^{\infty} \frac{\Gamma(n+\alpha_1)\Gamma(n+\alpha_2)}{\Gamma(n+\beta_1)\Gamma(n+1)} z^n.$$

(5.144)

If the parameters α_1, α_2 and β_1 are of the form $n + m\epsilon$, $n \in \mathbb{Z}$, we can compute the ϵ expansion of $_2F_1(\alpha_1, \alpha_2; \beta_1; z)$ as follows. Let us write $\alpha_i = a_i + c_i\epsilon$ and $\beta_1 = b_1 + d_1\epsilon$, where we assume that a_i and b_1 are integers. Let us concentrate on the following sum

$$\sum_{n=1}^{\infty} \frac{\Gamma(n+a_1+c_1\epsilon)\Gamma(n+a_2+c_2\epsilon)}{\Gamma(n+b_1+d_1\epsilon)\Gamma(n+1)} z^n.$$

(5.145)

First, we expand the gamma functions using Eq. (5.84). This expansion will introduce products of Z-sums of the form $Z_{11...1}(n + a_1 - 1)$, $Z_{11...1}(n + a_2 - 1)$, $Z_{11...1}(n + b_1 - 1)$, and $Z_{11...1}(n)$. Note that the expansion of the gamma functions in the denominator can be brought to this form using Eq. (5.88), followed by a conversion of S-sums to Z-sums. Next, we can synchronize the sums and use the algebra of Z-sums to write the products in terms of just single sums. After performing these manipulations, our problem reduces to the computation of sums of the form

$$\sum_{n=1}^{\infty} \frac{z^n}{(n+c)^m} Z(n-1, \ldots),\qquad(5.146)$$

where c is a non-negative integer. However, these are just the sums we have already encountered in Eq. (5.75)! We have seen there that they can be systematically reduced to Z-sums by recursively shifting $c \to c - 1$ until this offset is reduced to zero, at which point the sums trivially evaluate to Z-sums.

In fact, it is clear that the same algorithm (referred to as algorithm A in [4]) can be used to solve any sum of the form

$$\sum_{n=1}^{N} \frac{z^n}{(n+c)^m} \frac{\Gamma(n+a_1+c_1\epsilon)}{\Gamma(n+b_1+d_1\epsilon)} \cdots \frac{\Gamma(n+a_k+c_k\epsilon)}{\Gamma(n+b_k+d_k\epsilon)}$$

$$\times Z(n+o-1; m_1, \ldots, m_l; z_1, \ldots, z_l)\qquad(5.147)$$

in terms of Z-sums. The upper limit of summation is allowed to be infinity. This type of sum includes in particular the generalized Gauss hypergeometric function $_AF_B$.

Similar algorithms exist for other types of sums [4], which allow to perform the expansions of more general special functions. As an example, we briefly discuss one specific two-variable generalization of the Gauss hypergeometric function, the Appell function of the first kind, F_1. This function was introduced in Chap. 2 (see Eq. (2.189)). There it was defined by a double sum:

$$F_1(a, b_1, b_2; c, z_1, z_2) = \sum_{m=0}^{\infty} \sum_{n=0}^{\infty} \frac{(a)_{m+n}(b_1)_m(b_2)_n}{(c)_{m+n}m!n!} z_1^m z_2^n.\qquad(5.148)$$

This double series is absolutely convergent for $|z_1| < 1$ and $|z_2| < 1$. In order to make contact with the language of nested sums, we rewrite the double sum in the following way. First, we separate the terms where m or n is zero, and then we make the change of summation variable $n \to k = m + n$. Clearly the summation in k runs from 2 to infinity, while the positivity of $n = k - m$ implies that the summation in m runs from 1 to $k - 1$. Then we find

$$F_1(a, b_1, b_2; c, z_1, z_2) = 1$$

$$+ \frac{\Gamma(c)}{\Gamma(a)\Gamma(b_1)} \sum_{m=1}^{\infty} z_1^m \frac{\Gamma(m + a)\Gamma(m + b_1)}{\Gamma(m + c)\Gamma(m + 1)}$$

$$+ \frac{\Gamma(c)}{\Gamma(a)\Gamma(b_2)} \sum_{n=1}^{\infty} z_2^n \frac{\Gamma(n + a)\Gamma(n + b_2)}{\Gamma(n + c)\Gamma(n + 1)}$$

$$+ \frac{\Gamma(c)}{\Gamma(a)\Gamma(b_1)\Gamma(b_2)} \sum_{k=1}^{\infty} \frac{\Gamma(k + a)}{\Gamma(k + c)} \sum_{m=1}^{k-1} z_1^m \frac{\Gamma(m + b_1)}{\Gamma(m + 1)} z_2^{k-m} \frac{\Gamma(k - m + b_2)}{\Gamma(k - m + 1)}.$$

$$(5.149)$$

Let us assume that all parameters a, b_1, b_2, and c are of the form $i + j\epsilon$, where i is an integer. Then the expansion of the sums on the second and third lines of the above equation can be performed with the algorithm just discussed. However, the nested sum on the fourth line is new. In order to perform the expansion of this sum, let us first concentrate on the inner sum over m. We again use Eqs. (5.84) and (5.88), convert the S-sums to Z-sums, synchronize the subsums, and use the algebra of Z-sums to write products of sums with the same upper limit of summation in terms of just single sums. Then we obtain sums of the form

$$\sum_{m=1}^{k-1} \frac{z_1^m}{(m + o)^p} Z(m - 1; m_1, \ldots) \frac{z_2^{k-m}}{(k - m + o')^{p'}} Z(k - m - 1, m_1', \ldots). \quad (5.150)$$

After performing the partial fraction decomposition of the denominators (recall p and p' are fixed integers), the resulting expressions will involve either the denominator $(m + o)$ or $(k - m + o')$, but not both at the same time. Thus by changing the index of summation $m \rightarrow k - m$ in those terms that involve the denominator $(k - m + o')$, we can further reduce the above sum to sums of the type

$$\sum_{m=1}^{k-1} \frac{z^m}{(m + o)^p} Z(m - 1; m_1, \ldots) Z(k - m - 1; m_1', \ldots). \quad (5.151)$$

If the depth of $Z(k - m - 1; m_1', \ldots)$ is zero, we simply recover a sum of the form

$$\sum_{m=1}^{k-1} \frac{z^m}{(m + o)^p} Z(m - 1; m_1, \ldots), \quad (5.152)$$

with upper summation index $k - 1$. We have already discussed that this sum can be evaluated in terms of Z-sums of the form $Z(k - 1, \ldots)$. Thus, the double sum on the last line of Eq. (5.149) is reduced to a sum of the form of Eq. (5.147) which we can solve with algorithm A presented above. If the depth of $Z(k - m - 1; m'_1, \ldots)$ is greater than one, we can write Eq. (5.151) as

$$\sum_{n=1}^{k-1} \left[\sum_{m=1}^{n-1} \frac{z^m}{(m + o)^p} Z(m - 1; m_1, \ldots) \frac{(x'_1)^{n-m}}{(n - m)^{m'_1}} Z(n - m - 1; m'_2, \ldots) \right].$$

(5.153)

But this last sum is just of the form that we had started with, but the depth of the second Z-sum is reduced. Thus, we can use recursion to lower this depth to zero, when the recursion terminates. More generally, the algorithm just described (called algorithm B in [4]) can be used to evaluate sums of the form

$$\sum_{m=1}^{k-1} \frac{x^m}{(m + o)^p} \frac{\Gamma(m + a_1 + c_1\epsilon)}{\Gamma(m + b_1 + d_1\epsilon)} \cdots \frac{\Gamma(m + a_k + c_k\epsilon)}{\Gamma(m + b_k + d_k\epsilon)}$$

$$\times Z(m + r - 1; m_1, \ldots, m_l; z_1, \ldots, z_l)$$

$$\times \frac{y^{k-m}}{(k - m + o')^{p'}} \frac{\Gamma(k - m + a'_1 + c'_1\epsilon)}{\Gamma(k - m + b'_1 + d'_1\epsilon)} \cdots \frac{\Gamma(k - m + a'_k + c'_k\epsilon)}{\Gamma(k - m + b'_k + d'_k\epsilon)}$$

$$\times Z(k - m + r' - 1; m'_1, \ldots, m'_{l'}; z'_1, \ldots, z'_{l'}).$$

(5.154)

Thus, the expansion of the first Appell function can be computed with the help of algorithms A and B. To finish, we note that further algorithms exist to handle also certain sums in which binomial coefficients appear. We refer the interested reader to the original literature for further details [4]. Moreover, these algorithms have been implemented in the XSummer package [8] for the computer algebra system FORM. An adaptation of algorithm A described above, useful for the expansion of Gauss generalized hypergeometric functions $_AF_B$, has also been implemented in the HypExp package for Mathematica [9] (see Appendix A.3).

To finish, we reiterate that throughout we have assumed that the expansions are performed around integer values of the parameters of the hypergeometric functions and their generalizations. However, in practical calculations, sometimes special functions with half-integer parameters also appear. In this case, the algorithms described above cannot be applied in a straightforward manner, and new algorithms are needed. In particular, the expansion of generalized hypergeometric functions of the form $_PF_{P-1}$ around half-integer values can be performed with the methods described in [10], which have been implemented in the HypExp 2 package for Mathematica.

The Angular Integral with Two Denominators and One Mass

Consider the angular integral with two denominators and one mass, Eq. (3.172), with $j = k = 1$. This integral appears, e.g., when integrating over real soft radiation in a process with massive external legs. We can apply the techniques and tools described above to obtain the expansion of this integral around $\epsilon = 0$. Using the XSummer package, up to $O(\epsilon^2)$ accuracy, we find (see also the file MB_011_onemass_Springer.nb in the auxiliary material in [1]):

$$
\Omega_{1,1}(v_{11}, v_{12}, \epsilon) = -\Omega_{1-2\epsilon} \frac{\pi}{2v_{12}} \left\{ \frac{1}{\epsilon} - \mathrm{Li}_1(V_-) - \mathrm{Li}_1(V_+) \right.
$$
$$
- \left[2\mathrm{Li}_2(V_-) + 2\mathrm{Li}_2(V_+) + \mathrm{Li}_{1,1}(1, V_-) + \mathrm{Li}_{1,1}(1, V_+) \right.
$$
$$
\left. \left. - \mathrm{Li}_{1,1}\left(\frac{V_-}{V_+}, V_+\right) - \mathrm{Li}_{1,1}\left(\frac{V_+}{V_-}, V_-\right) \right] \epsilon + O(\epsilon^2) \right\},
$$

(5.155)

where we have set $V_\pm = \frac{2v_{12}-1\pm\sqrt{1-4v_{11}}}{2v_{12}}$. Furthermore, we have extracted a factor of $\Omega_{1-2\epsilon} = 2^{1-2\epsilon}\pi^{-\epsilon}\frac{\Gamma(1-\epsilon)}{\Gamma(1-2\epsilon)}$, whose expansion would introduce irrelevant constants like $\ln(4\pi)$ and γ into the final expression. In order to apply XSummer, we have used the nested sum representation of the Appell F_1 function as given in Eq. (5.149). We note in passing that the result in Eq. (5.155) can be expressed with just the logarithm and dilogarithm functions as follows:

$$
\Omega_{1,1}(v_{11}, v_{12}, \epsilon) = -\Omega_{1-2\epsilon} \frac{\pi}{2v_{12}} \left\{ \frac{1}{\epsilon} + \ln \frac{v_{11}}{v_{12}^2} - \left[\frac{1}{2}\ln^2\left(\frac{1-\sqrt{1-4v_{11}}}{1+\sqrt{1-4v_{11}}}\right) \right.\right.
$$
$$
\left.\left. + 2\mathrm{Li}_2\left(\frac{2v_{12}-1-\sqrt{1-4v_{11}}}{2v_{12}}\right) + 2\mathrm{Li}_2\left(\frac{2v_{12}-1+\sqrt{1-4v_{11}}}{2v_{12}}\right) \right] \epsilon + O(\epsilon^2) \right\}.
$$

(5.156)

5.1.7   More General Sums and Massive Propagators

Up to this point, we have been dealing with MB integrals where the integration variables enter the gamma functions only in the form $\Gamma(\ldots \pm az)$, $a = 1$, but not with more general a coefficients. Indeed, as alluded to at the beginning of this chapter, most of the tools that we have discussed are most useful in this case. However, more general MB integrals certainly arise in practical calculations ($a \neq 1$ appears in case of massive propagators and as we will see leads to series with binomial sums [11]). In order to illustrate the complications that arise in this case, consider the self-energy diagram of Fig. 3.5 in Chap. 3, where the internal propagators are massive, with mass m. The corresponding MB representation is given in Eq. (3.50). Let us consider the first (constant) term in the ϵ expansion of this integral:

$$G(1)_{SE2l2m} \equiv I_1 = \int_{-1/4-i\infty}^{-1/4+i\infty} \frac{dz_1}{2\pi i} \left(-\frac{m^2}{s}\right)^z \frac{\Gamma(1-z)^2\Gamma(-z)\Gamma(z)}{\Gamma(2-2z)}. \quad (5.157)$$

(This is just the function $aux(\epsilon = 0, \Re(z) = -1/4)$ of Eq. (4.7).) Notice that the gamma function in the denominator has the argument $2 - 2z$, i.e., the coefficient of the integration variable in the argument is -2.

Let us now apply Cauchy's residue theorem. For the sake of being explicit, we assume that $|m^2/s| < 1$ and close the contour to the right. The first few residues of the integrand read

$$\text{Res}_{z=0} \left(-\frac{m^2}{s}\right)^z \frac{\Gamma(1-z)^2\Gamma(-z)\Gamma(z)}{\Gamma(2-2z)} = -2 - \ln\left(-\frac{m^2}{s}\right), \quad (5.158)$$

$$\text{Res}_{z=1} \left(-\frac{m^2}{s}\right)^z \frac{\Gamma(1-z)^2\Gamma(-z)\Gamma(z)}{\Gamma(2-2z)} = \frac{2m^2}{s}\left[-1 + \ln\left(-\frac{m^2}{s}\right)\right] \quad (5.159)$$

$$\text{Res}_{z=2} \left(-\frac{m^2}{s}\right)^z \frac{\Gamma(1-z)^2\Gamma(-z)\Gamma(z)}{\Gamma(2-2z)} = \frac{m^4}{s^2}\left[1 + 2\ln\left(-\frac{m^2}{s}\right)\right]. \quad (5.160)$$

The residue for a general non-negative integer can be written as follows:[1]

$$\text{Res}_{z=n} \left(-\frac{m^2}{s}\right)^z \frac{\Gamma(1-z)^2\Gamma(-z)\Gamma(z)}{\Gamma(2-2z)} = 2\left(\frac{m^2}{s}\right)^n \frac{\Gamma(2n-1)}{\Gamma(n)\Gamma(n+1)}$$

$$\times \left[\ln\left(-\frac{m^2}{s}\right) - \psi^{(0)}(n) - \psi^{(0)}(n+1) + 2\psi^{(0)}(2n-1)\right], \quad n \in \mathbb{N}. \quad (5.161)$$

However, upon attempting to perform the summation over residues, we immediately face a problem: the summand cannot be reduced to a rational expression in n and so cannot be brought to the form of a Z-sum.

This is evident from appearance of the overall factor of $\frac{\Gamma(2n-1)}{\Gamma(n)\Gamma(n+1)}$. Using Eq. (2.107), it is easy to show that this factor grows exponentially as $n \to \infty$, and so it cannot correspond to a rational function of n. Thus, the solution cannot be expressed as a combination of polylogarithms of argument (m^2/s).

---

[1] For $n = 0$, the correct residue, $-2 - \ln\left(-\frac{m^2}{s}\right)$, is obtained by computing the limit as $n \to 0$ of the expression in Eq. (5.161).

In order to get an idea of the form of the solution, let us examine the sum proportional to $\ln(-m^2/s)$ in Eq. (5.161). This sum can be evaluated in closed form:

$$\sum_{n=0}^{\infty} 2\left(\frac{m^2}{s}\right)^n \frac{\Gamma(2n-1)}{\Gamma(n)\Gamma(n+1)} \ln\left(-\frac{m^2}{s}\right) = -\sqrt{1-\frac{4m^2}{s}} \ln\left(-\frac{m^2}{s}\right). \qquad (5.162)$$

Fractional and Inverse Binomial Sums

Let us evaluate the sum

$$\sum_{n=1}^{\infty} \frac{\Gamma(2n-1)}{\Gamma(n)\Gamma(n+1)} x^n. \qquad (5.163)$$

We begin by writing

$$\sum_{n=1}^{\infty} \frac{\Gamma(2n-1)}{\Gamma(n)\Gamma(n-1)} x^n = \sum_{n=1}^{\infty} \frac{(2n-2)!}{(n-1)!n!} x^n$$

$$= \sum_{n=1}^{\infty} \frac{1 \cdot 2 \cdot 3 \cdot 4 \cdot \ldots \cdot (2n-3) \cdot (2n-2)}{(n-1)!n!} x^n \qquad (5.164)$$

$$= \sum_{n=1}^{\infty} 2^{2n-2} \frac{\frac{1}{2} \cdot 1 \cdot \frac{3}{2} \cdot 2 \cdot \ldots \cdot \frac{2n-3}{2} \cdot (n-1)}{(n-1)!n!} x^n.$$

Written in this way, it is clear that we can cancel a factor of $(n-1)!$ between the numerator and denominator. Then we have

$$\sum_{n=1}^{\infty} 2^{2n-2} \frac{\frac{1}{2} \cdot 1 \cdot \frac{3}{2} \cdot 2 \cdot \ldots \cdot \frac{2n-3}{2} \cdot (n-1)}{(n-1)!n!} x^n$$

$$= \frac{1}{4} \sum_{n=1}^{\infty} (-1)^{n-1} \frac{\left(-\frac{1}{2}\right) \cdot \left(-\frac{3}{2}\right) \cdot \ldots \cdot \left(-\frac{2n-3}{2}\right)}{n!} (4x)^n$$

$$= -\frac{1}{2} \sum_{n=1}^{\infty} \frac{\frac{1}{2} \cdot \left(-\frac{1}{2}\right) \cdot \left(-\frac{3}{2}\right) \cdot \ldots \cdot \left(-\frac{2n-3}{2}\right)}{n!} (-4x)^n \qquad (5.165)$$

$$= -\frac{1}{2} \sum_{n=1}^{\infty} \frac{\frac{1}{2} \cdot \left(\frac{1}{2}-1\right) \cdot \left(\frac{1}{2}-2\right) \cdot \ldots \cdot \left(\frac{1}{2}-n+1\right)}{n!} (-4x)^n.$$

The sum on the last line can now be computed using the binomial theorem for fractional exponents (see Eq. (5.28)). Examining the sum in Eq. (5.165), we see that apart from missing the $n = 0$ term (which is equal to 1), it is just the generalized binomial sum of Eq. (5.28) with $\alpha = \frac{1}{2}$. Thus

$$
-\frac{1}{2} \sum_{n=1}^{\infty} \frac{\frac{1}{2} \cdot \left(\frac{1}{2} - 1\right) \cdot \left(\frac{1}{2} - 2\right) \cdot \ldots \cdot \left(\frac{1}{2} - n + 1\right)}{n!} (-4x)^n = -\frac{1}{2} \left(\sqrt{1 - 4x} - 1\right).
$$
(5.166)

Hence, we have established that

$$
\sum_{n=1}^{\infty} \frac{\Gamma(2n - 1)}{\Gamma(n)\Gamma(n + 1)} x^n = \frac{1}{2} - \frac{1}{2}\sqrt{1 - 4x}.
$$
(5.167)

A straightforward application of this result then leads immediately to Eq. (5.162).

We note in passing that if we choose to close the contour of integration to the left in Eq. (5.157), the residue of the integrand at a generic negative integer $z = -n$, $n \in \mathbb{N}^+$ reads

$$
\mathrm{Res}_{z=-n} \left(-\frac{m^2}{s}\right)^z \frac{\Gamma(1 - z)^2 \Gamma(-z)\Gamma(z)}{\Gamma(2 - 2z)} = \left(\frac{m^2}{s}\right)^{-n} \frac{\Gamma(n)\Gamma(n + 1)}{\Gamma(2n + 2)}
$$
$$
= \left(\frac{m^2}{s}\right)^{-n} \frac{1}{(2n + 1)n\binom{2n}{n}}, \qquad -n \in \mathbb{N}^+.
$$
(5.168)

After performing partial fractioning in n, we see that the solution involves *inverse binomial sums* such as

$$
\sum_{n=1}^{\infty} \frac{1}{n\binom{2n}{n}} x^n \quad \text{and} \quad \sum_{n=1}^{\infty} \frac{1}{(2n + 1)\binom{2n}{n}} x^n.
$$
(5.169)

Again, these sums are not of a form that we can directly treat with the algorithms presented above for nested sums.

In order to illustrate the one typical way of dealing with such sums, consider, e.g., the sum

$$
f(x) = \sum_{n=0}^{\infty} \frac{1}{(2n + 1)\binom{2n}{n}} x^n = \sum_{n=0}^{\infty} c_n x^n,
$$
(5.170)

which appeared already in Eq. (1.56). Let us denote the nth expansion coefficient as $c_n = \frac{1}{\binom{2n}{n}}$ and examine the $n + 1$st coefficient, c_{n+1}. It is easy to show using the definition of the binomial coefficient that $c_{n+1} = \frac{n+1}{2(2n+3)}c_n$, which implies the recursion relation

$$4(n+1)c_{n+1} + 2c_{n+1} = nc_n + c_n. \tag{5.171}$$

But then, multiplying this equation by x^n and summing over n form zero to infinity leads to the differential equation

$$4f'(x) + \frac{2}{x}[f(x) - 1] = xf'(x) + f(x). \tag{5.172}$$

Here we have used that sums involving the terms $(n+1)c_{n+1}$ and nc_n can be related to the derivative function $f'(x)$ and also that $c_0 = 1$. Equation (5.172) is an ordinary first-order inhomogeneous differential equation, and hence it can be reduced to integrations using the method of variation of constants. The integrals that appear are elementary, and we find

$$f(x) = \frac{2}{\sqrt{x(4-x)}}\left[\pi - 2\arctan\left(\frac{\sqrt{4-x}}{\sqrt{x}}\right)\right]. \tag{5.173}$$

Interestingly, the constant of integration is fixed by requiring that the expansion of the solution around $x = 0$ should only involve integer powers of x. The other sum in Eq. (5.169) can be handled in a similar fashion. We see that again square roots appear in the solution that can be simplified by using the conformal variable of Eq. (5.176).

Another approach to evaluating inverse binomial sums starts from the observation that the general inverse binomial coefficient admits the integral representation

$$\binom{n}{k}^{-1} = (n+1)\int_0^1 dt\, t^k(1-t)^{n-k}. \tag{5.174}$$

Then, e.g., for the particular sum in Eq. (5.170), we find

$$\sum_{n=0}^{\infty} \frac{1}{(2n+1)\binom{2n}{n}}x^n = \sum_{n=0}^{\infty} \frac{1}{(2n+1)}(2n+1)\int_0^1 dt\, t^n(1-t)^n x^n$$

$$= \int_0^1 \frac{dt}{1-t(1-t)x}, \tag{5.175}$$

where we have assumed that we can interchange the order of summation and integration. We can now try to perform the integration in t to obtain the solution. We will not pursue this example any further here, but we will have more to say on how such parametric integrals can be evaluated in Sect. 5.3.2.

Thus, we generally expect the expression $\sqrt{1 - \frac{4m^2}{s}}$ to appear in the solution. Let us then introduce (with some foresight) the so-called conformal variable

$$y = \frac{\sqrt{1 - 4m^2/s} - 1}{\sqrt{1 - 4m^2/s} + 1}. \tag{5.176}$$

Then $\left(-\frac{m^2}{s}\right)$ becomes $\frac{y}{(1-y)^2}$. We can get an idea for the structure of the final result by computing the sum of the first few residues, expressed as a series in y around zero. It is straightforward to perform this computation in Mathematica (recall we are closing the contour in the clockwise direction):

```
In[6]:= I1 = (y/(1-y)^2)^z Gamma[1-z]^2 Gamma[-z]
  ↪   Gamma[z]/Gamma[2-2z];
In[7]:= sum = Sum[-Residue[I1, {z, i}], {i, 0, 10}];
In[8]:= series = Normal[Series[sum, {y, 0, 10}]] // Simplify
Out[8]:= 2 + (1 + 2 y + 2 y^2 + 2 y^3 + 2 y^4 + 2 y^5 + 2 y^6 +
  ↪   2 y^7 + 2 y^8 + 2 y^9 + 2 y^10) Log[y]
```

Evidently part of the result builds a geometric series, which for $|y| < 1$ gives a final result

$$G(1)_{\text{SE2l2m}} = I_1 = 2 + \frac{1+y}{1-y} \ln y. \tag{5.177}$$

The condition $|y| < 1$ is satisfied for any values of m, s (see Fig. 5.1). This kind of change of variable is also very useful in calculation of Feynman integrals using the differential equations approach [12–17].

The ideas discussed above can be applied to other cases as well. For example, the the vertex function of Eq. (1.55)

$$V^{\epsilon^{-1}}_{V3l2m}(s) = -\frac{1}{2s} \int\limits_{-\frac{1}{2}-i\infty}^{-\frac{1}{2}+i\infty} \frac{dz}{2\pi i}(-s)^{-z} \frac{\Gamma^3(-z)\Gamma(1+z)}{\Gamma(-2z)}, \tag{5.178}$$

considered in Sect. 1.7, can be computed in a very similar way. Setting $s = -\frac{(1-y)^2}{y}$, which corresponds to the conformal transformation of Eq. (5.176) with $m = 1$, we again compute the first several residues and examine the expansion in y around zero. Once more, we are able to identify part of this expansion as a geometric series, and we find the result

$$V^{\epsilon^{-1}}_{V3l2m}(s) = -\frac{y \ln y}{1 - y^2}. \tag{5.179}$$

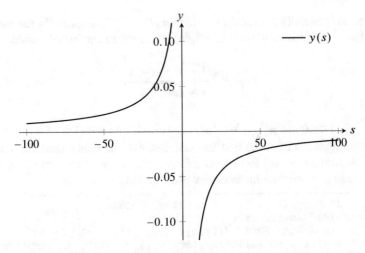

Fig. 5.1 The conformal transformation of Eq. (5.176). Since y only depends on the ratio of m^2/s, we have set $m = 1$ in the plot. The conformal variable y approaches ± 1 as $s \to 0^{\pm}$ and 0 as $s \to \infty$

Analytic Solutions with a Matching Procedure

In our examples above, we have relied in the last step on identifying parts of expansions with geometric series. There is however a more general way of approaching this problem, which can also be used in more complicated cases. This relies on building a proper ansatz for the solution. Then, we can "sum up" such series in an easy way by constructing a general solution, expanding it, and matching the coefficients of the specific sum with the coefficients of the general solution.

Choosing a proper ansatz requires some understanding of the structure of the result. For example, in the case of the self-energy integral which we discussed above, we must realize that the basic variable entering the ansatz should be the conformal variable of Eq. (5.176). Furthermore, we expect that a logarithmic function will play a pivotal role, since as a general rule, we know that such functions appear in one-loop computations. However, if we do not know the precise form of the solution, we should include more general functions as well, trying to guess the possible complete set of functions and coefficients, which can be monomials, polynomials, or rational functions in the conformal variables. Even if we choose a wide base of functions, the redundant ones should disappear when matching the series to the unknown sum, as we will see below.

So for the self-energy integral of Eq. (5.157), let us consider the minimal set S_1 of functions

$$S_1 = \{1, \ln(y), \ln(1 - y), \ln(1 + y)\}, \tag{5.180}$$

with y being the conformal variable of Eq. (5.176). We also define the minimal set of monomials

$$S_2 = \{1, y, \frac{1}{1-y}, \frac{1}{1+y}\}.$$ (5.181)

Taking the Cartesian product of S_1 and S_2, we obtain the following *basis*

$$
\begin{aligned}
S_B = S_1 \times S_2 = \Big\{ & 1, y, \frac{1}{1-y}, \frac{1}{1+y}, \ln(y), y\ln(y), \frac{\ln(y)}{1-y}, \frac{\ln(y)}{1+y}, \\
& \ln(1-y), y\ln(1-y), \frac{\ln(1-y)}{1-y}, \frac{\ln(1-y)}{1+y}, \\
& \ln(1+y), y\ln(1+y), \frac{\ln(1+y)}{1-y}, \frac{\ln(1+y)}{1+y} \Big\},
\end{aligned}
$$ (5.182)

which contains 16 terms, $i = 1,\ldots,16$. Our ansatz is then simply a linear combination of these 16 terms with undetermined coefficients $c[i]$:

$$\text{Ansatz} = \sum_{i=1}^{16} c[i] S_B[i].$$ (5.183)

In order to determine the $c[i]$, we first Taylor-expand our ansatz in y around zero. Since we must determine 16 unknowns, this expansion should include at least the first 16 terms. Second, we take the sum of residues[2] of the unknown function $G(1)_{\text{SE2l2m}}$ in Eq. (5.157) and Taylor-expand this too in y around zero. Next, we compare terms of the same order y in the two expansions. Since the function $\ln(y)$ does not have a Taylor expansion around zero, terms of the form $y^k \ln(y)$ will also appear in addition to just powers like y^k. The coefficients of terms of the same order in y that are also proportional to $\ln(y)$ can be matched separately from those terms that are free of the logarithm. In this way, we solve for the unknown coefficients $c[i]$. If the constructed basis S_B covers properly the space of functions which are necessary to build the solution of $G(1)_{\text{SE2l2m}}$, then the $c[i]$ will have unique solutions. Thus finally, we obtain a solution for the ansatz in Eq. (5.183) and so for $G(1)_{\text{SE2l2m}}$.

The above actions can be written in Mathematica as follows:

```
In[9] := eq = Array[c, Length[basis]].basis;
In[10] := serEq1 = Normal[Series[eq, {y, 0, 16}]] // Simplify;
In[11] := sumresid = Sum[-Residue[fun, {z, i}], {i, 0, 16}];
In[12] := serEq2 = Normal[Series[sumresid, {y, 0, 16}]] //
  ↪  Simplify;
```

---

[2] The sum of residues must include enough terms such that all terms in the Taylor expansion to the required order are generated in the following step.

```
In[13]:= nologs = serEq1 - serEq2; /. a_[] Log[y]^_[] -> 0;
In[14]:= logs = serEq1 - serEq2 - nologs // Simplify;
In[15]:= tab1 = Table[0 == Coefficient[nologs,y,i-1],{i,16}];
In[16]:= coeff1 = Solve[tab1, Array[c, 16]]
Out[16]:= {{c[1] -> 2, c[2] -> 0, c[3] -> 0, c[4] -> 0, c[9] ->
↪    0, c[10] -> 0, c[11] -> 0, c[12] -> 0, c[13] -> 0, c[14] ->
↪    0, c[15] -> 0, c[16] -> 0}}
In[17]:= rule1 = coeff1[[1]];
In[18]:= tab2 = Table[0 == Coefficient[logs,y,i-1],{i,16}];
↪    Solve[tab2, Array[c, 16]]
Out[18]:= {{c[5] -> -1, c[6] -> 0, c[7] -> 2, c[8] -> 0}}
In[19]:= rule2 = coeff2[[1]];
```

In the code above the "basis" refers to Eq. (5.182), and "fun" is (the integrand of) our unknown function $G(1)_{SE2l2m}$ in Eq. (5.157). The complete `Mathematica` file is available in the file `MB_SE2l2m_Springer.nb` in the auxiliary material in [1].

Finally, we can apply the rules derived above, `rule1` and `rule2`, to the ansatz in Eq. (5.183). Doing so, we obtain the solution:

```
In[20]:= eq /. Join[rule1, rule2] // Simplify
Out[20]:= 2 - ((1 + y) Log[y])/(-1 + y)
```

Clearly the solution obtained agrees with the one we found in Eq. (5.177).

To finish, let us make two final comments. First, after obtaining the solution in the way described above, it is of course possible to Taylor-expand both the ansatz and the unknown sum to some order in y that is higher than what was used to fix the solution. Verifying that the two expansions continue to match also at higher orders serves as a useful check of the result. Second, it can also happen that we do not find a solution for the $c[i]$ in Eq. (5.183). This indicates that our ansatz was incorrect or incomplete. In this case, we may try to modify or enlarge our basis and search for a solution with the new ansatz. Or we are unlucky, and the solution goes beyond the known classes of functions.

5.2   Decoupling Integrals Through a Change of Variables

In our case decoupling of `MB` representations or integrals means their decomposition into a product of two or more integrals with lower dimensionality.

One of the simplest examples can be found, for example, in Eq. (10) of [18]. It corresponds to an $1/\epsilon^2$ pole of a planar QED double box. The integral is only two-dimensional so decoupling is visible by eyes. One also should point out here that the decoupled one-dimensional integrals up to a coefficient have the same structure as the integral in Eq. (1.55). In its turn Eq. (1.55) corresponds to the $1/\epsilon$ pole of the one-loop QED vertex.

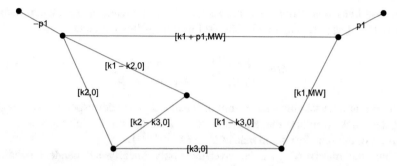

Fig. 5.2 A diagram corresponding to the integral in Eq. (5.184)

A more complicated example is given in Eq. (5.184) and corresponds to a diagram in Fig. 5.2.

$$I = -\frac{1}{(-s)^{1+3\epsilon}} \int_{-i\infty}^{+i\infty} \prod_{i=1}^{4} \frac{dz_i}{2\pi i} \left(-\frac{M_W^2}{s}\right)^{z_3} \frac{\Gamma(-\epsilon - z_1)\Gamma(-z_1)\Gamma(1 + 2\epsilon + z_1)}{\Gamma(1 - 2\epsilon)\Gamma(1 - 3\epsilon - z_1)}$$

$$\times \frac{\Gamma(-2\epsilon - z_{12})\Gamma(1 - \epsilon + z_2)\Gamma(1 + z_{12})\Gamma(1 + \epsilon + z_{12})\Gamma(1 + 3\epsilon + z_3)\Gamma(1 - \epsilon - z_4)}{\Gamma(1 - z_2)\Gamma(2 + \epsilon + z_{12})}$$

$$\times \frac{\Gamma(-\epsilon - z_2)\Gamma(-z_2)\Gamma(1 + z_3 - z_4)\Gamma(-z_4)\Gamma(-z_3 + z_4)\Gamma(-3\epsilon - z_3 + z_4)}{\Gamma(1 - 4\epsilon - z_3)\Gamma(2 + 2\epsilon + z_3 - z_4)}.$$

$$(5.184)$$

It is not clear from the diagram, but decoupling happens due to the presence of a massless two-loop self-energy subdiagram corresponding to the loop momenta k_1 and k_2. In this particular case, decoupling was obtained automatically by a proper choice of the order of integration in the LA approach.

In the general case, decoupled parts can be mixed up, and one needs to establish a procedure to check, if the decoupling is possible in principle. To answer this question, let us again consider the matrix representation used in Sect. 3.6 to study the possibility of MB integral decoupling. Decoupling means that the matrix representation should have a block form

$$M_\Gamma \longrightarrow \begin{pmatrix} A & 0 \\ 0 & B \end{pmatrix}. \qquad (5.185)$$

Transition to a block form is possible if it is possible to find a suitable transformation of the integration variables. For a faster search of the transformation, one can consider a smaller matrix $M_\Gamma \rightarrow \tilde{M}_\Gamma$ where we deleted all duplicated lines including lines that differ by overall sign. Now, elements of the transformation matrix $U = \{x_{ij}\}$ (see Eq. (3.53)) should fulfill the following conditions

$$\bar{M}_\Gamma = \begin{pmatrix} \bar{M}_\Gamma^{(1)} \\ \bar{M}_\Gamma^{(2)} \end{pmatrix}, \quad \begin{matrix} \bar{M}_\Gamma^{(1)} X^{(1)} = 0 \\ M_\Gamma^{(2)} X^{(2)} = 0 \end{matrix}, \tag{5.186}$$

where we assume that the first system of linear equations corresponds to the upper 0 in the block form and the second system to the bottom 0. Solutions will form columns of the transformation matrix $U = \left(X^{(2)} \ X^{(1)} \right)$.

Since the matrix U must be invertible, only linearly independent solutions should be taken. To have the total number of independent solutions equal to the dimensionality of matrix U, one can formulate the following condition

$$\mathrm{rank}(M_\Gamma^{(1)}) + \mathrm{rank}(M_\Gamma^{(2)}) = \dim \tag{5.187}$$

where dim is the dimensionality of the corresponding MB integral.

Now a test of decoupling possibility can be formulated in a simple algorithmic way: (i) construct matrix $\bar{M}_\Gamma$, (ii) divide it into two submatrices in all possible ways, and (iii) check the condition in Eq. (5.187). An instructive example for Eq. (5.184) is given in the file MB_Decoupling_Springer.nb in the auxiliary material in [1].

5.3   Solving via Integration

One can obtain analytic solutions for certain types of Mellin-Barnes integrals by transforming them to Euler-type integrals over real variables and evaluating the latter. In the present subsection, we explore this method.

5.3.1   From Mellin-Barnes to Euler Integrals

In this section, we will assume that the Mellin-Barnes integral we are evaluating is balanced, see section 5.1.1. As we have discussed above, this is not a real limitation for practical applications. Moreover it is straightforward to derive a real Euler-type integral representation for balanced Mellin-Barnes integrals as we will see in a moment. However, this representation will only be convergent if the arguments of all gamma functions have positive real parts. So in the following, let us assume that the integration contours are all straight lines parallel to the imaginary axis such that the real parts of the arguments of all gamma functions are positive. (We note that in dimensional regularization such contours may not exist for $\epsilon = 0$.) In this case, one can derive an Euler-type integral representation by noticing first of all that since the integral is balanced, its integrand can be expressed as a product of beta functions

$$B(a, b) = \frac{\Gamma(a)\Gamma(b)}{\Gamma(a + b)}. \tag{5.188}$$

For example, consider the integral

$$I = \int_{-i\infty-\frac{1}{2}}^{+i\infty-\frac{1}{2}} \frac{dz_1}{2\pi i} \int_{-i\infty-\frac{1}{4}}^{+i\infty-\frac{1}{4}} \frac{dz_2}{2\pi i} \Gamma(-z_1)\Gamma(-z_2)\Gamma(1 + z_1 + z_2)A^{z_1} B^{z_2} \,.$$

(5.189)

Clearly, the contours are such that the argument of each gamma function has a positive real part. Now, notice that, e.g.,

$$\Gamma(-z_2)\Gamma(1 + z_1 + z_2) = \Gamma(1 + z_1)B(-z_2, 1 + z_1 + z_2) \,,$$

(5.190)

so we have

$$I = \int_{-i\infty-\frac{1}{2}}^{+i\infty-\frac{1}{2}} \frac{dz_1}{2\pi i} \int_{-i\infty-\frac{1}{4}}^{+i\infty-\frac{1}{4}} \frac{dz_2}{2\pi i} \Gamma(-z_1)\Gamma(1 + z_1)B(-z_2, 1 + z_1 + z_2)A^{z_1} B^{z_2} \,.$$

(5.191)

Similarly, using

$$\Gamma(-z_1)\Gamma(1 + z_1) = \Gamma(1)B(-z_1, 1 + z_1),$$

(5.192)

we find

$$I = \int_{-i\infty-\frac{1}{2}}^{+i\infty-\frac{1}{2}} \frac{dz_1}{2\pi i} \int_{-i\infty-\frac{1}{4}}^{+i\infty-\frac{1}{4}} \frac{dz_2}{2\pi i} B(-z_1, 1 + z_1)B(-z_2, 1 + z_1 + z_2)A^{z_1} B^{z_2} \,.$$

(5.193)

The point is to "pair" gamma functions such that at least one integration variable cancels in the sum of arguments and replace such a pair by a corresponding beta function:

$$\Gamma(a \pm z)\Gamma(b \mp z) = \Gamma(a + b)B(a \pm z, b \mp z) \,.$$

(5.194)

Because the integral is balanced by assumption, this pairing can be continued until all gamma functions whose arguments contain integration variables have been exhausted. Notice that ratios of gamma functions can be replaced using

$$\frac{\Gamma(a \pm z)}{\Gamma(b \pm z)} = \frac{1}{\Gamma(b - a)}B(b - a, a \pm z) \,.$$

(5.195)

In this case, we are looking to pair a gamma function in the numerator with a gamma function in the denominator such that the difference of their arguments is free of at least one integration variable.

Next, we recall the real integral representation of $B(a, b)$:[3]

$$B(a, b) = \int_0^\infty dx\, x^{a-1}(1+x)^{-a-b}, \qquad \mathfrak{R}(a), \mathfrak{R}(b) > 0 \qquad (5.197)$$

and write all beta functions in the integrand of the Mellin-Barnes integral using this representation. The integral converges when $\mathfrak{R}(a), \mathfrak{R}(b) > 0$, and it is easy to check that these conditions are satisfied whenever the real parts of all arguments of gamma functions are positive in the original Mellin-Barnes integral. Because all integrals are convergent, we can exchange the Mellin-Barnes integrations and the real integrations coming from the beta functions. Returning to our example above, we obtain

$$
\begin{aligned}
I &= \int_{-i\infty-\frac{1}{2}}^{+i\infty-\frac{1}{2}} \frac{dz_1}{2\pi i} \int_{-i\infty-\frac{1}{4}}^{+i\infty-\frac{1}{4}} \frac{dz_2}{2\pi i} B(-z_1, 1+z_1) B(-z_2, 1+z_1+z_2) A^{z_1} B^{z_2} \\
&= \int_{-i\infty-\frac{1}{2}}^{+i\infty-\frac{1}{2}} \frac{dz_1}{2\pi i} \int_{-i\infty-\frac{1}{4}}^{+i\infty-\frac{1}{4}} \frac{dz_2}{2\pi i} \int_0^1 dx_1\, x_1^{-z_1-1}(1-x_1)^{z_1} \\
&\quad \times \int_0^1 dx_2\, x_2^{-z_2-1}(1-x_2)^{z_1+z_2} A^{z_1} B^{z_2} \\
&= \int_0^1 dx_1 \int_0^1 dx_2 \frac{1}{x_1 x_2} \int_{-i\infty-\frac{1}{2}}^{+i\infty-\frac{1}{2}} \frac{dz_1}{2\pi i} \int_{-i\infty-\frac{1}{4}}^{+i\infty-\frac{1}{4}} \frac{dz_2}{2\pi i} \left[\frac{(1-x_1)(1-x_2)A}{x_1} \right]^{z_1} \\
&\quad \times \left[\frac{(1-x_2)B}{x_2} \right]^{z_2}.
\end{aligned}
$$

$$(5.198)$$

In general, after this step, one finds an integral of the form

$$\int_0^1 \left(\prod_{k=1}^K dx_k \right) R_0(\mathbf{x}, \mathbf{v})\, [R_\epsilon(\mathbf{x}, \mathbf{v})]^\epsilon \int_{-i\infty}^{+i\infty} \prod_{l=1}^L \frac{dz_l}{2\pi i}\, [R_l(\mathbf{x}, \mathbf{v})]^{z_l}, \qquad (5.199)$$

where $\mathbf{x} = (x_1, \ldots, x_K)$ are K real integration variables, $\mathbf{v} = (v_1, \ldots, v_N)$ are the variables of the original Mellin-Barnes integral, and the R_l are ratios of products

---

[3] We note that in practice one can very often also use the alternative integral representation of the beta function

$$B(a, b) = \int_0^1 dx\, x^{a-1}(1-x)^{b-1}, \qquad \mathfrak{R}(a), \mathfrak{R}(b) > 0 \qquad (5.196)$$

with limits of integration zero and one. Most of the discussion that follows goes through without change in this case, but the appearance of boundary conditions can lead to certain technical subtleties.

of x_i, $(1 + x_i)$ and v_j. Although obvious, we note that R_ϵ is only present if the original Mellin-Barnes integral depends on ϵ. Hence, if we have already expanded the Mellin-Barnes integral in ϵ and are considering only the expansion coefficients, R_ϵ is absent.

Next, we need to perform the Mellin-Barnes integrations. This can be done using the formula

$$\int_{-i\infty+z_0}^{+i\infty+z_0} \frac{dz}{2\pi i} A^z = \delta(1 - A), \qquad A > 0. \tag{5.200}$$

Importantly, in the Euclidean regime, where all variables v_j are positive by definition, each R_l is clearly positive for $0 < x_i < \infty$, $i = 1, \ldots, K$. Thus, Eq. (5.199) takes the form

$$\int_0^\infty \left(\prod_{k=1}^K dx_k \right) R_0(\mathbf{x}, \mathbf{v}) \, [R_\epsilon(\mathbf{x}, \mathbf{v})]^\epsilon \prod_{l=1}^L \delta \, [1 - R_l(\mathbf{x}, \mathbf{v})] \,. \tag{5.201}$$

By solving the constraints implied by the δ functions, we arrive at the desired parametric integral representation. Continuing with our example in Eq. (5.198), we find

$$I = \int_0^1 dx_1 \int_0^1 dx_2 \, \frac{1}{x_1 x_2} \delta \left[1 - \frac{(1 - x_1)(1 - x_2)A}{x_1} \right] \delta \left[1 - \frac{(1 - x_2)B}{x_2} \right]. \tag{5.202}$$

The second δ function implies $x_2 = B/(1 + B)$, and so performing the integration over x_2, we obtain

$$I = \int_0^1 dx_1 \, \frac{1}{(1 + B)x_1} \delta \left[1 - \frac{(1 - x_1)A}{(1 + B)x_1} \right], \tag{5.203}$$

where we have used

$$\delta \left[1 - \frac{(1 - x_2)B}{x_2} \right] = \frac{B}{(1 + B)^2} \delta \left(x_2 - \frac{B}{1 + B} \right). \tag{5.204}$$

The remaining δ function now fixes $x_1 = A/(1 + A + B)$, and making use of

$$\delta \left[1 - \frac{(1 - x_1)A}{(1 + B)x_1} \right] = \frac{A(1 + B)}{(1 + A + B)^2} \delta \left(x_1 - \frac{A}{1 + A + B} \right), \tag{5.205}$$

we finally obtain a zero-dimensional integral representation

$$I = \int_0^1 dx_1 \, \frac{A}{(1 + A + B)^2 x_1} \delta \left(x_1 - \frac{A}{1 + A + B} \right) = \frac{1}{1 + A + B}. \tag{5.206}$$

Of course, in general, one obtains a finite-dimensional real integral representation, which must be solved to obtain the desired result. One way of imple-
menting the above manipulations in Mathematica is presented in the file
MB_Euler_Springer.nb in the auxiliary material in [1].

Finally, we mention that if the MB integral involves polygamma functions (this
will typically happen if we expand in ϵ first, and apply the above construction to the
MB integrals that appear in the expansion coefficients), we can use Eq. (2.137)

$$\psi(z) = \int_0^\infty [(1+x)^{-1} - (1+x)^{-z}]\frac{dx}{x} - \gamma, \qquad \mathfrak{R}(z) > 0, \qquad (5.207)$$

to express also the polygamma functions as real integrals.[4] The integral representation converges if the real part of the argument of the polygamma function is positive.

Mellin-Barnes Integrals to Euler-Type Integrals: An Overview

Given a balanced Mellin-Barnes integral such that the real parts of the arguments
of all gamma functions are positive, one can construct an Euler-type integral
representation for the Mellin-Barnes integral as follows:

1. Express the integrand as a product of beta functions. To do so, pair products
   (ratios) of two gamma functions such that at least one Mellin-Barnes integration
   variable cancels in the sum (difference) of their arguments. Continue this
   process until no gamma function remains whose argument contains an integration
   variable. Since the original Mellin-Barnes integral is balanced by assumption,
   this can always be achieved.
2. Replace all beta and polygamma functions by their integral representations as
   given in Eqs. (5.197) and (5.207). The real integrations introduced in this step
   are all convergent if the real parts of the arguments of all gamma and polygamma
   functions in the original Mellin-Barnes integral are positive.
3. Perform all Mellin-Barnes integrations using the result of Eq. (5.200). In the
   Euclidean region, the necessary condition is always fulfilled.
4. Finally, solve the constraints implied by the δ functions to obtain the desired real
   parametric representation.

---

[4] As with the beta function, the integral representation over the finite interval [0, 1] of Eq. (5.208),

$$\psi(z) = \int_0^1 \frac{t^{z-1} - 1}{t - 1}dt - \gamma, \qquad \mathfrak{R}(z) > 0, \qquad (5.208)$$

can also very often be used in practice, with the same caveat as for the beta function.

5.3.2 Symbolic Integration of Euler Integrals

Let us now turn to briefly describe how the parametric integrals that we obtain can be computed. More precisely, we consider an integral of the following form

$$I(\mathbf{v}; \epsilon) = \int_0^\infty \left(\prod_{k=1}^n dx_k \right) \prod_{l=1}^m P_l(\mathbf{x}, \mathbf{v})^{a_l + b_l \epsilon} \tag{5.209}$$

where again $\mathbf{x} = (x_1, \ldots, x_n)$ are n real integration variables and $\mathbf{v} = (v_1, \ldots, v_m)$ are m parameters fixed during the integration. We assume that a_l and b_l are integers and the $P_l(\mathbf{x}, \mathbf{v})$ are polynomials that cannot be factorized further into products of lower-degree polynomials. We assume that the limits of integration are zero and infinity. It is not very difficult to see that Eq. (5.201) can generally be written as a sum of such integrals. We assume that the integral is convergent for $\epsilon = 0$, so that we can expand the integrand in ϵ before integration. This is not always true, as in practical calculations the integrand sometimes develops non-integrable singularities when $\epsilon = 0$, typically when some of the x_i are zero (however, more elaborate structure of singularities can also show up). In such cases, we must first resolve the poles of the integral in ϵ, e.g., by subtracting the divergences. The detailed discussion of how this can be achieved for general integrals is beyond the scope of our discussion, and we point the reader to the literature [19, 20].

After performing the Taylor expansion of the integrand in Eq. (5.209) in ϵ, the expansion coefficients will involve rational functions (recall that the a_l can also be negative) and (powers of) logarithms of polynomials in the integration variables. Our goal is then to integrate out the x_i one after the other. This involves iterated integrations of rational functions and logarithms, and hence we expect that multiple polylogarithms, introduced in Chap. 2, will show up in the solution. We emphasize that of course not every integral can be solved in this way in terms of multiple polylogarithms; however this is still a powerful method for obtaining solutions in many situations. In fact, one can state a sufficient condition, first derived in [21], on the integrand, which if satisfied allows one to express the integral in terms of multiple polylogarithms. To do so, let us start by defining S as the set of all polynomials that are not monomials which appear in the integrand in Eq. (5.209):

$$S = \{P_l(\mathbf{x}, \mathbf{v})\}. \tag{5.210}$$

Recalling the definition of multiple polylogarithms as iterated integrals in Eq. (2.82),

$$G(a_1, \ldots, a_n; z) = \int_0^z \frac{dt}{t - a_1} G(a_2, \ldots, a_n; t), \tag{5.211}$$

we see that in order to begin the integration, we have to assume that there exists some integration variable, say x_a, such that all elements of S are linear in this variable. In this case, let us write each of the P_ls as

$$P_l(\mathbf{x}, \mathbf{v}) = Q_l^a(\mathbf{x'}, \mathbf{v})x_a + R_l^a(\mathbf{x'}, \mathbf{v}), \tag{5.212}$$

where $\mathbf{x'} = \{x_1, \ldots, x_{a-1}, x_{a+1}, \ldots, x_n\}$ denotes the set of integration variables *without* x_a. Of course, after the ϵ expansion, logarithms of P_l will also appear. These can be written in terms of multiple polylogarithms as

$$\ln P_l = \ln\left(Q_l^a x_a + R_l^a\right) = \ln R_l^a + \ln\left(1 + \frac{Q_l^a}{R_l^a}x_a\right)$$

$$= \ln R_l^a + G\left(-\frac{R_l^a}{Q_l^a}; x_a\right). \tag{5.213}$$

Moreover, products of multiple polylogarithms can be expressed in terms of functions of the form $G(a_1(\mathbf{x'}, \mathbf{v}), \ldots, a_l(\mathbf{x'}, \mathbf{v}); x_a)$ using the shuffle algebra of multiple polylogarithms, e.g.,

$$G\left(-\frac{R_k^a}{Q_k^a}; x_a\right) G\left(-\frac{R_l^a}{Q_l^a}; x_a\right) = G\left(-\frac{R_k^a}{Q_k^a}, -\frac{R_l^a}{Q_l^a}; x_a\right) + G\left(-\frac{R_l^a}{Q_l^a}, -\frac{R_k^a}{Q_k^a}; x_a\right). \tag{5.214}$$

Thus, the integration over x_a involves (sums of) integrals of the form

$$\int_0^\infty \frac{dx_a}{(Q_1^a x_a + R_1^a)^{-a_1} \ldots (Q_m^a x_a + R_m^a)^{-a_m}} G(\mathbf{a}; x_a). \tag{5.215}$$

In order to compute the integral, we partial fraction in x_a, e.g.,

$$\frac{1}{(Q_k^a x_a + R_k^a)(Q_l^a x_a + R_l^a)} = \frac{1}{Q_k^a R_l^a - Q_l^a R_k^a}\left(\frac{1}{x_a + R_k^a/Q_k^a} - \frac{1}{x_a + R_l^a/Q_k l^a}\right), \tag{5.216}$$

then simply apply the recursive definition of multiple polylogarithms in Eq. (5.211) to obtain a primitive. Finally, we must compute the limits of this primitive as $x_a \to 0$ and $x_a \to \infty$.

We would like to iterate this process and integrate over the rest of the variables one by one. However, to do this we must once again find an integration variable, say x_b in which all polynomials in the new integrand are linear. These polynomials are nothing but the Q_l^a and R_l^a introduced above, *as well as* the combinations $Q_k^a R_l^a - Q_l^a R_k^a$ that are introduced by partial fractioning. This last polynomial is not necessarily linear, even if the Q_k^a and R_l^a are. Thus, in order to proceed, it is necessary that the $Q_k^a R_l^a - Q_l^a R_k^a$ factor into polynomials that are linear in some integration variable. This analysis must be repeated after each step of the integration. In order to formalize this procedure, we can define $S_{(x_a)}$ as the set of all irreducible factors that appear inside the polynomials Q_l^a, R_l^a and $Q_k^a R_l^a - Q_l^a R_k^a$. Now, if we can find an integration variable, say x_b, such that all elements of $S_{(x_a)}$ are linear in x_b, we can restart the above procedure and integrate over x_b. If we can iterate this procedure and construct a sequence of sets of polynomials:

$$S_{(x_a)}, \ S_{(x_a,x_b)}, \ S_{(x_a,x_b,x_c)}, \ \ldots \tag{5.217}$$

such that all polynomials in each set are linear in at least one integration variable, then we can perform the integrations one by one and express the result in terms of multiple polylogarithms. *We stress three important facts:*

1. First, the existence of such a sequence of sets in general depends on the ordering of the integration variables. Of course, it is enough to find one ordering where all polynomials in each set are linear in at least one integration variable.
2. Second, it can be shown that this condition, called *linear reducibility*, is sufficient but not necessary for the integral to be expressible in terms of multiple polylogarithms. In fact, even if we fail to find a suitable sequence of sets, the result may still be given in terms of multiple polylogarithms, e.g., after a suitable change of variables. In addition, it is not necessary that the factors are linear for the last integration step, since in this case we can factor the polynomials into linear factors whose roots involve algebraic expressions of the variables **v**.
3. Third, notice that we can check if such a sequence of sets exists without actually performing any integrations. We remark though that in practical applications, it is rarely necessary to actually construct this sequence of sets: many times several orderings of variables can be chosen, some more convenient than others (e.g., leading to smaller intermediate expressions), and it is often quite simple to proceed by inspection.

We now have a criterion for determining if a given integral can be evaluated in terms of multiple polylogarithms; however in order to actually proceed and compute the result, we must address the following points. Let us assume that we have found a proper ordering of integration variables and have performed the (indefinite) integration over x_a. Then in order to continue and perform the next integration, we must:

1. Take the limits of the primitive with respect to x_a as $x_a \to 0$ and $x_a \to \infty$.
2. Find a way to write all multiple polylogarithms in the form $G(\mathbf{a}; x_b)$, where $\mathbf{a}$ is *independent* of x_b. This is nontrivial, as the arguments of multiple polylogarithms of the form $G\left(\ldots, -R_l^a / Q_l^a, \ldots\right)$ may well depend on x_b.

Regarding the first point, the limit at $x_a \to 0$ can easily be taken by using the shuffle algebra to extract pure logarithms in x_a (see the discussion in Chap. 2) and then using the sum representation of multiple polylogarithms, Eq. (2.88). In fact, in many situations, it is enough to consider that

$$\lim_{x_a \to 0} G(\mathbf{a}; x_a) = 0, \qquad \mathbf{a} \neq \mathbf{0}. \tag{5.218}$$

Turning to the limit at $x_a \to \infty$, we note that by introducing $\bar{x}_a = 1/x_a$, this problem can be reduced to taking the limit $\bar{x}_a \to 0$,

$$\lim_{x_a \to \infty} G(\mathbf{a}; x_a) = \lim_{\bar{x}_a \to 0} G\left(\mathbf{a}; \frac{1}{\bar{x}_a}\right). \tag{5.219}$$

If we can find a way of writing the multiple polylogarithms on the right-hand side as linear combinations of multiple polylogarithms of the form $G(\ldots; \bar{x}_a)$, then taking the $\bar{x}_a \to 0$ limit can be performed as discussed above. At this point, it is worth noticing that the problem of finding the appropriate inversion relations is formally equivalent to the second issue that we have not yet addressed, namely, how to bring the multiple polylogarithms entering the integrand to a form where x_b only appears in the last position, as $G(\mathbf{a}; x_b)$. In both cases we are looking for functional equations that bring the multiple polylogarithms into "standard form," where a certain variable only enters as the explicit argument but not in the parameters.

The framework necessary to address this problem involves the Hopf algebra structure of multiple polylogarithms. The description of this topic is beyond the scope of this book; however, we mention one main result. If the sufficient condition described above regarding the existence of the sequence of sets in Eq. (5.217) is met and *furthermore* the integration ranges are $[0, \infty]$, it can be shown that the multiple polylogarithms that appear can always be brought to a form

$$\sum_i c_i G(\mathbf{a}_i; x), \tag{5.220}$$

for some variable x such that $\mathbf{a}_i$ is independent of x and the coefficients c_i involve only multiple polylogarithms that are independent of x. Furthermore, this rewriting can be performed in a constructive algorithmic way. The appropriate algorithms are implemented, e.g., in the `Mathematica` package `PolyLogTools` and the `Maple` program `HyperInt` (see Appendix A.3).

We end our discussion with a brief comment about generic limits of integration. Although many statements described above remain true for generic integration boundaries, nevertheless in certain cases, some algorithms do break down due to the appearance of boundary terms. However, a generic region of integration $x \in [a, b]$ can always be mapped to $y \in [0, \infty]$ by the change of variables

$$y = \frac{x - b}{x - a}. \tag{5.221}$$

Thus from a purely formal point of view, when constricting Euler integral representations of MB integrals, we should use the representation of the beta function as an integral between zero and infinity. However, from a practical standpoint, we can very often proceed with the representation based on the integration over $[0, 1]$ as well.

5.3.3 Merging MB Integrals with Euler Integrals and Symbolic Integration

Finally, as an illustration of the above procedure, let us discuss a simple example. Consider the phase space integral

$$I = \int d\Omega_{d-1}(k) \, d\Omega_{d-1}(l) \, \frac{1}{p_1 \cdot (k + l)}, \tag{5.222}$$

where p_1 is fixed massless four-vectors, while k and l are massless vectors whose directions we integrate over with the measures $d\Omega_{d-1}(k)$ and $d\Omega_{d-1}(k)$ given in Eq. (3.137). Such integrands appear, for example, in factorization formulae describing the double soft limits of QCD real-emission matrix elements. We can easily relate this integral to the phase space integrals studied in Chap. 2 by applying the basic MB formula to the denominator:

$$I = \int d\Omega_{d-1}(k) \, d\Omega_{d-1}(l) \int_{-i\infty}^{+i\infty} \frac{dz}{2\pi i} \Gamma(-z)\Gamma(1+z)(p_1 \cdot k)^z (p_1 \cdot l)^{-1-z}. \tag{5.223}$$

Now, let us exchange the order of integrations and then apply the Eq. (3.163):

$$\int d\Omega_{d-1}(q) \frac{1}{(p_1 \cdot q)^j} = \Omega_j(0, \epsilon) = 2^{2-j-2\epsilon} \pi^{1-\epsilon} \frac{\Gamma(1-j-\epsilon)}{\Gamma(2-j-2\epsilon)}, \tag{5.224}$$

to perform the angular integrations.[5] We obtain

$$I = \int_{-i\infty}^{+i\infty} \frac{dz}{2\pi i} \Gamma(-z)\Gamma(1+z) \int d\Omega_{d-1}(k) \frac{1}{(p_1 \cdot k)^{-z}} d\Omega_{d-1}(l) \frac{1}{(p_1 \cdot l)^{1+z}}$$

$$= 2^{3-4\epsilon} \pi^{2-2\epsilon} \int_{-i\infty}^{+i\infty} \frac{dz}{2\pi i} \Gamma(-z)\Gamma(1+z) \frac{\Gamma(1+z-\epsilon)}{\Gamma(2+z-2\epsilon)} \frac{\Gamma(-z-\epsilon)}{\Gamma(1-z-2\epsilon)}. \tag{5.225}$$

Now we must find a contour for the z integration, such that the poles of the gamma functions with $\Gamma(\ldots + z)$ and those with $\Gamma(\ldots - z)$ are separated. In this case, it is

---

[5] Equation (5.224) is valid provided that all vectors in the integrand are normalized such that their zeroth components are one. It is obviously trivial to rescale the integrand in Eq. (5.222) to achieve this, and we assume that this has been done.

quite straightforward to find a suitable straight-line contour parallel to the imaginary axis. By simple inspection, we see that for $\epsilon = 0$, we may choose any $z_0 \in (-1, 0)$ and the contour running from $z_0 - i\infty$ to $z_0 + i\infty$ will separate the poles as required. As we have found a suitable contour directly at $\epsilon = 0$, there is no need to analytically continue to $\epsilon \to 0$.

Having derived the MB representation, let us now turn to obtaining an analytic solution by converting it into an Euler-type integral and integrating that. To do this, we can combine the factors of $\Gamma(-z)$ and $\Gamma(1 + z)$ in the numerator using[6]

$$\Gamma(-z)\Gamma(1 + z) = \Gamma(1) \int_0^\infty dx_1 \, x_1^{-1-z}(1 + x_1)^{-1}, \tag{5.226}$$

while the two ratios of gamma functions can be written as

$$\frac{\Gamma(1 + z - \epsilon)}{\Gamma(2 + z - 2\epsilon)} = \frac{1}{\Gamma(1 - \epsilon)} \int_0^\infty dx_2 \, x_2^{z-\epsilon}(1 + x_2)^{-2+2\epsilon-z}, \tag{5.227}$$

$$\frac{\Gamma(-z - \epsilon)}{\Gamma(1 - z - 2\epsilon)} = \frac{1}{\Gamma(1 - \epsilon)} \int_0^\infty dx_3 \, x_3^{-1-\epsilon-z}(1 + x_3)^{-1+2\epsilon+z}. \tag{5.228}$$

Then we obtain

$$I = \frac{2^{3-4\epsilon}\pi^{2-2\epsilon}}{\Gamma^2(1-\epsilon)} \int_{-i\infty}^{+i\infty} \frac{dz}{2\pi i} \int_0^\infty dx_1 \int_0^\infty dx_2 \int_0^\infty dx_3$$
$$\times \frac{x_2^{-\epsilon}(1 + x_2)^{2\epsilon-2}x_3^{-\epsilon-1}(1 + x_3)^{2\epsilon-1}}{x_1(1 + x_1)} \left(\frac{x_2(1 + x_3)}{x_1(1 + x_2)x_3} \right)^z. \tag{5.229}$$

Next, we perform the MB integration using Eq. (5.200). Notice that the quantity being raised to the power z, $\frac{x_2(1+x_3)}{x_1(1+x_2)x_3}$, is positive in the integration region, essentially by construction. Then we find

$$I = \frac{2^{3-4\epsilon}\pi^{2-2\epsilon}}{\Gamma^2(1-\epsilon)} \int_0^\infty dx_1 \int_0^\infty dx_2 \int_0^\infty dx_3 \, \frac{x_2^{-\epsilon}(1 + x_2)^{2\epsilon-2}x_3^{-\epsilon-1}(1 + x_3)^{2\epsilon-1}}{x_1(1 + x_1)}$$
$$\times \delta\left(1 - \frac{x_2(1 + x_3)}{x_1(1 + x_2)x_3} \right). \tag{5.230}$$

Finally, we must resolve the Dirac δ function by performing one integration. Solving the constraint for x_1 yields

$$x_1 = \frac{x_2(1 + x_3)}{(1 + x_2)x_3} \tag{5.231}$$

---

[6] The choice of contour is such that the real parts of the arguments of all gamma functions in Eq. (5.225) are positive; thus all integral representations we write will automatically converge.

and we find the Euler-type integral representation

$$I = \frac{2^{3-4\epsilon}\pi^{2-2\epsilon}}{\Gamma^2(1-\epsilon)} \int_0^\infty dx_2 \int_0^\infty dx_3 \frac{x_2^{-\epsilon}(1+x_2)^{-1+2\epsilon}x_3^{-\epsilon}(1+x_3)^{-1+2\epsilon}}{x_2 + x_3 + 2x_2x_3}.$$

(5.232)

Obviously this representation is not unique: it depends on the particular way in which we combine the gamma functions into beta functions as well as on the choice of which variable we solve the Dirac δ constraint for.

Of course, we may choose to solve the Dirac δ constraint for any of the variables. However, some choices are less ideal than others. To see why, consider solving for x_2 in our example instead of x_1. We find

$$x_2 = \frac{x_1 x_3}{1 + x_3 - x_1 x_3}.$$

(5.233)

But x_2 must be non-negative in the integration region, which leads to the constraint $1 + x_3 - x_1 x_3 \geq 0$. If $0 \leq x_1 \leq 1$, this is clearly satisfied for any non-negative x_3; however for $x_1 > 1$, it implies $x_3 \leq \frac{1}{x_1-1}$. We must then be careful to implement this constraint in the subsequent integrations, in this case, e.g., by splitting the integration in x_1 at one. Notice that the choice of solving for x_1 instead completely circumvents such issues.

Now we are ready to solve the remaining real integrations. To do so, we first note that the result is finite as $\epsilon \to 0$, and we may simply expand in ϵ under the integral sign. Let us consider the leading term of this expansion, i.e., the finite part:

$$I = 8\pi^2 \int_0^\infty dx_2 \int_0^\infty dx_3 \frac{1}{(1+x_2)(1+x_3)(x_2+x_3+2x_2x_3)} + O(\epsilon).$$

(5.234)

It can be checked that this integral is linearly reducible and hence expressible with multiple polylogarithms (at some specific arguments). The order of integrations clearly cannot matter in this example, since the integrand is symmetric under $x_2 \leftrightarrow x_3$ exchange. Thus, let us partial fraction the integrand in x_2:

$$\frac{1}{(1+x_2)(1+x_3)(x_2+x_3+2x_2x_3)} = -\frac{1}{(1+x_3)^2(1+x_2)}$$
$$+ \frac{1}{(1+x_3)^2[x_3/(1+2x_3)+x_2]}.$$

(5.235)

It is straightforward to compute a primitive with respect x_2 now, and we find

$$
\int dx_2 \left[-\frac{1}{(1+x_3)^2(1+x_2)} + \frac{1}{(1+x_3)^2[x_3/(1+2x_3)+x_2]} \right]
$$

$$
= -\frac{1}{(1+x_3)^2} G(-1; x_2) + \frac{1}{(1+x_3)^2} G\left(-\frac{x_3}{1+2x_3}; x_2\right).
$$

(5.236)

Although the integration is of course elementary, we have expressed the result with multiple polylogarithms in order to illustrate the general case. Next, we must evaluate this primitive in the limits $x_2 \to 0$ and $x_2 \to \infty$. The limit at $x_2 = 0$ is trivial: using Eq. (5.218) we see that in the limit, the primitive is simply zero. In order to find the limit at $x_2 \to \infty$, let us set $x_2 = 1/\bar{x}_2$ and consider $\bar{x}_2 \to 0$. This requires writing the multiple polylogarithms $G\left(-1; \frac{1}{\bar{x}_2}\right)$ and $G\left(-\frac{x_3}{1+2x_3}; \frac{1}{\bar{x}_2}\right)$ in terms of multiple polylogarithms of the form $G(\ldots; \bar{x}_2)$. As discussed above, the procedures for how this can be done in general go beyond the scope of this book; however for the specific case at hand, we may proceed simply by recalling that a weight 1 multiple polylogarithm is simply an ordinary logarithm in disguise, $G(a; z) = \ln\left(1 - \frac{z}{a}\right)$, if $a \neq 0$. So we have

$$
G\left(-1; \frac{1}{\bar{x}_2}\right) = \ln\left(1 + \frac{1}{\bar{x}_2}\right) = \ln(1 + \bar{x}_2) - \ln(\bar{x}_2) = G(-1; \bar{x}_2) - G(0; \bar{x}_2)
$$

(5.237)

and similarly

$$
G\left(-\frac{x_3}{1+2x_3}; \frac{1}{\bar{x}_2}\right) = \ln\left(1 + \frac{1+2x_3}{\bar{x}_2 t_3}\right)
$$

$$
= \ln(1 + 2x_3) + \ln\left(1 + \frac{x_3}{1+2x_3}\bar{x}_2\right) - \ln(x_3) - \ln(\bar{x}_2)
$$

(5.238)

$$
= G\left(-\frac{1}{2}; x_3\right) + G\left(-\frac{1+2x_3}{x_3}; \bar{x}_2\right) - G(0; x_3) - G(0; \bar{x}_2).
$$

Using the above, it is straightforward to compute the limit of the primitive at $\bar{x}_2 \to 0$. Notice in particular that terms involving $G(0; \bar{x}_2) = \ln(\bar{x}_2)$ cancel and the limit is finite. Thus we obtain

$$
\int_0^\infty dx_2 \left[-\frac{1}{(1+x_3)^2(1+x_2)} + \frac{1}{(1+x_3)^2[x_3/(1+2x_3)+x_2]} \right]
$$

$$
= \frac{G\left(-\frac{1}{2}; x_3\right) - G(0; x_3)}{(1+x_3)^2}.
$$

(5.239)

Finally, we must evaluate

$$\int_0^\infty dx_3 \frac{G\left(-\frac{1}{2}; x_3\right) - G(0; x_3)}{(1 + x_3)^2}. \tag{5.240}$$

Notice that the denominator involves $(1 + x_3)$ to the second power, so the primitive can be computed using integration by parts and will not involve multiple polylogarithms of weight higher than one. Computing the limits then proceeds in the same way as above, using the functional identities for the logarithm, and we find the above integral is simply equal to $2G(0; 2) = 2 \ln 2$. Recalling the overall factor of $8\pi^2$ in Eq. (5.234), we have finally (see the file MB_symbolicint_Springer.nb in the auxiliary material in [1])

$$I = 16\pi^2 \ln 2 + O(\epsilon). \tag{5.241}$$

In a similar fashion, it is possible to compute the result at higher orders in the ϵ expansion, and the linear reducibility of the integrand guarantees that the expansion coefficients at higher orders are still expressible in terms of multiple polylogarithms (at specific arguments). However, already at $O(\epsilon)$, the full machinery of the Hopf algebra of multiple polylogarithms comes into play. The method discussed here has been used in the literature to compute real radiation integrals, e.g., in Higgs boson production at N^3LO accuracy [22].

Choosing Proper Methods of MB Integration

Sometimes it is beneficial to combine the methods presented above. For example, it may happen that after replacing only some of the gamma functions in the integrand of a Mellin-Barnes integral by beta functions, the Mellin-Barnes integrations simplify. One way this may happen is that multi-dimensional Mellin-Barnes integrals decouple into products of one-dimensional integrals, but this is not the only option as we will see shortly. In such situations, it is not necessary to then continue to replace further gamma functions by beta functions until we arrive at a proper Euler integral representation, but one may go ahead and evaluate the simplified Mellin-Barnes integrals directly, e.g., using the methods of symbolic summation. This way of proceeding can sometimes be more efficient than if we had derived a full Euler integral representation. As an application of these ideas, here we derive the solution of the two-denominator angular integral with one mass, Eq. (3.172).

To start, recall from Eq. (3.171) that the Mellin-Barnes representation of the angular integral $\Omega_{j,k}(v_{11}, v_{12}, \epsilon)$ involves the two-dimensional integral

$$I = \int_{-i\infty}^{+i\infty} \frac{dz_1}{2\pi i} \frac{dz_2}{2\pi i} \Gamma(-z_1)\Gamma(-z_2)\Gamma(a + 2z_1 + z_2)\Gamma(b + z_2)$$
$$\times \Gamma(c - z_1 - z_2)x^{z_1}y^{z_2}. \tag{5.242}$$

Notice that this integral involves a gamma function of the form $\Gamma(\cdots + 2z_1)$ and hence it cannot easily be computed using the summation methods presented above. Throughout we tacitly assume that all parameters lie in a strip of the convex plane such that each integral we write is convergent. Then let us begin by writing the product of the second and fourth gamma functions in the integrand as a beta function. Here we choose the representation of the beta function where the domain of integration is $[0, 1]$ and write

$$\Gamma(-z_2)\Gamma(b + z_2) = \Gamma(b)\int_0^1 dt\, t^{-1-z_2}(1-t)^{b-1+z_2}\,. \tag{5.243}$$

Then, Eq. (5.242) becomes

$$I = \Gamma(b)\int_{-i\infty}^{+i\infty} \frac{dz_1}{2\pi i}\frac{dz_2}{2\pi i}\int_0^1 dt\, t^{-1-z_2}(1-t)^{b-1+z_2}$$
$$\times\,\Gamma(-z_1)\Gamma(a + 2z_1 + z_2)\Gamma(c - z_1 - z_2)x^{z_1}y^{z_2}\,. \tag{5.244}$$

Next, we perform the change of variables $z_2 \to -a - 2z_1 - z_2$, which makes the following manipulations more straightforward to follow. Then we find

$$I = \Gamma(b)\int_{-i\infty}^{+i\infty} \frac{dz_1}{2\pi i}\frac{dz_2}{2\pi i}\int_0^1 dt\, t^{a-1+2z_1+z_2}(1-t)^{b-a-1-2z_1-z_2}$$
$$\times\,\Gamma(-z_1)\Gamma(-z_2)\Gamma(a + c + z_1 + z_2)x^{z_1}y^{-a-2z_1-z_2}\,. \tag{5.245}$$

After exchanging the order of integrations and rearranging some factors, I can be written in the following form:

$$I = y^{-a}\Gamma(b)\int_0^1 dt\, t^{a-1}(1-t)^{a+b+2c-1}\int_{-i\infty}^{+i\infty} \frac{dz_1}{2\pi i}\frac{dz_2}{2\pi i}$$
$$\times\,\Gamma(-z_1)\Gamma(-z_2)\Gamma(a + c + z_1 + z_2)\left(\frac{xt^2}{y^2}\right)^{z_1}\left[\frac{t(1-t)}{y}\right]^{z_2}[(1-t^2)]^{-a-c-z_1-z_2}\,. \tag{5.246}$$

At this point we could continue to express products of gamma functions as real Euler-type integrals. However, a more clever way to proceed presents itself. Indeed, the Mellin-Barnes integrations are now formally easy to perform, since the integral we are left with is just a special case of the general formula

$$\frac{1}{(A + B + C)^\nu} = \frac{1}{\Gamma(\nu)}\int_{-i\infty}^{+i\infty} \frac{dz_1\, dz_2}{(2\pi i)^2}\Gamma(-z_1)\Gamma(-z_2)\Gamma(\nu + z_1 + z_2)A^{z_1}B^{z_1}C^{-\nu-z_1-z_2}, \tag{5.247}$$

where of course the appropriate choice of contour is understood. Thus, Eq. (5.246) becomes

$$I = y^{-a}\Gamma(b)\Gamma(a+c) \int_0^1 dt\, t^{a-1}(1-t)^{a+b+2c-1} \left[(1-t)^2 + \frac{t(1-t)}{y} + \frac{xt^2}{y^2}\right]^{-a-c}.$$
(5.248)

The last factor of the integrand can be factored in the integration variable t as follows:

$$(1-t)^2 + \frac{t(1-t)}{y} + \frac{xt^2}{y^2} = \left(1 - \frac{2y - 1 - \sqrt{1-4x}}{2y}t\right)\left(1 - \frac{2y - 1 + \sqrt{1-4x}}{2y}t\right).$$
(5.249)

Note that the appearance of square roots is not an issue, since only the parameter x appears under the root, the variable of integration t does not. (Recall also the discussion below Eq. (5.217) regarding the last, and in the present case only, integration step.) Finally, we obtain the one-dimensional real integral representation

$$I = y^{-a}\Gamma(b)\Gamma(a+c) \int_0^1 dt\, t^{a-1}(1-t)^{a+b+2c-1}$$
$$\times \left(1 - \frac{2y - 1 - \sqrt{1-4x}}{2y}t\right)^{-a-c}\left(1 - \frac{2y - 1 + \sqrt{1-4x}}{2y}t\right)^{-a-c}.$$
(5.250)

The last integral can now be performed in terms of the Appell function of the first kind (see Eq. (2.194)), and we find the final result

$$I = y^{-a} \frac{\Gamma(b)\Gamma(b)\Gamma(a+c)\Gamma(a+b+2c)}{\Gamma(2a+b+2c)}$$
$$\times F_1\left(a, a+c, a+c, 2a+b+2c, \frac{2y - 1 - \sqrt{1-4x}}{2y}, \frac{2y - 1 + \sqrt{1-4x}}{2y}\right).$$
(5.251)

We note that if the parameters a, b, and c are all of the form $n + m\epsilon$ with $n \in \mathbb{Z}$, the integral representation can serve as a starting point for the computation of the ϵ expansion of the result. Alternatively, the Appell function can be represented as nested sum of depth two and the techniques of symbolic summation discussed above may also be applied to evaluate the ϵ expansion.

5.3.4   More General Integrals

We have seen in Sect. 5.1.7 that for MB integrals that contain the integration variable
also in the form $\Gamma(\ldots \pm az)$ with $a \neq 1$ (as is typical, e.g., for problems with
massive propagators), obtaining analytic solutions by applying Cauchy's theorem
and summing over residues become problematic even for one-dimensional integrals.
Indeed, as demonstrated there, already in such simple cases, we can encounter sums
that we are no longer able to write as S- or Z-sums and thus the systematic way of
obtaining the solution given in Sect. 5.1.4 is no longer applicable.

Here we want to give a brief description of how these issues manifest themselves
in the integration approach to obtaining analytic solutions. To do so, let us consider
the vertex function of Eq. (1.55), which we have also seen in Eq. (5.252):

$$V_{V3l2m}^{\epsilon^{-1}}(s) = -\frac{1}{2s} \int\limits_{-\frac{1}{2}-i\infty}^{-\frac{1}{2}+i\infty} \frac{dz}{2\pi i} (-s)^{-z} \frac{\Gamma^3(-z)\Gamma(1+z)}{\Gamma(-2z)}. \tag{5.252}$$

Noting that the contour runs parallel to the imaginary axis with real part $\Re(z) = -\frac{1}{2}$, we see immediately that the real parts of the arguments of all gamma functions
are positive. Thus, we can express the integral as a convergent Euler-type integral as
described in Sect. 5.3.1. For example, using

$$\frac{\Gamma^2(-z)}{\Gamma(-2z)} = \int_0^\infty dx_1 \, x_2^{-1-z} (1+x_1)^{2z} \tag{5.253}$$

and

$$\Gamma(-z)\Gamma(1+z) = \int_0^\infty dx_2 \, x_2^{-1-z} (1+x_2)^{-1}, \tag{5.254}$$

we find the integral representation

$$V_{V3l2m}^{\epsilon^{-1}}(s) = -\frac{1}{2s} \int_0^\infty dx_1 \int_0^\infty dx_2 \frac{1}{x_1 x_2 (1+x_2)} \delta\left(1 - \frac{(1+x_1)^2}{(-s)x_1 x_2}\right). \tag{5.255}$$

To proceed, we must solve the Dirac delta constraint for one of the variables. Solving
for x_2, we find

$$x_2 = \frac{(1+x_1)^2}{(-s)x_1}, \tag{5.256}$$

and the following integral representation for the vertex function

$$V_{V3l2m}^{\epsilon^{-1}}(s) = \frac{1}{2}\int_0^\infty dx_1 \frac{1}{1 + (2-s)x_1 + x_1^2}. \tag{5.257}$$

For the sake of simplicity, let us assume that we are working in the Euclidean region where $s < 0$ and thus there are no singularities in the integration domain. Examining the above integral, we see the following issue: the denominator is not a product of factors that are linear in the integration variable x_1, so we cannot perform the integration in a straightforward manner in terms of MPLs. However, in this particular case, it is easy to see how to proceed. Let us simply factor the denominator by solving for the roots of the quadratic:

$$1 + (2-s)x_1 + x_1^2 = \left(x_1 - \frac{s-2+\sqrt{(s-4)s}}{2}\right)\left(x_1 - \frac{s-2-\sqrt{(s-4)s}}{2}\right). \tag{5.258}$$

Using this factorization, the integral in Eq. (5.257) can be computed easily after performing the partial fraction decomposition. *Evidently, logarithms of the roots* $\frac{s-2\pm\sqrt{(s-4)s}}{2}$ *will appear in the result.* The final expression can be simplified by introducing once again the conformal variable y, given in Eq. (5.176) (with $m = 1$). In fact, setting $s = -\frac{(1-y)^2}{y}$, the quadratic expression in the denominator of Eq. (5.257) factors immediately

$$1 + (2-s)x_1 + x_1^2 = \frac{(x_1+y)(1+x_1y)}{y}, \tag{5.259}$$

and the integration yields the same result that we found in Eq. (5.179):

$$V_{V3l2m}^{\epsilon^{-1}}(s) = -\frac{y\ln y}{1-y^2}. \tag{5.260}$$

Thus, once more we find that a suitable change of variable allows to compute the solution in terms of MPLs.

This is, however, not true generally. In order to get a feeling for why this is the case, let us return to Eq. (5.255) and solve the Dirac delta constraint for x_1. In this case we find

$$x_1 = \frac{-2 - sx_2 \pm \sqrt{sx_2(4+sx_2)}}{2}. \tag{5.261}$$

Recalling that

$$\delta(f(x)) = \sum_i \frac{1}{|f'(x_{0,i})|}\delta(x - x_{0,i}), \tag{5.262}$$

where $x_{0,i}$ are the zeros of $f(x)$, we find the following integral representation in this case

$$V_{V3l2m}^{\epsilon^{-1}}(s) = \frac{1}{2} \int_{-\frac{4}{s}}^{\infty} dx_2 \, \frac{1}{(1+x_2)\sqrt{sx_2(4+sx_2)}}. \tag{5.263}$$

The nontrivial lower limit of integration appears because the solutions for x_1 in Eq. (5.261) must be non-negative and hence real. But recall that we are working in the Euclidean region where $s < 0$. Thus, the expression under the square root in Eq. (5.261) must be non-negative, which implies $-\frac{4}{s} \le x_2$ (recall that $x_2 > 0$). Now we are faced with the following problem: the integrand includes a polynomial of the integration variable under a square root. Such integrals are genuine generalizations of multiple polylogarithms and in general cannot be evaluated in terms of MPLs. Notice too that introducing the conformal variable y does not solve this problem. Setting $x_2 \to x_2 - \frac{4}{s}$ to transform the limits of integration to zero and infinity and using $s = -\frac{(1-y)^2}{y}$ together with the fact that in the Euclidean region $0 < y < 1$, we find

$$V_{V3l2m}^{\epsilon^{-1}}(s) = \frac{1}{2} \int_0^{\infty} dx_2 \, \frac{y(1-y)}{[(1+y)^2 + (1-y)^2 x_2]\sqrt{x_2(4y + (1-y)^2 x_2)}}. \tag{5.264}$$

Evidently, the integration variable x_2 still appears under the square root. In order to proceed, we must find some transformation of the integration variable which *rationalizes the root*, i.e., a transformation which converts the integrand into a rational function of x_2. There are some systematic procedures known which can be used to search for such transformations, see, e.g., the algorithm of [23], which has also been implemented in the Mathematica package RationalizeRoots. However, in general such transformations are not guaranteed to exist. In our particular case, the transformation

$$x_2 \to -\frac{4y}{(1-y)^2 - 16y^2 t_2^2} \tag{5.265}$$

does the trick. Note that the lower limit of integration after this change of variables becomes $\frac{1-y}{4y}$, while the upper limit is still infinity. Finally, we shift the integration variable, $t_2 \to t_2 + \frac{1-y}{4y}$, in order to set the lower limit of integration to zero and obtain

$$V_{V3l2m}^{\epsilon^{-1}}(s) = \frac{1}{2} \int_0^{\infty} dt_2 \, \frac{2y(1-y)}{[1 - y + 2(1+y)t_2][1 - y + 2y(1+y)t_2]}. \tag{5.266}$$

This final integral is easy to perform and once more gives the result of Eq. (5.260).

However, the fact that we were able to find a transformation which rationalized the integrand is not generic. In fact, we emphasize that in the general case, one can encounter square roots that cannot be rationalized (with a rational transformation). Such integrands lead to genuinely new functions, such as the complete elliptic integrals

$$K(z) = \int_0^1 dt \, \frac{1}{\sqrt{(1 - t^2)(1 - zt^2)}} \quad \text{and} \quad E(z) = \int_0^1 dt \, \frac{\sqrt{1 - zt^2}}{\sqrt{(1 - t^2)}}. \tag{5.267}$$

Moreover, in complicated examples, one typically encounters several different square roots in the integrand. In such cases, the variable transformations must rationalize all square roots that appear together simultaneously. The complete elliptic integrals already showcase this scenario. In fact, we can easily find transformations that rationalize either $\sqrt{1 - t^2}$ or $\sqrt{1 - zt^2}$, but not both roots together. Hence, in the general case, we are left with some square roots and cannot proceed to compute the integral in terms of MPLs. Instead, the set of functions must be enlarged to include these more general integrals. The study of the appropriate generalizations, such as elliptic polylogarithms, is an active area of research, and their discussion is beyond the scope of this book. For a summary of the current state of the art, we refer the reader to [24].

5.4 Approximations

5.4.1 Expansions in the Ratios of Kinematic Parameters

Expansions of the MB integrals depend on the kinematic variables and physical regimes. As an example we start from the integrals where two variables, say m and s, are present, as in Eq. (3.50), which can be cast in the form $ms = -m^2/s$

$$I_1 = \int_{-1/4-i\infty}^{-1/4+i\infty} \frac{dz_1}{2\pi i} \, (ms)^{z_1} \frac{\Gamma(1 - z_1)^2 \Gamma(-z_1) \Gamma(z_1)}{\Gamma(2 - 2z_1)}. \tag{5.268}$$

We can approximate the integral and expand it in the ms parameter by closing the contour on the right half in a complex plane, which implies $s \gg m^2$. This has been done in Sect. 5.1.2, and the Mathematica file with derivations can be found in the file MB_SE212m_Springer.nb in the auxiliary material in [1]. There you can also find how the result can be obtained directly with the package MBasymptotics.m (see Appendix A.1). If another kinematical variable exists, we can work our approximations systematically in regions where the ratios of them are small. For instance, in the case of the scattering processes like Bhabha, we have s and t Mandelstam variables, we can expand in the limit $s, t \gg m^2$ by considering ratios

$$\ln\left(-\frac{m^2}{s}\right) \longrightarrow \ln\left(-\frac{m^2}{t}\right) = \ln\left(-\frac{m^2}{s}\right) - \ln\left(\frac{t}{s}\right), \tag{5.269}$$

and

$$\frac{t}{s} \longrightarrow \frac{s}{t} = 1 / \left(\frac{t}{s} \right). \tag{5.270}$$

The above transformations may imply some algebra in transforming the arguments of the polylogarithms to a unique form or basis (see HPLs Sect. 2.4).

In general, integrals of the form

$$I = (m^2)^{-2\epsilon} \int_{-i\infty}^{i\infty} \frac{dz}{2\pi i} \left(-\frac{m^2}{s} \right)^z f \left(\frac{t}{s}, z \right) \tag{5.271}$$

have been considered in kinematic expansions in [25]. Here the f function contains, among others, a product of gamma or possibly polygamma functions, which have poles in z. The f function is given in general by a multi-dimensional MB integral. To expand such integrals, we take a trick since it is difficult to directly take residues in this form, and we change the order of integration and close the z contour to the right. This procedure is subsequently applied recursively, until no further poles at the required order of expansion occur. We can check resulting approximations numerically with existing MB software (Appendix A.4) and other methods and packages (Appendix A.5).

For more complicated cases, we consider 4PF integrals with three kinematic Mandelstam variables s, t, u, we follow [25].

To summarize, we would like to evaluate MB integrals in Eq. (5.271) using the following series variables

$$x = \frac{t}{s}, \quad L = \ln \left(-\frac{m^2}{s} \right). \tag{5.272}$$

The advantage of such choice is that the results are explicitly real in the Euclidean domain, $s, t < 0$. We should note that in the limit $s, t \gg m^2$ we have $s + t + u = 0$, which simplifies possible analytic solutions of MB integrals involved. In fact, we know already that without this approximation the analytic solutions goes beyond HPLs [26].

That approximations of MB integrals lead to analytic solutions with simpler classes of functions is a typical observation. This fact can be certainly explored systematically in frontier studies of evaluation of multiloop and multileg FI integrals.

Saying that, in the file MB_B74m1_Springer.nb in the auxiliary material in [1], we derive six-dimensional MB representation for the scalar integral given in

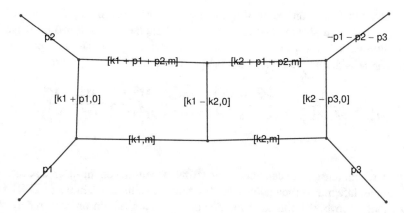

Fig. 5.3 The seven-propagator planar two-loop topology B714m1 with the momentum distribution defined in the file MB_B714m1_Springer.nb (see text for details)

Fig. 5.3, which was studied for the first time in [27]. Also, MBasymptotic.m package is used there to expand the integral in the ms variable. After expansion and additional manipulations, we are left with a bunch of one-dimensional MB integrals which can be solved by one of the summation packages described in Appendix A.2. We leave it as Problem 5.5.

After summation over the x variable in Eq. (5.272), we get [25]

$$
\begin{aligned}
\text{B714m1} = & + \frac{1}{\epsilon^2} \frac{2L^2}{s^2 t} \\
& - \frac{1}{\epsilon} \frac{1}{3s^2 t} \left[-10\, L^3 + 6\, L\, \zeta_2 + 6\, \zeta_3 + 12\, L^2\, \ln(x) \right] \\
& - \frac{1}{6s^2 t} \Big\{ -12\, L^4 + 28\, L^3\, \ln(x) - 4\, L^2\, (-30\, \zeta_2 + 3\, \ln^2(x)) \\
& - 4\, L\, (9\, \zeta_3 + 30\, \zeta_2\, \ln(x) + \ln^3(x) - 18\, \zeta_2\, \ln(1+x) \\
& - 3\, \ln^2(x)\, \ln(1+x) - 6\, \ln(x)\, \text{Li}_2(-x) + 6\, \text{Li}_3(-x)) \\
& - 15\zeta_4 - 24\zeta_3 \ln(x) \Big\} .
\end{aligned}
\tag{5.273}
$$

5.4.2 Taylor and Region Expansions with MB

Another interesting idea to evaluate difficult multi-dimensional MB integrals, which is different from direct kinematic expansions just discussed in the previous section, is to decrease the number of MB variables by direct propagator expansion in the first step in some ratios of the propagator masses. A simple way to do that is to Taylor-expand, as discussed and applied in more general contexts in [28, 29]. In [30] an example has been shown. The idea is simple. Imagine we have two propagators with different masses, M_W and M_Z, say. Then the integrand can be expanded at, e.g., $M_W = M_Z$, and the expanded propagator takes the following form

$$\frac{1}{k^2 - M_W^2} = \frac{1}{k^2 - M_Z^2} + \frac{(M_W^2 - M_Z^2)}{(k^2 - M_Z^2)^2} + \frac{(M_W^2 - M_Z^2)^2}{(k^2 - M_Z^2)^3} + \frac{(M_W^2 - M_Z^2)^3}{(k^2 - M_Z^2)^4} + \dots .$$
(5.274)

As a result, one gets to calculate more terms (depending on the desired accuracy) but simpler integrals (propagators with M_W mass are eliminated in Eq. (5.274)). For more masses involved, this idea can be applied recursively in order to reduce the number of scales in the integral.

In [31] the method of expansions by regions [32] has been merged with the MB method. The contribution from each region is expressed in terms of MB integrals, and the integrals have been solved systematically for the Higgs pair production cross section at two-loop order, in the high energy limit. However, the method can have wider applications. In work, the generalized Barnes lemmas have been considered. In general, using 1BL, 2BL, and generalized Barnes lemmas, four-dimensional MB integrals were reduced up to two-dimensional cases which were solved analytically in terms of generalized hypergeometric functions discussed in Sect. 2.7 or summed (in the case of one dimensional MB integrals). For essential points of the MB techniques used, see [33].

5.4.3 MB: Other Directions

Another strategy useful for getting analytical results for FI expanded in large or small limits has been developed in [34]. The starting point is the Feynman parametrization of Eq. (3.13) rewritten to the form

$$\mathcal{F}(\rho_j) = \int_0^1 d\alpha_1 \int_0^1 d\alpha_2 \cdots \int_0^1 d\alpha_F \frac{N(\alpha_i)}{\left[\sum_j D_j(\alpha_i)\rho_j\right]^{n+(p/q)\epsilon}},$$
(5.275)

where the ρ_j denote the scalar products of the external momenta and squared masses normalized to a fixed mass scale in the given Feynman diagram, so that $\rho_0 = 1$, and $n + (p/q)\epsilon > 0$ with n, p, and q some positive integers, in general. For instance, in the simplest case, there can be only one ρ–parameter $\sum_j D_j(\alpha_i)\rho_j = D_0(\alpha_i) + D_1(\alpha_i)\rho$, and the behavior of the integral $\mathcal{F}(\rho)$ for $\rho \ll 1$ can be examined (for details see [34] and related works by the authors). Using the basic MB relation, we get the relation between the ρ real space and the complex Mellin s-plane

$$\mathcal{F}(\rho) = \frac{1}{2\pi i} \int\limits_{c-i\infty}^{c+i\infty} ds \, \rho^{-s} \, M[\mathcal{F}](s), \tag{5.276}$$

$$M[\mathcal{F}](s) = \int_0^1 d\alpha_1 \int_0^1 d\alpha_2 \cdots \int_0^1 d\alpha_F \frac{N(\alpha_i)}{[D_0(\alpha_i)]^\nu} \left(\frac{D_0(\alpha_i)}{D_1(\alpha_i)}\right)^s \frac{\Gamma(s)\Gamma(\nu - s)}{\Gamma(\nu)}.$$

The relation in Eq. (5.276) constitutes the so-called Mellin transform. *The asymptotic behaviors of $\mathcal{F}(\rho)$, both for $\rho \ll 1$ and $\rho \gg 1$, are encoded in the so-called converse mapping theorem [35] which establishes a relation between the singularities in the Mellin s–plane and the asymptotic behavior(s) one is looking for.* The method has been applied to the calculation of the vacuum polarization contributions to the muon anomaly in [34].

The problem of the convergence of MB sums is difficult, either in asymptotic or full analytic form [36, 37]. The public programs are gathered in Appendix A.2.

In this chapter we discussed possible ways in which MB representations are used in analytic studies of FI. Certainly the subject is not "out of print." For instance, as shown in [38], Mellin-Barnes representations can be used to derive linear systems of homogeneous differential equations for the original Feynman integrals with arbitrary powers of propagators, without the need for IBP relations. This method in addition can be used to deduce extra relations between master integrals, beyond the IBP reduction [39–41].

Problems

Problem 5.1 Discuss the behavior of the term $\Gamma(a + z)\Gamma(b - z)x^z$ in Eq. (5.5) for $z \to \infty$.
Hint: Notice that the large-z behavior is controlled by the exponential function $e^{z \ln x}$, since this grows or vanishes faster than any power-law function of z. Thus, the limit at $z \to \infty$ is zero or infinity, depending on the sign of the real part of the product $z \ln x$.

Problem 5.2 Derive Barnes' first lemma in Eq. (2.179).
Hint: Consider

$$I = \int_{-i\infty}^{+i\infty} \frac{dz}{2\pi i} \Gamma(a + z)\Gamma(b + z)\Gamma(c - z)\Gamma(d - z) \tag{5.277}$$

and use the methods discussed above to derive a real integral representation for this onefold Mellin-Barnes integral. You should end up with a one-dimensional real integral that can be immediately performed in terms of a beta function. See also discussion in [42].

Problem 5.3 Show that

$$I = \int_{-i\infty}^{+i\infty} \frac{dz}{2\pi i} \frac{\Gamma(a+z)\Gamma(b+z)\Gamma(-z)}{\Gamma(c+z)} x^z \tag{5.278}$$

can be evaluated in terms of Gauss' hypergeometric function $_2F_1$. Note that $_2F_1(a, b; c; x)$ has the following standard integral representation

$$_2F_1(a, b; c; x) = \frac{\Gamma(c)}{\Gamma(b)\Gamma(b-c)} \int_0^1 dt \, t^{b-1}(1-t)^{c-b-1}(1-tx)^{-a} . \tag{5.279}$$

Hint: See discussion in Sect. 2.7, Eq. (2.180), and Problem 2.10.

Problem 5.4 Analyze construction and expansion of the MB integral given in Fig. 5.3, which is given in the file MB_B714m1_Springer.nb in the auxiliary material in [1]. Note that the original MB integral is six-dimensional; after analytic expansion in ϵ up to the ϵ^0 term it is four-dimensional; and after ms expansion (2BL must be applied afterward) the final result is a set of zero- and one-dimensional MB integrals.

Problem 5.5 Taking the set of MB integrals obtained in Problem 5.4, which are expanded in the ms parameter, derive the analytic result and check it against Eq. (5.273).
Hint: Use, e.g., MBsums.m and one of the packages for summing series, e.g., Xsummer (see Appendix A.2).

References

1. https://github.com/idubovyk/mbspringer, http://jgluza.us.edu.pl/mbspringer.
2. J. Gluza, T. Jelinski, D.A. Kosower, Efficient evaluation of massive Mellin-Barnes integrals. Phys. Rev. **D95**(7), 076016 (2017). arXiv:1609.09111, https://doi.org/10.1103/PhysRevD.95.076016
3. B. Ananthanarayan, S. Banik, S. Friot, S. Ghosh, Multiple series representations of N-fold Mellin-Barnes integrals. Phys. Rev. Lett. **127**(15), 151601 (2021). arXiv:2012.15108, https://doi.org/10.1103/PhysRevLett.127.151601
4. S. Moch, P. Uwer, S. Weinzierl, Nested sums, expansion of transcendental functions and multiscale multiloop integrals. J. Math. Phys. **43**, 3363–3386 (2002). arXiv:hep-ph/0110083, https://doi.org/10.1063/1.1471366
5. L. Lewin, *Polylogarithms and Associated Functions* (North-Holland, Amsterdam, 1981)

6. R. Kellerhals, Volumesinhyperbolic5-space. Geometric Funct. Anal. **5**(4), 640–667 (1995). http://eudml.org/doc/58206

7. A. Goncharov, Volumes of hyperbolic manifolds and mixed Tate motivesarXiv:alg-geom/9601021

8. S. Moch, P. Uwer, XSummer: Transcendental functions and symbolic summation in form. Comput. Phys. Commun. **174**, 759–770 (2006). arXiv:math-ph/0508008

9. T. Huber, D. Maitre, HypExp: A mathematica package for expanding hypergeometric functions around integer-valued parameters. Comput. Phys. Commun. **175**, 122–144 (2006). arXiv:hep-ph/0507094, https://doi.org/10.1016/j.cpc.2006.01.007

10. T. Huber, D. Maitre, HypExp 2, expanding hypergeometric functions about half-integer parameters. Comput. Phys. Commun. **178**, 755–776 (2008). arXiv:0708.2443, https://doi.org/10.1016/j.cpc.2007.12.008

11. A.I. Davydychev, M.Y. Kalmykov, Massive Feynman diagrams and inverse binomial sums. Nucl. Phys. B **699**, 3–64 (2004). arXiv:hep-th/0303162, https://doi.org/10.1016/j.nuclphysb.2004.08.020

12. A. Kotikov, Differential equations method: New technique for massive Feynman diagrams calculation. Phys. Lett. **B254**, 158–164 (1991). https://doi.org/10.1016/0370-2693(91)90413-K

13. A. Kotikov, Differential equations method: The Calculation of vertex type Feynman diagrams. Phys. Lett. **B259**, 314–322 (1991)

14. A.V. Kotikov, Differential equation method: The Calculation of N point Feynman diagrams. Phys. Lett. **B267**, 123–127 (1991)

15. E. Remiddi, Differential equations for Feynman graph amplitudes. Nuovo Cim. **A110**, 1435–1452 (1997). arXiv:hep-th/9711188

16. J.M. Henn, Multiloop integrals in dimensional regularization made simple. Phys. Rev. Lett. **110**(25), 251601 (2013). arXiv:1304.1806, https://doi.org/10.1103/PhysRevLett.110.251601

17. M. Czakon, J. Gluza, T. Riemann, Master integrals for massive two-loop Bhabha scattering in QED. Phys. Rev. **D71**, 073009 (2005). arXiv:hep-ph/0412164

18. M. Czakon, Automatized analytic continuation of Mellin-Barnes integrals. Comput. Phys. Commun. **175**, 559–571 (2006). arXiv:hep-ph/0511200, https://doi.org/10.1016/j.cpc.2006.07.002

19. G. Heinrich, Sector decomposition. Int. J. Mod. Phys. A **23**, 1457–1486 (2008). arXiv:0803.4177, https://doi.org/10.1142/S0217751X08040263

20. E. Panzer, Feynman integrals and hyperlogarithms. Ph.D. thesis, Humboldt U. (2015). arXiv:1506.07243, https://doi.org/10.18452/17157

21. F. Brown, The Massless higher-loop two-point function. Commun. Math. Phys. **287**, 925–958 (2009). arXiv:0804.1660, https://doi.org/10.1007/s00220-009-0740-5

22. C. Anastasiou, C. Duhr, F. Dulat, B. Mistlberger, Soft triple-real radiation for Higgs production at N3LO. JHEP **07**, 003 (2013). arXiv:1302.4379, https://doi.org/10.1007/JHEP07(2013)003

23. M. Besier, P. Wasser, S. Weinzierl, RationalizeRoots: Software package for the rationalization of square roots. Comput. Phys. Commun. **253**, 107197 (2020). arXiv:1910.13251, https://doi.org/10.1016/j.cpc.2020.107197

24. J.L. Bourjaily, et al., Functions beyond multiple polylogarithms for precision collider physics (2022). arXiv:2203.07088

25. M. Czakon, J. Gluza, T. Riemann, The planar four-point master integrals for massive two-loop Bhabha scattering. Nucl. Phys. B **751**, 1–17 (2006). arXiv:hep-ph/0604101, https://doi.org/10.1016/j.nuclphysb.2006.05.033

26. J.M. Henn, V.A. Smirnov, Analytic results for two-loop master integrals for Bhabha scattering I. JHEP **1311**, 041 (2013). arXiv:1307.4083, https://doi.org/10.1007/JHEP11(2013)041

27. V.A. Smirnov, Analytical result for dimensionally regularized massive on-shell planar double box. Phys. Lett. B **524**, 129–136 (2002). arXiv:hep-ph/0111160, https://doi.org/10.1016/S0370-2693(01)01382-X

28. V.A. Smirnov, Asymptotic expansions in momenta and masses and calculation of Feynman diagrams. Mod. Phys. Lett. A **10**, 1485–1500 (1995). arXiv:hep-th/9412063, https://doi.org/10.1142/S0217732395001617

29. M. Misiak, M. Steinhauser, Three loop matching of the dipole operators for $b \rightarrow s\gamma$ and $b \rightarrow sg$. Nucl. Phys. B **683**, 277–305 (2004). arXiv:hep-ph/0401041, https://doi.org/10.1016/j.nuclphysb.2004.02.006

30. I. Dubovyk, J. Usovitsch, K. Grzanka, Toward three-loop Feynman massive diagram calculations. Symmetry **13**(6), 975 (2021). https://doi.org/10.3390/sym13060975

31. G. Mishima, High-energy expansion of two-loop massive four-point diagrams. JHEP **02**, 080 (2019). arXiv:1812.04373, https://doi.org/10.1007/JHEP02(2019)080

32. M. Beneke, V.A. Smirnov, Asymptotic expansion of Feynman integrals near threshold. Nucl. Phys. **B522**, 321–344 (1998). arXiv:hep-ph/9711391, https://doi.org/10.1016/S0550-3213(98)00138-2

33. J. Davies, G. Mishima, M. Steinhauser, D. Wellmann, Double-Higgs boson production in the high-energy limit: planar master integrals. JHEP **03**, 048 (2018). arXiv:1801.09696, https://doi.org/10.1007/JHEP03(2018)048

34. S. Friot, D. Greynat, E. De Rafael, Asymptotics of Feynman diagrams and the Mellin-Barnes representation. Phys. Lett. B **628**, 73–84 (2005). arXiv:hep-ph/0505038, https://doi.org/10.1016/j.physletb.2005.08.126

35. P. Flajolet, X. Gourdon, P. Dumas, Mellin transforms and asymptotics: Harmonic sums. Theoret. Comput. Sci. **144**(1-2), 3–58 (1995)

36. S. Friot, D. Greynat, On convergent series representations of Mellin-Barnes integrals. J. Math. Phys. **53**, 023508 (2012). arXiv:1107.0328, https://doi.org/10.1063/1.3679686

37. J. Blumlein, I. Dubovyk, J. Gluza, M. Ochman, C.G. Raab, T. Riemann, C. Schneider, Nonplanar Feynman integrals, Mellin-Barnes representations, multiple sums. PoS **LL2014**, 052 (2014). arXiv:1407.7832

38. M.Y. Kalmykov, B.A. Kniehl, Mellin-Barnes representations of Feynman diagrams, linear systems of differential equations, and polynomial solutions. Phys. Lett. B **714**, 103–109 (2012). arXiv:1205.1697, https://doi.org/10.1016/j.physletb.2012.06.045

39. M.Y. Kalmykov, B.A. Kniehl, Counting master integrals: Integration by parts versus differential reduction. Phys. Lett. B **702**, 268–271 (2011). arXiv:1105.5319, https://doi.org/10.1016/j.physletb.2011.06.094

40. M.Y. Kalmykov, B.A. Kniehl, Counting the number of master integrals for sunrise diagrams via the Mellin-Barnes representation. JHEP **07**, 031 (2017). arXiv:1612.06637, https://doi.org/10.1007/JHEP07(2017)031

41. T. Bitoun, C. Bogner, R.P. Klausen, E. Panzer, Feynman integral relations from parametric annihilators. Lett. Math. Phys. **109**(3), 497–564 (2019). arXiv:1712.09215, https://doi.org/10.1007/s11005-018-1114-8

42. B. Jantzen, New proofs for the two Barnes lemmas and an additional lemma. J. Math. Phys. **54**, 012304 (2013). arXiv:1211.2637, https://doi.org/10.1063/1.4775770

MB Numerical Methods

6

Abstract

We discuss the main issues which lead to numerical instabilities of multiloop MB integrals in the Minkowskian region. We then present practical procedures for overcoming the obstacles such as the transformation of variables to finite intervals, the shifting and deformation of integration contours, and the construction of MB representations that take kinematic thresholds into account. Steepest descent and Lefschetz thimbles are applied to find optimal integration contours. The numerical evaluation of phase space MB integrals is also briefly discussed.

6.1 Introduction

Since the very beginning, the computer language FORTRAN (1957, John Backus, IBM) has been used for numeric scientific computing. For the purpose of algebraic evaluations, a special software Reduce (started in 1963 by Anthony Hearn), Schoonschip (started in 1967 by Martinus J. G. Veltman), and its descendant, the Form package (initially released in 1989 by Jos Vermaseren), has been also developed. Nowadays Form, C++, Mathematica and python are the main environments used in particle physics. More on the history of numerical and algebra systems and software development can be read in [1].

In the 1980s, many multiloop methods to calculate higher-order corrections in particle physics have been developed, based on generalized unitarity, tree-duality, simultaneous numerical integration of amplitudes over the phase space and the loop momentum, reductions of the integrals at the integrand level, improved diagrammatic approach and recursion relations applied to higher-rank tensor integrals, contour deformations, expansions by regions, sector decomposition, dispersion relations, differential equations, summation of series, and Mellin-Barnes representations. Many methods aim at direct Feynman integral calculations. For general

© The Author(s), under exclusive license to Springer Nature Switzerland AG 2022
I. Dubovyk et al., *Mellin-Barnes Integrals*, Lecture Notes in Physics 1008,
https://doi.org/10.1007/978-3-031-14272-7_6

Table 6.1 Strong and weak points of analytic versus numerical evaluation of loop integrals as discussed in [8]

	Analytic	Numerical
Pole cancellation	Exact	With numerical uncertainty
Control of integrable singularities	Analytic continuation	Less straightforward
Fast and stable evaluation	Yes (mostly)	Depends
Extension to more scales/loops	Difficult	Promising
Automation	Difficult	Less difficult

reviews see [2–6]. Due to experimental requirements defined by precision physics at present and future colliders, notably HL-LHC and Tera-Z physics at FCC-ee, there is a large activity in the field, and the methods are in permanent development [7, 8]. See also a recent workshop at CERN on the subject (June 2022) [9].

To solve Feynman integrals, analytical methods can be used, though they exhibit natural limitations when sophisticated integrals with many parameters appear. Some analytical methods connected with MB integrals have been discussed in previous sections, and some useful software is summarized in the Appendix. Thus, with going to higher and higher loop levels and growing complexity (number of parameters, dimensionality of integrals), it is natural that numerical methods become more and more relevant. The trade-off between analytical and numerical or semi-numerical methods has been nicely summarized in [8] (see Table 6.1).

So far a lot has been done for precise analytical and numerical calculations at the one-loop order (which is also called "next to leading," NLO). Many NLO programs allow to consider processes automatically and in a numerical way (see, for instance, [8, 10]). Going beyond the one-loop calculations in Minkowskian regions, the situation is quite different. There are technical obstructions in the calculation of FI in this setting due to threshold effects, singularities, on-shellness, or several mass parameters involved. With the increasing number of loops and increasing number of mass and momentum scales, it becomes more difficult to compute these corrections analytically or even semi-analytically. In 2014 the only advanced automatic numerical two-loop method in Minkowskian kinematics was SD [11, 12], where a complex contour deformation of the Feynman parameter integrals is implemented in the publicly available numerical packages FIESTA 3 [13] (since 2013) and SecDec 2 [14] (since 2012), followed by pySecDec [15].

Evaluation of the multiloop, multiscale integrals is in general very challenging. For instance, for the two-loop Z-decay (representative Feynman diagram is shown in Fig. 6.1), the typical evaluation problems which we meet [16] are connected with (i) up to four dimensionless scales at $s = M_Z^2$ with a variance of masses M_Z, M_W, m_t, M_H involved and (ii) intricate threshold and on-shell effects. To tackle these problems, the semi-numerical approach to the calculation of Feynman integrals based on Mellin-Barnes representations has been developed. In calculations the MB method is used with many suitable packages gathered in what we call the MB-suite. An important part of the MB-suite is the AMBRE project (see

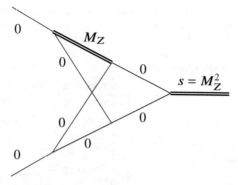

Fig. 6.1 The non-planar vertex diagram V611m which is present, for instance, in a calculation of the Z-decay process to two massless fermions. It is a difficult case for the SD method due to a single massive propagator, whose mass M_Z coincides with external invariant energy. On the other hand, it can be nicely integrated with high accuracy with the MB method (see Table 6.2). This MB integral has been discussed initially in [29]

Chap. 3). It includes among others the construction of MB representations [17–20]) and the recognition of planarity of Feynman diagrams [21, 22] with the corresponding packages AMBRE.m and PlanarityTest.m. For the extraction of ϵ-singularities in dimensional regularization of multiloop integrals, the MB.m [23] and MBresolve.m [24] packages are used (see Chap. 4). As they offer the possibility of numerical integrations in Euclidean kinematics, they are used also as a numerical cross-check of analytical results for multiloop integrals. Working on the completion of two-loop corrections to the Z-boson decay, it has been observed that serious convergence problems for some classes of integrals both with SecDec and FIESTA appear and the new numerical Mellin-Barnes (MB) approach to the calculation of multiloop and massive Feynman integrals is used, with the MB package MBnumerics.m [25, 26] (see Section E6 in [7] and [27]). This package is able to evaluate MB integrals directly in Minkowskian kinematics, solving so far untouchable regions of precision physics. Most of the mentioned packages can be found at web pages [20, 28] (see also Appendix A).

In Table 6.2 we show the problem that meets the SD method and the laborious progress made in recent years to overcome it. It appears that for such cases where most of the propagators are massless, the MB method may serve as a complementary tool. In this way, using SD and MB approaches, the calculation of the two-loop SM FI has been completed [16, 30, 31].

The analytical result can be found in [32]. The MB running file MB_V611m_Springer.sh for the scalar integral which corresponds to Fig. 6.1 evaluated with MBnumerics.m is available at [33]. Similarly configuration files SD_V611m_Springer_generate.py and SD_V611m_Springer_integrate.py for the pySecDec evaluation are given in [33]. The discussion of the integral by the SecDec team (2017) can be

Table 6.2 Minkowskian results for the diagram given in Fig. 6.1, for the constant part in ϵ. As discussed in [16] (see also Sect. 2.2), the integral V611m which corresponds to the Feynman diagram in Fig. 6.1 is evaluated at $s = M_Z^2 \equiv 1 + i\delta'$ where δ' is a small parameter. In numerical evaluations $\delta' = i \cdot 10^{-7}$. To get the correct sign of the imaginary part of FI in calculations made in this chapter, the same $i\delta'$ prescription is assumed everywhere

Analytical	$-\mathbf{0.77859960897968} - \mathbf{4.12351259339631} \cdot i$
MBnumerics	$-\mathbf{0.77859960}832476 - \mathbf{4.12351260}051601 \cdot i$
MB+thresholds	$-\mathbf{0.77854}282426410 - \mathbf{4.12349}826423109 \cdot i$
SecDec	big error [2016]
	$-\mathbf{0.77} - i \cdot \mathbf{4.1}$ [2017]
	$-\mathbf{0.778} - i \cdot \mathbf{4.123}$ [2019]
pySecDec+rescaling	$-\mathbf{0.778598} - i \cdot \mathbf{4.123512}$ [2020]

found in [15]. In Table 6.2 the new pySecDec result obtained with the so-called rescaling is also given.

The Idea of Rescaling

Let us briefly discuss the rescaling of Feynman parameters in the context of numerical integration with the SD approach, though the method can be used elsewhere. The basic idea is that we can rescale each Feynman parameter by an arbitrary factor $x_i \to \kappa_i x_i, \kappa > 0$ resulting in a new integral $I = \prod \kappa_i^{n_i} I'$ where each term in U and F polynomials will get new coefficients. The transformation is possible due to the Cheng-Wu theorem discussed in Chap. 3. One can sequentially extract Feynman parameters from the common delta function and integrate them from 0 to ∞. Rescaling, in this case, doesn't change integration boundaries and doesn't affect the delta function. After this manipulation, one can reverse the theorem and put the corresponding parameter back into the delta function, making changes only in coefficients of graph polynomials. Starting from Eq. (3.13) and omitting the coefficient in front of the integral, it works as follows

$$\int_0^1 \prod_{j=1}^N dx_j \, x_j^{n_j-1} \delta\left(1 - \sum_{i=1}^N x_i\right) \frac{U(\mathbf{x})^{N_\nu - d(L+1)/2}}{F(\mathbf{x})^{N_\nu - dL/2}}$$

$$= \kappa_{i_1}^{n_{i_1}} \int_0^\infty dx_{i_1} \int_0^1 \prod_{j\neq i_1}^N dx_j \prod_{j=1}^N x_j^{n_j-1} \delta\left(1 - \sum_{i\neq i_1}^N x_i\right) \frac{U(\ldots, \kappa_{i_1} x_{i_1}, \ldots)^{N_\nu - d(L+1)/2}}{F(\ldots, \kappa_{i_1} x_{i_1}, \ldots)^{N_\nu - dL/2}}$$

$$= \prod_{j=1}^N \kappa_i^{n_i} \int_0^1 \prod_{j=1}^N dx_j \, x_j^{n_j-1} \delta\left(1 - \sum_{i=1}^N x_i\right) \frac{U(\mathbf{x}, \boldsymbol{\kappa})^{N_\nu - d(L+1)/2}}{F(\mathbf{x}, \boldsymbol{\kappa})^{N_\nu - dL/2}}. \qquad (6.1)$$

After rescaling, one can proceed further in the usual way with the SD algorithm.

In the SD method, a deformation of integration contours is performed to avoid a problem with the so-called Landau singularity for physical kinematic points. Integrations start at 0 and end at 1 but go in a complex plane. In the case shown in Table 6.2, we do the calculation at a specific point for which coefficients in the F polynomial are equal to 1 or -1. In such a situation, the Landau singularity is located at some corner of the integration domain and cannot be fixed by the contour's transformation. The main purpose of rescaling, in this case, is to move the singularity to the interior of the integration domain. In principle, rescaling may also improve the behavior of the integrand and increase the accuracy of results.

At the end of this introductory section, we should acknowledge that very recently numerical calculations have been pushed forward independently by DEs [34–37]. For other exploratory methods in FI computation, see a review [8]. Now we will consider in more detail the main problems and features of the MB numerical approach to evaluating the FI.

6.2 MB Numerical Evaluation Using Bromwich Contours

We begin from a discussion of numerical integrations over the Bromwich contour introduced in Sect. 2.6.

6.2.1 Straight-Line Contours and Their Limitations

As shown in Sect. 4, the general form of the MB representation in Eq. (3.34) is well-defined if real parts of all gamma functions are positive (equivalent to the separation of all right poles of gamma functions from all left ones). That is accomplishes by the appropriate choice of points where integration contours cross the real axis. The final form of MB integrals suited to numerical integration is the following:

$$I = \frac{1}{(2\pi i)^r} \int_{-i\infty+z_{10}}^{+i\infty+z_{10}} \cdots \int_{-i\infty+z_{r0}}^{+i\infty+z_{r0}} \prod_i^r dz_i \, \mathbf{F}(Z, S) \frac{\prod_{j=1}^{N_n} \Gamma(\Lambda_j)}{\prod_{k=1}^{N_d} \Gamma(\Lambda_k)} f_\psi(Z). \quad (6.2)$$

Notations are the same as in Eq. (3.34), but now it doesn't depend on ϵ. The positions of contours are fixed by z_{i0}. The part $f_\psi(Z)$ may depend on polygamma functions and constants like Euler's constant γ, or it is equal to 1 if the corresponding Feynman integral has no ϵ poles.

To understand the problems appearing during numerical integrations of MB integrals, one has first to study the asymptotic behavior of integrands. The main building blocks of MB integrals are gamma and polygamma functions which have the following asymptotics for large arguments $|z| \to \infty$

$$\Gamma(z)_{|z|\to\infty} = \sqrt{2\pi} e^{-z} z^{z-\frac{1}{2}} \left[1 + \frac{1}{12z} + \frac{1}{288z^2} + \dots \right], \tag{6.3}$$

$$\psi(z)_{|z|\to\infty} = \ln z - \frac{1}{2z} - \frac{1}{12z^2} + \dots$$

$$\psi'(z)_{|z|\to\infty} = \frac{1}{z} + \frac{1}{2z^2} + \frac{1}{6z^3} + \dots$$

$$\psi''(z)_{|z|\to\infty} = -\frac{1}{z^2} - \frac{1}{z^3} + \dots \tag{6.4}$$

$$\psi^{(3)}(z)_{|z|\to\infty} = \frac{2}{z^3} + \dots$$

$$\dots \ .$$

We are interested only in the leading term of the expansion for large $|z|$. It is evident that the polygamma functions, as well as the linear fractional part of the gamma function asymptotics, do not contribute, and we are looking only at the $e^{-z} z^{z-\frac{1}{2}}$ part of the Stirling formula in Eq. (6.3). It is easy to see that e^{-z} parts from different functions cancel each other. That is a general property of MB representations and extends to any multi-dimensional integral. Based on that the asymptotic behavior of MB integrals is determined by the $z^{z-\frac{1}{2}} = e^{(z-\frac{1}{2})\ln z}$ part of Eq. (6.3).

Let's now consider a simple one-dimensional example

$$I_{5,\epsilon^{-2}}^{0h0w} = \frac{1}{2s} \frac{1}{2\pi i} \int\limits_{-i\infty-\frac{1}{2}}^{+i\infty-\frac{1}{2}} dz \left(\frac{M_Z^2}{-s} \right)^z \frac{\Gamma^3(-z)\Gamma(1+z)}{\Gamma^2(1-z)}, \tag{6.5}$$

where the ratio of gamma functions, which we will call a *core* of the MB representation, in the limit $|z| \to \infty$ are

$$\frac{\Gamma^3(-z)\Gamma(1+z)}{\Gamma^2(1-z)} \xrightarrow{|z|\to\infty} e^{z(\ln z - \ln(-z)) + \frac{1}{2}\ln z - \frac{5}{2}\ln(-z)}. \tag{6.6}$$

Here one should notice that independently of the contour of integration, the following relation holds: $\ln z - \ln(-z) = i\pi \, \mathrm{sign}(\Im(z))$ (see discussion in Sect. 2.2 and Eq. (2.27)).

In case of practical applications, the integration contour is the Bromwich contour (a straight line parallel to the imaginary axis), so $z = z_0 + it$, $t \in (-\infty, \infty)$ and the *core* of the MB integral in Eq. (6.5) in the limit $|z| \to \infty \Leftrightarrow t \to \pm\infty$ is

$$\frac{\Gamma^3(-z)\Gamma(1+z)}{\Gamma^2(1-z)} \longrightarrow e^{-\pi|t|}\frac{1}{|t|^2}. \tag{6.7}$$

It is a well-behaving, non-oscillating function.

We can extend our reasoning to any multi-dimensional integral. The asymptotic of the representation core in generalized spherical coordinates has the following form

$$\frac{\prod_j \Gamma(\Lambda_j)}{\prod_k \Gamma(\Lambda_k)} \xrightarrow[|z_i| \to \infty]{r \to \infty} \frac{e^{-\beta r}}{r^\alpha}, \quad \beta = \beta(\theta) \geq \pi, \quad \alpha = \alpha(z_{i0}) \tag{6.8}$$

where the coefficient β depends on direction and the expression is valid only for contours parallel to the imaginary axis.

Now let's look at the kinematic term $\left(\frac{M_Z^2}{-s}\right)^z$. In the Euclidean case $s < 0$, this factor gives oscillations which are well damped by the factor $e^{-\pi|t|}$. In the Minkowskian case $s \to s + i\delta$ ($s > 0$), where $i\delta$ comes from the propagator definition and is needed to choose the proper branch of the logarithm:

$$\left(\frac{M_Z^2}{-s}\right)^z = e^{z\ln\left(-\frac{M_Z^2}{s}+i\delta\right)} \longrightarrow e^{it\ln\frac{M_Z^2}{s}}e^{-\pi t}, s > 0. \tag{6.9}$$

As one can see, the *core* factor $e^{-\pi|t|}$ cancels with $e^{-\pi t}$ only when $t \to -\infty$ and oscillations are not damped in general. Moreover, it may happen that an exponent α in the fractional part $\frac{1}{|t|^\alpha}$ of the limit in Eq. (6.7) (in our example $\alpha = 2$) is not large enough to guarantee the convergence of the integral (see, e.g., Sec. (3.4) in [23]). That is the main problem of numerical integration of MB integrals for physical kinematics.

One of the ways to avoid the described problem is to perform a deformation of the integration path in a way that the overall exponential damping factor is restored. This will be discussed in Sect. 6.2.3.

6.2.2 Transforming Variables to the Finite Integration Range

In practice, the integration over infinite intervals requires their transformation into finite ones (e.g., in CUBA numerical library [38], it is the interval [0, 1]). In the package MB.m this transformation is done in the following way:

$$t_i \to \ln\left(\frac{x_i}{1-x_i}\right), \quad dt_i \to \frac{dx_i}{x_i(1-x_i)}. \tag{6.10}$$

In case of the example in Eq. (6.5), the limit $t \to -\infty$ is equivalent to $x \to 0$, and in this limit the integrand behaves like

$$\frac{1}{x \, \ln^2 x} \xrightarrow{x \to 0} \infty. \tag{6.11}$$

This singularity is integrable but prevents reaching a high accuracy result. As an alternative one can transform the integration interval $(-\infty, \infty)$ into $[0, 1]$ differently:

$$t_i \to \tan\left(\pi\left(x_i - \frac{1}{2}\right)\right), \quad dt_i \to \frac{\pi \, dx_i}{\cos^2\left(\pi\left(x_i - \frac{1}{2}\right)\right)}. \tag{6.12}$$

The corresponding limit now is

$$\frac{1}{\sin^2\left(\pi\left(x_i - \frac{1}{2}\right)\right)} \xrightarrow{x \to 0} 1, \tag{6.13}$$

and the integration can be easily performed. One should stress that with the new type of transformation imaginary parts of arguments of gamma functions grow much faster than with ln-type transformation. At some moment gamma functions in the denominator become equal to 0, numerically. To avoid this problem, we compute the *core* of MB integral in the following way:

$$\frac{\prod\limits_{j=1}^{N_n} \Gamma(\Lambda_j)}{\prod\limits_{k=1}^{N_d} \Gamma(\Lambda_k)} = \mathrm{Exp}\left(\sum_{j=1}^{N_n} \ln\Gamma(\Lambda_j) - \sum_{k=1}^{N_d} \ln\Gamma(\Lambda_k)\right), \tag{6.14}$$

where $\ln\Gamma$ denotes the log-gamma function (see, e.g., `CernLib` documentation and Appendix A.4).

Let's consider a three-dimensional MB integral

$$I_{2,11}^{0h0w} = \frac{1}{s^2} \frac{1}{(2\pi i)^3} \int\limits_{-i\infty-\frac{47}{37}}^{i\infty-\frac{47}{37}} dz_1 \int\limits_{-i\infty-\frac{139}{94}}^{i\infty-\frac{139}{94}} dz_2 \int\limits_{-i\infty-\frac{176}{235}}^{i\infty-\frac{176}{235}} dz_3 \left(-\frac{s}{m^2}\right)^{-z_1} \Gamma(-1-z_1)$$

$$\Gamma(2+z_1)\Gamma(-1-z_2)\Gamma(z_1-z_2)\Gamma(1+z_2-z_3)^2\Gamma(-z_3)\Gamma(1+z_3)$$

$$\Gamma(-z_1+z_3)^2\Gamma(-z_2+z_3)/\Gamma(-z_1)\Gamma(1+z_1-z_2)\Gamma(1-z_1+z_3). \tag{6.15}$$

Table 6.3 Numerical results for the integral Eq. (6.15) for $s = m^2 = 1$. AB - analytical solution [40]. MB1 to MB5—numerical integration of the MB integrals with different integration routines and transformations of the infinite integration region as described in the text

AB	-1.199526183135	$+5.567365907880i$	
MB1	-1.199525259137	$+5.567367419371i$	Cuhre, 10^7, 10^{-8}
MB2	-1.199526183168	$+5.567365907904i$	Cuhre, 10^7, 10^{-8}
MB3	-1.204597845834	$+5.567518701898i$	Vegas, 10^7, 10^{-3}
MB4	-1.199516455248	$+5.567376681167i$	QMC, 10^6, 10^{-5}
MB5	-1.199527580305	$+5.567367345229i$	QMC, 10^7, 10^{-6}

Its derivation, corresponding diagram and files needed for calculations are in the file MB_3dimNum_Springer.nb in the auxiliary material in [33].

The integral in Eq. (6.15) has the cancellation of the overall damping factor along the $z_1(t_1)$-axis ($t_1 = t$, $t_2 = t_3 = 0$) or in spherical coordinates along the direction $\theta = (\theta = \pi/2, \phi = 0)$. Numerical results for this integral obtained with different combinations of transformations in Eqs. (6.10) and (6.12) are compared with an analytical solution in Table 6.3. This example is taken from [39].

In the table, the label MB1 corresponds to the numerical integration of Eq. (6.15), where the mapping into the integration interval [0, 1] is done by the tan-type of transformation of Eq. (6.12) for all variables. MB2 denotes a tan-mapping for t_1 and ln-mapping for the remaining variables. Integrations are done by the CUHRE routine of the CUBA library. The maximum number of integrand evaluations allowed was set to 10^7. The absolute error reported by the routine is at the level of 10^{-8}. In MB3 the integration is done by the VEGAS routine [41,42] and the ln-type of transformation is used for all variables. Corresponding error estimation is of the order of $\sim 10^{-3}$. The last two rows MB4 and MB5 show results for the numerical integration of Eq. (6.15) and tan-mapping for all variables with the quasi-Monte Carlo library QMC [43]. Numbers in the last column give the maximum number of integrand evaluations and the absolute error.

Integration of MB integrals in the Minkowskian region with only logarithmic mapping and a deterministic algorithm implemented in CUHRE leads to a NaN result. However, the Monte Carlo algorithm in VEGAS can handle integrable singularities and give a few correct digits for relatively simple integrals. As one can see from Table 6.3, the highest real accuracy was obtained in the case MB2. Here one should point out that for both results MB1 and MB2 an absolute error returned by the integration routine is at the same level $\sim 10^{-8}$. *This shows that the direct integration of MB integrals is possible but limited by several factors.* First, the best accuracy is achieved when the cancellation of the exponential damping factor happens along with one of the axes in the integration space. In the case of integrals with more than one scale, a cancellation takes place in multiple directions or in some sector of the integration space, and getting a high accuracy result becomes much more complicated. Second, the exponent α in $\frac{1}{|t|^\alpha}$ in the asymptotical expansion along the direction of the damping factor cancellation may not be big enough for a good convergence. In a multi-dimensional case in contrast to the one-dimensional case, this problem can be solved by shifts of integration contours $z_i \to z_{i0} + s_i + it_i$, $s_i \in Z$ which is the subject of the next section. An appropriate set of $\{s_i\}$ allows to increase α. For example, in case of the integral in Eq. (6.15), taking $z_3 \to -176/235 - 1 + it_3$, one can get damping factor $\alpha = 881/235 \simeq 3.75$ instead of $646/235 \simeq 2.75$. In addition to the shifted integral, contributions from the residues must be added. The additional terms are MB integrals with one integration less (see Chap. 4) and hence simpler to evaluate.

6.2.3   Shifting and Deforming Contours of Integration

Let's come back to the example from Sect. 1.7 and Eq. (1.55), with a solution in Eq. (1.56) (see also Sect. 5.1.7):

$$
V_{V3l2m}^{\epsilon^{-1}}(s) = -\frac{1}{2s} \int\limits_{-\frac{1}{2}-i\infty}^{-\frac{1}{2}+i\infty} \frac{dz}{2\pi i} \left(\frac{-s}{M_Z^2}\right)^{-z} \frac{\Gamma^3(-z)\Gamma(1+z)}{\Gamma(-2z)}. \tag{6.16}
$$

Doing the same asymptotic analysis as in Sect. 6.2.1, one can find that the integrand behaves like $\frac{1}{\sqrt{t}}$ and this is not enough for a fast convergence. Moreover, shifts in the integration contour in one-dimensional cases do not affect the asymptotic (see Problem 6.2). The analytical solution for this integral is well-defined, so numerical integration should also be possible.

One of the methods for overcoming difficulties connected with slow convergence is the deformation of the integration contours. It can be written in general from like

$$
z_i = z_{i0} + f_i(t_1, \ldots, t_n) + it_i \tag{6.17}
$$

where we add some real-valued function to our standard contour parallel to the imaginary axis. This function f_i must fulfill two basic conditions. First, it must restore the exponential damping factor in the integration region. Second, it must prevent the crossing of poles of gamma functions. Briefly speaking, when the imaginary part of arguments of gamma functions is equal to zero, the real part should not be equal to zero or negative integers. We automatically have the second condition with the standard contour, but in general, this is a highly non-trivial task. In addition, this function can be chosen to improve the general behavior of the integrand by removing oscillations and making it smoother (see Sect. 6.3 and [44]).

The simplest ansatz for the function f_i which fulfills the second condition is a linear function $f_i(t_1, \ldots, t_n) = \theta t_i$ with the same coefficient θ for all integration variables, and we have

$$z_i = z_{i0} + (i + \theta)t_i. \qquad (6.18)$$

Formally we rotate all integration contours at the same angle. This approach is described in [45] and works well for certain integrals, but it is not general. The main problem is that we have only one parameter to fix asymptotic behavior in the whole integration domain. Let's look at Eq. (6.9) and use a new contour:

$$\left(\frac{M_Z^2}{-s}\right)^z \longrightarrow e^{t(i+\theta)(\ln \frac{M_Z^2}{s} + i\pi)} \longrightarrow e^{t(\theta \ln \frac{M_Z^2}{s} - \pi)}, s > 0. \qquad (6.19)$$

Keeping in mind the new asymptotic for gamma functions, one can choose θ in such a way as to control the overall exponential damping everywhere (see Problem 6.3). The situation becomes more complicated when we go to the integrals with more scales. From Eq. (6.19) we see that even in the Euclidean case, rotated contours give some exponential factor, and if we have more than one kinematic coefficient, it can be in general not possible to find θ which will provide exponential damping in all directions.

In Fig. 6.2 a parabolic deformation is shown in addition to the parallel and rotated contours. It is a special case of transformations described in Sect. 6.3. The function θt^2 doesn't fulfill the condition of not crossing poles of gamma

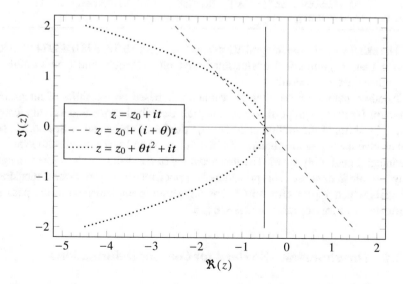

Fig. 6.2 Contour deformations discussed in [19]

functions, but this type of deformation can be beneficial for one-dimensional integrals. Numerical examples for the contours in Eq. (1.55) are given in the file `MB_1dimNum_Springer.nb` in the auxiliary material in [33].

Contour deformations can be easily implemented in `Mathematica` as seen below:

```
In[1]:= V1 =
 ↪   MBint[-((Gamma[-z1]^3*Gamma[1+z1])/(-s)^z1)/(2*s*Gamma[-2*z1]),
 ↪   {{eps->0},{z1->-(1/2)}}];
In[2]:= {int,z0} = V1/.MBint[mb_,{{___},{Rule[z0_,val_],___}}]->{mb,val};
```

```
In[3]:= f1 = z0 + I t;     (* straight contour *)
In[4]:= J1 = D[f1, t];     (* Jacobian *)
In[5]:= intobj1 = J1/(2 Pi I) int /. z1 -> f1;
In[6]:= res1 = NIntegrate[intobj1 /. s -> 2, {t, -Infinity,
 ↪   Infinity}, Method -> DoubleExponential]
Out[6]:= 0.572124 - 0.0364984 I
```

```
In[7]:= f2 = z0 + theta t + I t;  (* rotation *)
In[8]:= J2 = D[f2, t];            (* Jacobian *)
In[9]:= intobj2 = J2/(2 Pi I) int /. z1 -> f2;
In[10]:= res2 = NIntegrate[intobj2 /. s -> 2 /. theta -> -1,
 ↪   {t, -Infinity, Infinity}, Method -> DoubleExponential]
Out[10]:= 0.785398
```

```
In[11]:= f3 = z0 + theta t^2 + I t;  (* parabolic deformation *)
In[12]:= J3 = D[f3, t];              (* Jacobian *)
In[13]:= intobj3 = J3/(2 Pi I) int /. z1 -> f3;
In[14]:= res3 = NIntegrate[intobj3 /. s -> 2 /. theta -> -1,
 ↪   {t, -Infinity, Infinity}, Method -> DoubleExponential]
Out[14]:= 0.785398
```

The exact value of the integral V1 at $s = 2$ is $\pi/2 \simeq 0.78539816339744830962$. Notice that the numerical evaluation based on a straight-line contour fails to reproduce the correct result.

Another method of MB integral evaluation is based on the shifts of integration contours. It relies on properties of MB integrals described above, and its main feature is that by shifts of integration contours $\{s_i\}$ discussed at the end of Sect. 6.2.2, one can change the asymptotic behavior of the integrand and make the absolute value of the integral negligibly small. The low accuracy of the result for the shifted integral plays no role in this case. This procedure is applied recursively to lower-dimensional integrals which appear after shifts. The algorithm is implemented in the package `MBnumerics.m` (for more details see [26]).

6.2.4   Thresholds and No Need for Contour Deformations

Here we would like to discuss an approach to the construction of MB representations which allows for numerical integration without contour deformations and shifts.

Fig. 6.3 The two-loop
planar vertex diagram

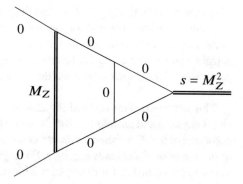

Table 6.4 Numerical results for the diagram in Fig. 6.3 for two different values of s and $M_Z^2 = 1$. AB—analytical solution [46]. MB—numerical integration of the corresponding representation is performed by the MB.m package

	$s = 1$	
AB	−5.**46318633201975**	+3.**83353434712221**i
MB	−5.**46318633201924**	+3.**83353434712**128i
	$s = 2$	
AB	−2.**01402077672107**	+2.**06704179425209**i
MB	−2.**01402077672170**	+2.**06704179425242**i

For some Feynman integrals, the MB-suite works without MBnumerics and any additional tricks because even in the Minkowskian kinematic case, cancellation of the damping factor does not happen, and the integrand of the corresponding MB integral has always Euclidean-like asymptotic by default. As an example, Table 6.4 gives a result for the integral which corresponds to the Feynman diagram shown in Fig. 6.3. The numerical evaluation has been obtained just with the MB.m package.

As shown in Sect. 3.2, Eq. (3.22), the F polynomial for a given Feynman diagram can be written as a sum of two parts

$$F = F_0 + U \sum_{i=1}^{n} x_i m_i^2. \qquad (6.20)$$

The first one, denoted as $F_0(x)$, corresponds to a diagram with all massless propagators. It depends on kinematic invariants. The second term $U(x) \sum_i m_i^2 x_i$ depends on masses of internal particles. To minimize the dimensionality of the representation within the GA approach, *we do not expand the second term*, and we construct the representation as follows

$$G(X) \sim \frac{U(x)^{N_v - d(L+1)/2}}{\left(F_0(x) + U(x) \sum_i m_i^2 x_i \right)^{N_v - dL/2}} \sim \prod_i \left(m_i^2 x_i \right)^{z_i} \frac{U(x)^{N_v - d(L+1)/2 + \sum_i z_i}}{F_0(x)^{N_v - dL/2 + \sum_i z_i}}. \qquad (6.21)$$

This way, we effectively get a massless diagram plus as many additional integrations as many massive propagators we have. In practice, the number of additional integrations equals to the number of different masses in each chain of the diagram. Equal masses in one chain can be collected, giving a linear combination of Feynman parameters. Later such a combination can be simplified by 1BL as discussed in Sect. 3.6.

This approach gives optimal dimensionality, but at a price, we lose information about physical and pseudo-thresholds. *By a physical threshold, we mean a kinematic point where the F polynomial starts to be negative. For the pseudo-threshold, one of the terms in F becomes negative.* For a general two-loop diagram with a single kinematic invariant s, for example, a $Zb\bar{b}$ vertex, we can rearrange terms in F and write it schematically in the following way:

$$F(x, s) = -s \sum_{l,n,k \in \Omega_{F_0}} x_l x_n x_k + \sum m_i^2 \sum_{l,n,k \in \Omega_{m_i}} x_l x_n x_k \qquad (6.22)$$

$$= -s \sum_{l,n,k \in \Omega_{F_0} \setminus \Omega_{m_i}} x_l x_n x_k + \sum \left(m_i^2 - s \right) \sum_{l,n,k \in \Omega_{F_0} \cap \Omega_{m_i}} x_l x_n x_k$$

$$+ \sum m_i^2 \sum_{l,n,k \in \Omega_{m_i} \setminus \Omega_{F_0}} x_l x_n x_k, \qquad (6.23)$$

where in Eq. (6.22) the first term corresponds to F_0 in Eq. (6.20) and the second is an expanded $U \sum_{i=1}^{n} x_i m_i^2$ term. In Eq. (6.23) we separated and collected $x_l x_n x_k$ terms with a common $\sum \left(m_i^2 - s \right)$ dependency. There is a physical threshold at $s = 0$ and pseudo-thresholds at points $s = \sum m_i^2$.

According to our observations, sometimes it is better to construct the MB representation using the expanded version of the F polynomial where all pseudo-thresholds are separated and collected explicitly. This leads to representations with higher dimensionality. However, in this case, we have a chance to get the exponential damping of the integrand in all kinematic regions. More precisely, in this way of construction, the coefficient $\beta(\theta)$ in Eq. (6.8) fulfills a new condition $\beta(\theta) \geq 2\pi$, and the overall damping factor remains for Minkowskian kinematical points. In this case, we don't need to perform a contour deformation or other tricks. Integration can be done straightforwardly with the logarithmic mapping of Eq. (6.10) and MB.m package.

The integral in Fig. (6.3) has no intersection between Ω_{F_0} and the massive part Ω_{m_i}, so explicit threshold separation is unnecessary, and the integral can be easily integrated without additional tricks. For this example, see the file MB_TH1_Springer.nb in the auxiliary material in [33]. In the file, the representation was obtained with the help of the LA method. To collect all the peculiarities of the F polynomial, in general, using the GA method is mandatory, even for planar diagrams.

Now let's consider a more complicated example for the diagram shown in Fig. 6.1 with results in Table 6.2. The diagram has one massive propagator. In addition to the two-dimensional representation in Sect. 3.9.2, Eq. (3.104), we have one more integration, and the minimal dimensionality for this integral is three:

```
In[1]:= MBreprNP[{1}, {PR[k1, 0, n1] PR[k1 - k2, 0, n2] PR[k2,
 ↪  0, n3] PR[k1 - k2 + p1, m, n4] PR[k2 + p2, 0, n5] PR[k1 +
 ↪  p1 + p2, 0, n6]}, {k1, k2}];
...
In[2]:= Fauto[0];
...
In[3]:= fupc = m^2 x[1] x[2] x[4] + m^2 x[1] x[3] x[4] + m^2
 ↪  x[2] x[3] x[4] + m^2 x[1] x[4]^2 + m^2 x[3] x[4]^2 (* + m^2
 ↪  x[1] x[4] x[5]-s x[1] x[4] x[5]*) + m^2 x[2] x[4] x[5] +
 ↪  m^2 x[4]^2 x[5] - s x[1] x[2] x[6] - s x[1] x[3] x[6] - s
 ↪  x[2] x[3] x[6] - s x[1] x[4] x[6] + m^2 x[2] x[4] x[6] +
 ↪  m^2 x[3] x[4] x[6] + m^2 x[4]^2 x[6] - s x[1] x[5] x[6] +
 ↪  m^2 x[4] x[5] x[6];
```

In the frame above, we show a partial output from AMBREv3 package for this integral. The program allows manual manipulation with the F polynomial. The pseudo-threshold terms m^2 x[1] x[4] x[5] - s x[1] x[4] x[5] are commented. We can keep the threshold terms in the form m^2 x[1] x[4] x[5] - s x[1] x[4] x[5] → m2s x[1] x[4] x[5]. In this case rescaling of Feynman parameters as in Eq. (3.93) reduces dimensionality to 7. For the specific kinematic point $s = m^2$, the term $m^2 - s = $ m2s can be dropped out, reducing dimensionality by one. The final result after ϵ-expansion and simplifications with BLs is four-dimensional, which is only one dimension higher than the optimal one. As in the case of the diagram in Fig. 6.3, the MB integrand now does not cause numerical problems, and the integration can be done with MB.m package. The result is given in Table 6.2. All the manipulations for the F polynomial discussed here are given in the file MB_TH2_Springer.nb in the auxiliary material in [33].

At this point, we should stress that at the pseudo-threshold point $s = m^2$, the integral is continuous, and we can drop out the related term in the F polynomial. Calculations at physical thresholds, in our case $s = 0$, should be done differently, by analytical expansion around the point $s \ll 1$. After that, we can take the limit $s = 0$ and compute left and right limits $s \to 0^\pm$. For analytical expansion of MB integrals, see Sect. 5.4.1.

6.3 MB Numerical Evaluation by Steepest Descent

The method of steepest descent has already been used in works by Riemann, who applied it to estimate the hypergeometric function, and by Cauchy, Debye, and Nekrasov. For historical records and references to their original works, see [47]. The method is often used for the asymptotic evaluation of integrals and is based on the idea that many functions in the complex plane have a stationary point. There is

one direction at that stationary point in which the function decreases rapidly, and
there is an orthogonal direction in which the function increases rapidly. In other
words, the stationary point is a saddle point; the function decreases in one direction,
while it increases in another direction. By deforming the integration path so that it
goes through the stationary point in the direction in which the function decreases,
one can evaluate the integral asymptotically.

Here we will connect stationary points with MB through the Lefschetz thimbles
(LT), by searching for a stationary phase contours C as solutions of properly defined
differential equations.

LT are applied in research of mathematics, crossing many issues like behavior of
LT in the presence of poles, singularities and branch cuts, behavior at complex infinity, and Stokes phenomenon, relation to relative homology of a punctured Riemann
sphere. In physics, it can be applied to the analytic continuation of 3D Chern-Simons
theory, QCD with chemical potential, resurgence theory, counting master integrals,
or the repulsive Hubbard model. Applying this method to the numerical evaluation
of MB integrals is at the exploratory stage and has been discussed in [44, 48]. Here
we will present the main idea for the lowest one-dimensional MB integrals, in both
Euclidean ($s < 0$) and Minkowski ($s > 0$) regions. These cases have been explored
in fine details in [44]. For higher dimensions, even solving two-dimensional MB
integrals is not fully understood and explored and can be a potential subject of a
nice research work by the reader (see Problem 6.5).

6.3.1   General Idea

Let us write then a general MB integrand $F(z)$, transformed into exponential form.
For brevity, we suppress the dependence on s and shall use $F(z)$ instead of $F(s, z)$

$$I(s) = \frac{1}{2\pi i} \int_{C_0} dz \, F(z) = \frac{1}{2\pi i} \int_{c_0 - i\infty}^{c_0 + i\infty} dz \, e^{-f(z)}. \qquad (6.24)$$

C_0 is a contour defined by $\Re(z) = c_0$, while $f(z) = -\ln F(z)$.

The core of the problem with integration over C_0 is highly oscillatory behavior
of the integrand $F(z)$. For such a class of integrands, standard methods of numerical
integration are often not adequate.

One of possible ways to get rid of numerical problems with the MB integrand
$F(z)$ which is of highly oscillatory behavior (see Sect. 1.7 and [19]) is to integrate
Eq. (6.24) over a new contour $C = \mathcal{J}_1 + \mathcal{J}_2 + \mathcal{A}$.

A typical example is sketched in Fig. 6.4 where C is a sum of three contours $\mathcal{J}_1$,
$\mathcal{J}_2$, and $\mathcal{A}$ along which the behavior of f is under control.

Let us describe in details this decomposition.

As $\mathcal{J}_{1,2}$ we choose such stationary phase contours which start at saddle points
$z_*^{(1,2)}$ and go toward infinity without hitting other poles. Both contours are chosen

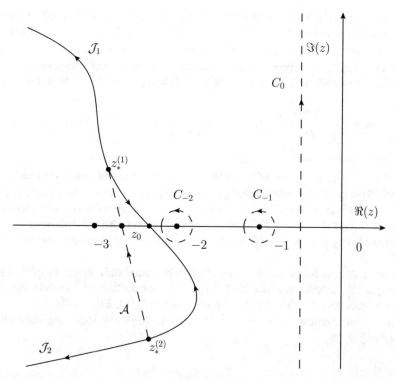

Fig. 6.4 A deformation of the integration contour C_0 defined by $\Re(z) = c_0$ to a contour $C = \mathcal{J}_1 + \mathcal{J}_2 + \mathcal{A}$. $\mathcal{J}_{1,2}$ are two Lefschetz thimbles which start at saddle points $z_*^{(1,2)}$ and go toward infinity. The compact contour $\mathcal{A}$ (interval) connects the two saddle points $z_*^{(1)}$ and $z_*^{(2)}$. When there is an obstruction in deriving the parametrization of $\mathcal{J}_{1,2}$ around some point, e.g., z_0, one can bypass that region using the contour $\mathcal{A}$. Note that here a deformation $C_0 \to C$ requires taking into account integrals over two "small" contours, C_{-2} and C_{-1} around poles $z = -2$ and $z = -1$ which contribute to $\sum \operatorname{Res} F$ in (6.25)

such that $\Im(f)$ is constant along them and function $\Re(f)$ is stricly increasing when one moves away from $z_*^{(1,2)}$. Varieties defined in such a way are called steepest descent contours [49, 50] or Lefschetz thimbles [51–56]. Usage of $\mathcal{J}_{1,2}$ allows to control the behavior of $f(z)$ when $z \to \infty$. Because $\Re(f)$ is strictly increasing, the integrand e^{-f} decreases rapidly at the ends of $\mathcal{J}_{1,2}$. That transforms the integral (6.24) into a form which is more suitable for numerical treatment.

In this method, $\mathcal{A}$ is an auxiliary compact contour with an explicit analytical parametrization. It connects two saddle points on different Lefschetz thimbles $\mathcal{J}_k$.[1] $\mathcal{A}$ is chosen such that F does not have singularities along $\mathcal{A}$. Then the integral

---

[1] More generally, the contour $\mathcal{A}$ can connect any two points z_1 and z_2 which belongs to $\mathcal{J}_1$ and $\mathcal{J}_2$, respectively. They do not necessarily have to be saddle points of f.

$\int_{\mathcal{A}} dz\, F$ can be calculated with high precision with the help of standard numerical methods. The simplest choice of $\mathcal{A}$ is an interval spanned between two saddle points $z_*^{(1,2)}$. In some cases of integrands F, it is convenient to introduce such an additional contour to improve convergence and accuracy of the numerical integration.

After deformation $C_0 \to C$, the original MB integral (6.24) can be written as

$$I(s) = \sum_{k=1,2} \frac{e^{-i\phi_k}}{2\pi i} \int_{\mathcal{J}_k} dz\, e^{-\Re(f)} + \frac{1}{2\pi i} \int_{\mathcal{A}} dz\, F + \sum_{C_0 \to C} \text{Res}\, F, \qquad (6.25)$$

where $\phi_k = \Im(f)\big|_{\mathcal{J}_k}$. According to the Cauchy theorem, the sum over residues in Eq. (6.25) is necessary when deformation from C_0 to C encircles any poles of $F = e^{-f}$. The additional condition is that during transition from C_0 to C one surrounds only a finite set of poles of F. In other words, C is such that the difference $C - C_0$ consists of only a finite sum of "small" closed contours encircling respective poles of F.

The analytical formula describing $\mathcal{J}_k$ can be found only in the simplest cases by explicitly solving the equation $\Im(f) = \text{const}$. Instead, we use the fact that the function $\Re(f)$ defines a Morse flow [57, 58]. Such a flow is realized by a parametrization $t \mapsto z(t)$ of $\mathcal{J}_k$ which obeys the following differential equation [52, 55]:

$$\frac{dz}{dt} = (\partial_z f)^* \qquad (6.26)$$

with an initial condition:

$$\lim_{t \to -\infty} z(t) = z_*. \qquad (6.27)$$

With respect to various methods known in the literature [4, 23, 45, 59] which shift/rotate contours or use approximate forms thereof, this method relies on deriving numerically a parametrization $z(t)$ of $\mathcal{J}_k$ as a solution of differential equation (6.26) and then, again numerically, integrating the function $e^{-\Re(f)}$ along contour C composed of Lefschetz thimbles $\mathcal{J}_k$ (and compact contour $\mathcal{A}$ if necessary). This purely numerical approach is complementary to the Padé approximation [44].

We choose z_* as a starting point for differential equation (6.26). It is worth mentioning that other initial points are also possible. However, from the practical point of view, it is easier to find a point z_* which satisfies $\partial_z f(z_*) = 0$ and starting from that point properly construct $\mathcal{J}_k$, than to find a point z_{in} lying on some stationary phase contour and check if $\Re(f)$ is increasing along that contour.

Let us note that at first sight the presented method seems to be not appropriate for finding a parametrization of a Lefschetz thimble around a saddle point z_* which is, at the same time, a zero of the integrand F. The reason is that in this region the differential equation (6.26) is not well-defined because $\partial_z f = -\partial_z F/F$ diverges at the saddle point z_*. However, one can shift F by a holomorphic function, e.g., $F \to \widetilde{F} = F + 1$, without changing the value of the integral $I(s) = \int_C dz\, \widetilde{F} = \int_C dz\, F$. Now, $\widetilde{F}(z_*) = 1$ and $\partial_z \widetilde{F}(z_*) = 0$. Hence one can use $\widetilde{f} = -\ln \widetilde{F}$ in the differential equation (6.26) to derive the parametrization of a Lefschetz thimble related to a saddle point z_*.

6.3.2   Implementation of the Method

Let us now describe how to numerically derive parametrizations $z_k(t)$ of Lefschetz thimbles $\mathcal{J}_k$ using the differential equation (6.26). Below we present all crucial steps of the method and enumerate the routines used from the `Mathematica` language.

To solve (6.26) one has to specify three ingredients:

(a) The saddle point z_* from which Lefschetz thimble starts
(b) A line $l(t) = z_* + te^{i\beta}$ tangent to the curve $z(t)$ at point z_*
(c) A distortion

$$z_*(\epsilon) = z_* + \epsilon e^{i\beta} \tag{6.28}$$

from z_* along $l(t)$ which is controlled by a small parameter $\epsilon \ll 1$, the smaller ϵ is the more accurate solution of (6.26) one gets.

In practice, the distorted point $z_*(\epsilon)$ plays a role of the initial point at $t = 0$ for the `NDSolve`. Such a small deviation is necessary to start off the abovementioned numerical routine. At z_* the derivative dz/dt is zero so setting z_* as the initial point, i.e., $z(0) = z_*$, would only generate static solution $z(t) \equiv z_*$.

The phase β in (6.28) is related to the slope of $l(t)$. To find possible values of β, one uses the Laurent series of f around z_* and solves the condition for stationary phase $\Im(f) = \phi$ keeping only leading terms. The easiest way to decide which value of β corresponds to Lefschetz thimble(s) $\mathcal{J}$ is to solve (6.26) for all possible distortions $z_*(\epsilon)$ in some small range of t and then compare along which solution $\Re(f)$ is increasing. In cases when $\partial_z^2 f(z_*) \neq 0$, one can use eigenvectors of the Hessian matrix of $\Re(f)$ at z_* to find directions along which $\Re(f)$ is increasing.

To find saddle points, we use `FindRoot` which looks for a solution of an equation $\partial_z f = 0$ in the vicinity of a chosen point. It is convenient to first start looking for z_* within the original fundamental region $-1 < \Re(z) < 0$. If this fails then one jumps to another region $-2 < \Re(z) < -1$, etc. In the Minkowskian region, the position of a saddle point is not restricted at all, while in the Euclidean region, a saddle point has to be real, or it can be complex, but then it must come in pair with its complex conjugate $(z_*)^*$.

After finding a saddle point $z_*^{(1)}$, one constructs all its possible distortions (6.28) and uses them in `NDSolve`. If `NDSolve` returns two Lefschetz thimbles $\mathcal{J}_{1,2}$

starting from $z_*^{(1)}$ and such that the sum $\mathcal{J}_1 + \mathcal{J}_2$ is well-defined deformation of C_0 (see Fig. 6.4), then this step of the method is accomplished and one can go further.

In cases in which NDSolve is not able to find two Lefschetz thimbles starting from one saddle point, one can use the already mentioned trick which relies on using auxiliary compact contour $\mathcal{A}$. It allows to bypass, in a fully controlled way, regions in which NDSolve does not work well or returns warnings/errors. In such a situation, one constructs only one Lefschetz thimble $\mathcal{J}_1$ which starts at $z_*^{(1)}$ and goes to ∞, and then one finds another saddle point $z_*^{(2)}$ and applies the same procedure as for $z_*^{(1)}$. If for $z_*^{(2)}$ it is possible to construct at least one Lefschetz thimble $\mathcal{J}_2$ such that $z_*^{(1)}$ and $z_*^{(2)}$ can be connected via a compact contour $\mathcal{A}$ and the sum $\mathcal{J}_1 + \mathcal{J}_2 + \mathcal{A}$ is well-defined deformation of C_0 (see Fig. 6.4), then this step of the method is accomplished, and one can go to the next step. If one fails, e.g., because the obtained contours are closed or they hit poles, then one has to choose another saddle point of f and repeat the whole procedure.

The final output of NDSolve is the InterpolatingFunction which is further used to integrate numerically the integrand F using NIntegrate.

To monitor whether the solution of (6.26) parametrizes a contour which goes toward ∞ without hitting a pole, one can make use of the slope of the line which is tangent to that contour:

$$\frac{\Im \frac{dz}{dt}}{\Re \frac{dz}{dt}} = -\tan\left[\arg\left(\partial_z f\right)\right]. \tag{6.29}$$

If $\mathcal{J}$ asymptotically approaches a line $\Im(f(z \to \infty)) = \text{const.}$, then along $\mathcal{J}$, in the limit $t \to \infty$, one gets

$$\Delta\theta = \theta_{\pm\infty} + \arg(\partial_z f)|_{z=z(t)} \to n\pi, \quad n \in \mathbb{Z}. \tag{6.30}$$

In other words, $\Delta\theta$ measures whether $\mathcal{J}$ approaches a line of constant phase $\Im(f(z \to \infty)) = \text{const.}$ when $t \to \infty$.

$\theta_{\pm\infty}$ for the stationary phase contour are defined as follows:

$$\begin{aligned}
z &\xrightarrow{t \to +\infty} z_\infty + i e^{i\theta_{+\infty}} t, \quad t > 0, \\
z &\xrightarrow{t \to -\infty} z_\infty + i e^{-i\theta_{-\infty}} t, \quad t < 0,
\end{aligned} \tag{6.31}$$

their general solutions are discussed in [44].

Let us now discuss numerical integration of F over contour C. Taking into account the change of the integration measure given by (6.26), the integral (6.25) can be written as

$$
I(s) = \frac{e^{-i\phi_1}}{2\pi i} \int_{-\infty}^{+\infty} dt \; (\partial_z f)^* e^{-\mathcal{R}(f)} \Big|_{z=z_1(t)} - \frac{e^{-i\phi_2}}{2\pi i} \int_{-\infty}^{+\infty} dt \; (\partial_z f)^* e^{-\mathcal{R}(f)} \Big|_{z=z_2(t)}
$$

$$
+ \frac{1}{2\pi i} \int_0^1 dt \; F(z_{\mathcal{A}}(t)) \frac{dz_{\mathcal{A}}(t)}{dt} + \sum_{C_0 \to C} \text{Res} \, F,
\tag{6.32}
$$

where $\phi_{1,2} = \Im(f)|_{\mathcal{J}_{1,2}}$ take into account the fact that the value of $\Im(f)$ can be different on each Lefschetz thimble $\mathcal{J}_k$. The minus sign in front of the second term in (6.32) corresponds to the opposite orientation of $\mathcal{J}_2$ with respect to the orientation of C (see Fig. 6.4). $z_k = z_k(t)$ are parametrizations of $\mathcal{J}_k$ derived from NDSolve as discussed above. In both cases the flow starts from the saddle points $z_*^{(k)} = z_k(t = -\infty)$, but because the distorted point $z_*(\epsilon) = z(t = 0)$ is close to $z_* = z(t = -\infty)$, due to the small parameter ϵ, one can integrate over $t \in (0, +\infty)$ instead of $t \in (-\infty, +\infty)$. In practice one integrates over $(0, t_{\max})$ where $t_{\max}$ is chosen such that all contributions from the integrand, up to declared precision, are taken into account. Varying $t_{\max}$ provides additional test whether the method works properly. When one makes $t_{\max}$ bigger and bigger, then $I(s)$ should asymptotically approach a finite value. Finally, $z_{\mathcal{A}}(t)$ used in (6.32) is an explicit analytical parametrization of the contour $\mathcal{A}$. When $\mathcal{A}$ is an interval, then its parametrization can be of the following form:

$$
t \mapsto z_{\mathcal{A}}(t) = z_2 + t(z_1 - z_2).
\tag{6.33}
$$

MBDE Solutions in Euclidean and Minkowskian Kinematics

Let us consider the badly behaving integrand Eq. (1.55), considered in Sects. 1.7 and 5.1.7 (see Eqs. (5.178) and (5.179) [23, 44]):

$$
F_1(z) = (-s)^{-z} \frac{\Gamma^3(-z)\Gamma(1+z)}{\Gamma(-2z)}
\tag{6.34}
$$

at three different kinematic points $s = -1/20, 1 + i\delta, 5$. Functional behavior of the function is given in Figs. 6.5, 6.6, and 6.7 with the aid of a function designed to compress the vertical scale (left-side figures):

$$
\underset{m}{\text{sln}} \, x \equiv \text{sign}(x) \, \ln(1 + |x|e^m).
\tag{6.35}
$$

The file `MBDE_Springer.nb` with a complete solution can be found in the book auxiliary file repository [33]. Based on a discussion in Sect. 6.3.1, the easy-to-understand major steps for the integral evaluation of the function in Eq. (6.34) are easy to follow there.

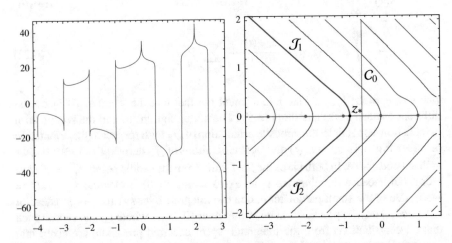

Fig. 6.5 A plot of $\mathrm{sln}_{20}\,\mathfrak{R}(F_1)$ for $s = -1/20$ (left). $\mathfrak{I}(f_1) = 0$ contours for $s = -1/20$. Thick blue line corresponds to the solution of (6.26) with $z_* = -0.8256$ (right)

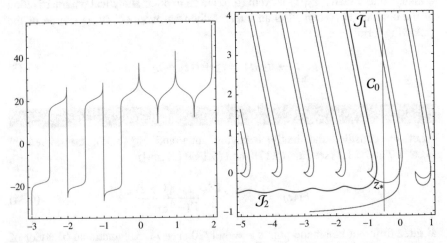

Fig. 6.6 A plot of $\mathrm{sln}_{20}\,\mathfrak{R}(F_1)$ for $s = 1 + i\delta$ (left). $\mathfrak{I}(f_1) = 2.289$ contours for $s = 1 + i\delta$. Thick blue line corresponds to the solution of (6.26) with $z_* = -0.7893 - 0.1745i$ (right)

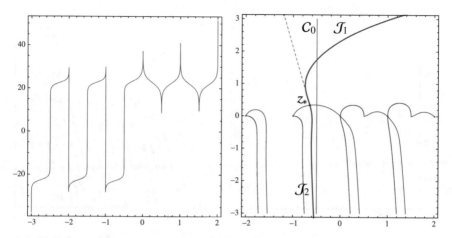

Fig. 6.7 A plot of $\mathrm{sln}_{20}\, \mathfrak{R}(F_1)$ for $s = 5$ (left). $\mathfrak{I}(f_1) = -1.895$ contours for $s = 5$. Thick blue line corresponds to the solution of (6.26) with $z_* = -0.6470 + 0.3366i$ (right)

6.3.3 Tricks and Pitfalls for "Vanishing Derivatives"

Let us consider the integrand $F_4(z) = F_1(z)\psi^4(1 - z)$, $F_4^{(0-3)}(z_*) = 0$ for $z_* \in (-1, 0)$, $s = -2$. In some cases, for example, F_4 in the Minkowskian region, it is hard to find a saddle point z_* through which a contour goes to infinity (see Fig. 6.8). Then one can use the following trick. One looks for a new contour C which can be decomposed into three pieces $\mathcal{J}_1 + \mathcal{A} + \mathcal{J}_2$, where $\mathcal{J}_{1,2}$ are two $\mathfrak{I}(F_4) = 0$ contours ending at ∞, while $\mathcal{A} = \{z \in \mathbb{C} : z = z_2 + t(z_1 - z_2), t \in \langle 0, 1 \rangle\}$ is an interval connecting z_1 and z_2. These two points lie on $\mathcal{J}_1$ and $\mathcal{J}_2$, respectively, and are such that $F_4^{(k)}(z_{1,2}) \neq 0$. The additional requirement is that C_0 can be deformed into C by encircling only the finite number of poles of F_4. For example, $z_{1,2}$ can be chosen as intersection points of two $\mathfrak{I}(F_4) = 0$ contours with the line $\mathfrak{R}(z) = -1/2$ (see Fig. 6.8 (right)).

From the practical point of view, there is a way to control if numerically derived $\mathcal{J}_{1,2}$ really go to ∞ and do not hit a zero or pole. Namely, one can use (6.26) to monitor the asymptotic behavior of $\mathcal{J}_{1,2}$:

$$\frac{dy}{dx} = \frac{\mathfrak{I}\frac{dz}{dt}}{\mathfrak{R}\frac{dz}{dt}} = -\tan\left[\arg\left(\partial_z f_4\right)\right]. \tag{6.36}$$

So, for $t \to \infty$ the value of dy/dx along $\mathcal{J}_{\pm}$ should asymptotically approach the slope of the lines defined by $\mathfrak{I}(f_4(z \to \infty)) = 0$.

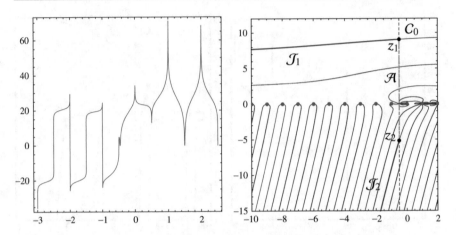

Fig. 6.8 A plot of $\mathrm{sln}_{20} F_4$ for $s = 2$ (left). $\mathfrak{I}(f_4) = -2.793$ contours for $s = 2$. Thick blue lines correspond to contours $\mathcal{J}_{1,2}$ which go to ∞. They intersect $\mathfrak{R}(z) = -1/2$ line at $z_1 = -1/2 + 9.057i$ and $z_2 = -1/2 - 5.169i$, respectively. The orginal integration contour C_0 is deformed to $\mathcal{J}_1 + \mathcal{A} + \mathcal{J}_2$, where $\mathcal{J}_1$ is an upper part of the contour C, starting at z_1; $\mathcal{J}_2$ is a lower part of the contour C, starting at z_2; and $\mathcal{A} = \{z \in \mathbb{C} : z = z_2 + t(z_1 - z_2), t \in \langle 0, 1 \rangle\}$ is an interval connecting z_1 and z_2, orange, dot-dashed line (right)

6.3.4   Beyond One-Dimensional Cases

There is so far not much development within the LT method beyond the one-dimensional case. Eq. (6.26) can be generalized to

$$\frac{dz_i}{dt} = -\left(\frac{\partial f}{\partial z_i}\right)^*, \quad i = 1, \ldots, n. \tag{6.37}$$

This allows us to define Lefschetz thimbles associated with a given critical point as

$$\mathcal{J}_\alpha = \{z \in \mathbb{C}^n : \lim_{t \to \infty} z(t) = z_\alpha\}, \tag{6.38}$$

where $z(t)$ is the solution of the flow Eq. (6.37). The Lefschetz thimble is a collection of curves flowing in the direction of the steepest descent that asymptotically reach the corresponding critical point. Note that the real part of the integrand for critical points is at the same time saddle points. There exists an alternative direction in which the real part changes the most, i.e., the direction of the steepest ascent. With this direction are associated geometrical structures similar to the Lefschetz thimbles, called anti-thimbles, which satisfy the system of upward flow differential equations

$$\mathcal{K}_\alpha = \{z \in \mathbb{C}^n : \lim_{t \to \infty} \bar{z}(t) = z_\alpha\}, \tag{6.39}$$

where $\bar{z}(t)$ are curves that satisfy the following system of upward flow equations

$$\frac{dz_i}{dt} = \left(\frac{\partial f}{\partial z_i}\right)^*, \quad i = 1, \ldots, n. \tag{6.40}$$

The idea of thimbles as a surface of steepest descent appeared in [51]. It has been shown there that within the homology theory, an initial contour of integration C can be changed into a finite sum of thimbles, each associated with one critical point

$$C = \sum_\alpha N_\alpha \mathcal{J}_\alpha. \tag{6.41}$$

The sum in the above equation goes over critical points and $N_\alpha \in \mathbb{Z}$. This way, we can change the integration over one surface into the sum of integrals over thimbles. See W. Flieger's contribution to [7] for more details.

6.4 Numerical Evaluation of Phase Space MB Integrals

The tools described above for the numerical evaluation of MB integrals can also be very effective for obtaining numerical solutions for phase space integrals. The power of the MB approach is nicely demonstrated in the following example. Let us consider an angular integral with three denominators:

$$I_{j,k,l} = \int d\Omega_{d-1}(q) \frac{1}{(p_1 \cdot q)^j (p_2 \cdot q)^k (p_3 \cdot q)^l}, \tag{6.42}$$

where the fixed momenta p_1, p_2, and p_3 are massless and we are integrating over the angular variables of the (massless) momentum q. Such integrals appear, e.g., when integrating over soft radiation beyond NLO accuracy [60–62]. Let us attempt to evaluate the above integral numerically. One obvious choice is to use a parametrization in terms of angular variables as described in Sect. 3.14. So let us choose a Lorentz frame such that

$$p_1^\mu = (1, \mathbf{p}_1) = (1, \mathbf{0}_{d-2}, 1),$$

$$p_2^\mu = (1, \mathbf{p}_2) = (1, \mathbf{0}_{d-3}, \sin \chi_2^{(1)}, \cos \chi_2^{(1)}), \tag{6.43}$$

$$p_3^\mu = (1, \mathbf{p}_3) = (1, \mathbf{0}_{d-4}, \sin \chi_3^{(2)} \sin \chi_3^{(1)}, \cos \chi_3^{(2)} \sin \chi_3^{(1)}, \cos \chi_3^{(1)}).$$

In this frame, the vector q^μ has the form

$$
q^\mu = \left(1, ..\text{``angles''}.., \cos\vartheta_n \prod_{k=1}^{n-1} \sin\vartheta_k, \cos\vartheta_{n-1} \prod_{k=1}^{n-2} \sin\vartheta_k, \ldots, \cos\vartheta_2 \sin\vartheta_1, \cos\vartheta_1 \right),
$$

(6.44)

where .."angles".. denotes those angular variables that can be integrated trivially (since the integrand does not depend on them). In this frame, our integral $I_{j,k,l}$ becomes

$$
I_{j,k,l} = \int d\Omega_{d-4} \int_{-1}^{1} d(\cos\vartheta_1) d(\cos\vartheta_2) d(\cos\vartheta_3)(\sin\vartheta_1)^{-2\epsilon}(\sin\vartheta_2)^{-1-2\epsilon}
$$

$$
\times (\sin\vartheta_3)^{-2-2\epsilon}(1-\cos\vartheta_1)^{-j}\left(1 - \cos\chi_2^{(1)}\cos\vartheta_1 - \sin\chi_2^{(1)}\sin\vartheta_1\cos\vartheta_2\right)^{-k}
$$

$$
\times \left(1 - \cos\chi_3^{(1)}\cos\vartheta_1 - \cos\chi_3^{(2)}\sin\chi_3^{(1)}\sin\vartheta_1\cos\vartheta_2\right.
$$

$$
\left. - \sin\chi_3^{(2)}\sin\chi_3^{(1)}\sin\vartheta_1\sin\vartheta_2\cos\vartheta_3\right)^{-l}.
$$

(6.45)

However, the straightforward evaluation of this integral runs into the following difficulty. If at least two of the exponents j, k, and l are positive integers (at $\epsilon = 0$), already the integrand has a line singularity inside the integration region. Indeed, if say $j = 1$ and $k = 1$, the integrand is singular not only at $\cos\vartheta_1 = 1$ but also along the line $\cos\vartheta_1 = \cos\chi_2^{(1)}$ whenever $\cos\vartheta_2 = 1$. Obviously these singularities lead to poles in ϵ which must be made explicit before any actual numerical computation can be attempted. However, the presence of the line singularity means that the usual procedure of making these poles explicit, *sector decomposition* (see, e.g., [63] and references therein) is not directly applicable to Eq. (6.45), as it relies on the assumption that the only singularities are on the integration boundaries. If all three of the exponents are positive integers, the last factor in Eq. (6.45) develops singularities inside the integration region as well. These divergences must then be treated on a case-by-case basis before numerical integration can proceed.

Removing Line Singularities by Partial Fractioning

In certain cases the partial fraction decomposition of the integrand can be applied to eliminate line singularities. To see this in its simplest form, consider the case $j = 1$, $k = 1$, and $l = 0$. Clearly we are in the situation described above, where in addition to the singularity on the integration boundary at $\cos\vartheta_1 = 1$, the line singularity at $\cos \vartheta_1 = \cos \chi_2^{(1)}$, $\cos \vartheta_2 = 1$ is also present. However, we can remove this singularity as follows. Let us perform partial fractioning of the denominator:

$$\frac{1}{(p_1 \cdot q)(p_2 \cdot q)} = \left[\frac{1}{(p_1 \cdot q)} + \frac{1}{(p_2 \cdot q)} \right] \frac{1}{(p_1 + p_2) \cdot q}. \qquad (6.46)$$

Now, it is easy to check that the denominator $(p_1 + p_2) \cdot q$ is no longer singular. Hence, all singularities are now associated with the two separate denominators in the bracket. The first term is only divergent at $\cos \vartheta_1 = 1$ in the chosen frame, and we can proceed to apply sector decomposition to resolve the ϵ pole. The second term has the line singularity *in this particular frame*. However, we are free to use rotational invariance to compute this second integral in another Lorentz frame, where the roles of p_1 and p_2 are interchanged! In this rotated frame, the line singularity would appear in the denominator $(p_1 \cdot q)$, while $(p_2 \cdot q)^{-1}$ is singular only at $\cos \vartheta_1' = 1$, where $\cos \vartheta'$ is the variable in the rotated Lorentz frame. Thus, we do not have to deal with the line singularity explicitly.

Another approach to dealing with the line singularity in this particular case would be to split the region of integration in $\cos \vartheta_1$ as

$$\int_{-1}^{1} d(\cos \vartheta_1) = \int_{-1}^{\cos \chi_2^{(1)}} d(\cos \vartheta_1) + \int_{\cos \chi_2^{(1)}}^{1} d(\cos \vartheta_1). \qquad (6.47)$$

Obviously the two integrals only have singularities on the boundaries of the integration region in this case. Note however that this approach requires the precise knowledge of the geometry of the singularities inside the integration region, which in general can be quite elaborate.

However, all of these complications can be sidestepped immediately by using the MB representation of the angular integral, discussed in Sect. 3.14.

Indeed, for general exponents j, k, and l, we find [64]

$$I_{j,k,l} = 2^{2-j-k-l-2\epsilon} \pi^{1-\epsilon} \frac{1}{\Gamma(j)\Gamma(k)\Gamma(l)\Gamma(2-j-k-l-2\epsilon)} \int_{-i\infty}^{+i\infty} \frac{dz_1 \, dz_2 \, dz_3}{(2\pi i)^3}$$

$$\times \Gamma(-z_1)\Gamma(-z_2)\Gamma(-z_3)\Gamma(j+z_1+z_2)\Gamma(k+z_1+z_3)\Gamma(l+z_2+z_3)$$

$$\times \Gamma(1-j-k-l-\epsilon-z_1-z_2-z_3)v_{12}^{z_1}v_{13}^{z_2}v_{23}^{z_3},$$

$$(6.48)$$

where

$$v_{12} = \frac{p_1 \cdot p_2}{2}, \qquad v_{13} = \frac{p_1 \cdot p_3}{2} \qquad \text{and} \qquad v_{23} = \frac{p_2 \cdot p_3}{2}. \tag{6.49}$$

Thus, for any specific exponents j, k, and l, we can directly proceed with resolving the poles, performing the ϵ-expansion and evaluating the resulting MB integrals numerically. As an example, consider, e.g., $j = 1$, $k = 1 + \epsilon$, and $l = \epsilon$. This particular set of exponents appears when integrating over soft-gluon emission from the interference of tree-level and one-loop matrix elements, as happens in QCD computations at NNLO accuracy [60, 62]. Then we must evaluate the integral

$$I_{1,1+\epsilon,\epsilon} = 2^{-4\epsilon} \pi^{1-\epsilon} \frac{1}{\Gamma(1+\epsilon)\Gamma(\epsilon)\Gamma(-4\epsilon)} \int_{-i\infty}^{+i\infty} \frac{dz_1\, dz_2\, dz_3}{(2\pi i)^3} \Gamma(-z_1)$$

$$\times \Gamma(-z_2)\Gamma(-z_3)\Gamma(1 + z_1 + z_2)\Gamma(1 + \epsilon + z_1 + z_3)\Gamma(\epsilon + z_2 + z_3)$$

$$\times \Gamma(-1 - 3\epsilon - z_1 - z_2 - z_3)v_{12}^{z_1} v_{13}^{z_2} v_{23}^{z_3}. \tag{6.50}$$

In this particular case, we must address one further subtlety: it is relatively straightforward to see that we cannot find straight-line contours running parallel to the imaginary axis such that the real parts of the arguments of all gamma functions involving integration variables are positive, for any real value of ϵ. Indeed, clearly on the one hand, we must have $\Re(z_i) < 0$ from the first three gamma functions in the integrand. But then, looking at the last gamma function on the second line, $\Gamma(\epsilon + z_2 + z_3)$, we deduce that $\epsilon > 0$. However, the sum of the arguments of the last three gamma function in the integrand, $\Gamma(1 + \epsilon + z_1 + z_3)$, $\Gamma(\epsilon + z_2 + z_3)$, and $\Gamma(-1 - 3\epsilon - z_1 - z_2 - z_3)$ is $-\epsilon + z_3$. If the real parts of all arguments were positive, this would imply $\Re(-\epsilon + z_3) > 0$, and given that $\Re(z_3) < 0$, we derive $\epsilon < 0$, in contradiction to our previous observation! One way to proceed is to introduce an auxiliary regulator, say we set $l = \epsilon$ to $l = \epsilon + \delta$. Now it becomes possible to find straight-line contours, and we may then proceed to analytically continue the integral first to $\delta \to 0$ and then to $\epsilon \to 0$. This strategy has been also discussed in Chap. 4. It is imperative to check that this procedure does not introduce any poles or logarithms of the regulator δ, so that the $\delta \to 0$ limit exists and is finite (in δ). The algorithmic procedure for performing the analytic continuations was detailed in Chap. 4, and we can implement the computation in Mathematica using the MB.m package as follows (see also the file MB_RealPSNum_Springer.nb in the auxiliary material in [33]):

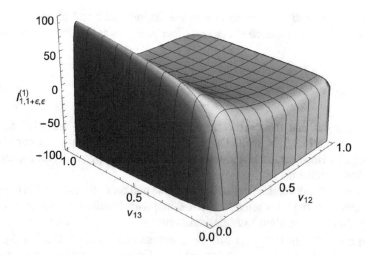

Fig. 6.9 Numerical result for the third expansion coefficient $I^{(1)}_{1,1+\epsilon,\epsilon}$ defined in Eqs. (6.50) and (6.51) for the co-planar case $\cos\chi^{(2)}_3 = -1$. In this case, we have simply $v_{12} = (1 - \cos\chi^{(1)}_2)/2$ and $v_{13} = (1 - \cos\chi^{(1)}_3)/2$

```
In[1]:= Ijkl = 2^(2-j-k-1-2ep) Pi^(1-ep) * 1/(Gamma[j] Gamma[k]
↪    Gamma[1] Gamma[2-j-k-1-2ep]) Gamma[-z1] Gamma[-z2]
↪    Gamma[-z3] Gamma[j+z1+z2] Gamma[k+z1+z3] Gamma[1+z2+z3]
↪    Gamma[1-j-k-1-ep-z1-z2-z3] v12^z1 v13^z2 v23^z3 /. {j->1,
↪    k->1+ep, 1->ep+delta};
In[2]:= contour = MBoptimizedRules[Ijkl, delta->0, {},
↪    {delta}];
In[3]:= Ijkl = MBcontinue[Ijkl, delta->0, contour];
In[4]:= Ijkl = MBexpand[Ijkl, 1, {delta, 0, 0}];
In[5]:= FreeQ[Ijkl /. MBint[int_, rule_]->int, delta]
Out[5]:= True
In[6]:= Ijkl = Ijkl /. {MBint[int_,rule_] :> MBint[int,
↪    {Select[rule[[2]], !FreeQ[#,ep]&], Select[rule[[2]],
↪    FreeQ[#,ep]&]}]};
In[7]:= Ijkl = Ijkl/. {MBint[int_, rule_]:>MBcontinue[int,
↪    ep->0, rule];
```

The fifth input checks that indeed the $\delta \to 0$ limit is free of δ (and hence does not involve poles of logarithms of δ), while the sixth input brings the rule in MBint to a form where the analytic continuation can then performed in ϵ in the next step. (For further details, consult the documentation of the MB.m package [23].)

Now we are ready to expand in ϵ to the desired order and obtain a numerical solution for the expansion coefficients. In fact, in this example, the expansion starts at $1/\epsilon$:

$$I_{1,1+\epsilon,\epsilon} = \frac{1}{\epsilon}I^{(-1)}_{1,1+\epsilon,\epsilon} + I^{(0)}_{1,1+\epsilon,\epsilon} + \epsilon I^{(1)}_{1,1+\epsilon,\epsilon} + O(\epsilon^2) \tag{6.51}$$

and it turns out that the coefficient of $\frac{1}{\epsilon}$ and the finite part are given by zero-dimensional MB integrals, while the coefficient of ϵ involves only one-dimensional MB integrals. However, three-dimensional MB integrals appear for all higher-order expansion coefficients.

In terms of the numerical evaluation, we note that for real angles, all variables v_{ij} are positive, and so in this sense, we are dealing with "Euclidean" kinematics. Thus, we may simply use straight-line integration contours. In Fig. 6.9 we show the order ϵ expansion coefficient $I^{(1)}_{1,1+\epsilon,\epsilon}$, while the complete numerical solution for Eq. (6.51) can be found in the file MB_RealPSNum_Springer.nb in the auxiliary material in [33]. In the numerical calculation, we have specialized to the case where the three vectors $\mathbf{p}_1$, $\mathbf{p}_2$, and $\mathbf{p}_3$ are co-planar. Then the result depends only on the angles $\chi^{(1)}_2$ and $\chi^{(1)}_3$, and we can represent the functions as surfaces in three dimensions.

Problems

Problem 6.1 Calculate the asymptotics for one-dimensional integrals Eqs. (6.5) and (1.55), and check that they do not depend on the position of the contour z_0.
Hint: See derivation above Eq. (6.7).

Problem 6.2 For Eq. (6.15) check the direction of the damping factor cancellation, and calculate the value of the parameter α in the fractional part $\frac{1}{|t|^\alpha}$ of the asymptotic for a contour position before and after the shift $z_2 \to z_2 + 2$.
Hint: For checking the direction, keep in mind that the it term in the gamma function leads to $e^{-\pi|t|/2}$ in the limit. For calculating α, put all irrelevant t_i equal to 0.

Problem 6.3 Find optimal θ parameter values for the rotated and parabolic contours illustrated in Fig. 6.2 for the integral in Eq. (1.55).
Hint: The relation $\ln z - \ln(-z) = i\pi \operatorname{sign}(\Im(z))$ doesn't depend on parametrization. Compare asymptotics of gamma function ratio with the kinematic part, and θ should restore exponential damping in both directions $t \to \pm\infty$. The result should have the form of an inequality.

Problem 6.4 Install and run examples for numerical integration of MB integrals using the MBnumerics package given in [20].
Hint: For installation, follow the Appendix and HOWTO discussed there.

Problem 6.5 Concerning the MBDE approach described in Sect. 6.3, it is a very efficient method for one-dimensional MB integrals. However, already two-dimensional cases are not fully understood (see, for instance, W. Flieger's contribution to [7] and the PhD thesis by Z. Peng [65]). In the case of MB integrals, we have to face the following issues:

- Since the gamma function has infinitely many critical points, we have to consider how many of them are necessary to match the assumed numerical precision.
- The parametrization of thimbles by solving flow equations for each relevant critical point. This stage requires also calculation of the Hessian matrix and the associated eigensystem.
- The knowledge of singularities of the MB integrand is necessary to avoid them by thimbles.
- The determination of the inverse function for the cut approach.
- The summation of thimbles with suitable coefficients in (6.41) to reconstruct the whole integral.

You are invited to make a research in this direction!

References

1. S. Weinzierl, Computer algebra in particle physics, in *Proceedings of the 11th National Seminar of Theoretical Physics (In Italian)* (2002). arXiv:hep-ph/0209234
2. V.A. Smirnov, Evaluating Feynman integrals. Springer Tracts Mod. Phys. **211**, 1–244 (2004)
3. V. Smirnov, *Feynman Integral Calculus* (Springer, Berlin, 2006)
4. C. Anastasiou, A. Daleo, Numerical evaluation of loop integrals. JHEP **10**, 031 (2006). arXiv: hep-ph/0511176. https://doi.org/10.1088/1126-6708/2006/10/031
5. J. Gluza, T. Riemann, Massive Feynman integrals and electroweak corrections. Nucl. Part. Phys. Proc. **261–262**, 140–154 (2015). arXiv:1412.3311. https://doi.org/10.1016/j.nuclphysbps.2015.03.012
6. A. Freitas, Numerical multi-loop integrals and applications. arXiv:1604.00406
7. A. Blondel, et al., Standard model theory for the FCC-ee Tera-Z stage, in *Mini Workshop on Precision EW and QCD Calculations for the FCC Studies: Methods and Techniques*. CERN Yellow Reports: Monographs, Vol. 3/2019 (CERN, Geneva, 2018). arXiv:1809.01830. https://doi.org/10.23731/CYRM-2019-003
8. G. Heinrich, Collider Physics at the Precision Frontier. Phys. Rept. **922**, 1–69 (2021). arXiv: 2009.00516. https://doi.org/10.1016/j.physrep.2021.03.006
9. Precision calculations for future e+e– colliders: targets and tools. Workshop, CERN 7–17 June 2022. https://indico.cern.ch/event/1140580/
10. I. Dubovyk, J. Gluza, T. Jelinski, T. Riemann, J. Usovitsch, New prospects for the numerical calculation of Mellin-Barnes integrals in Minkowskian kinematics. Acta Phys. Polon. **B48**, 995 (2017). arXiv:1704.02288. https://doi.org/10.5506/APhysPolB.48.995

11. K. Hepp, Proof of the Bogolyubov-Parasiuk theorem on renormalization. Commun. Math. Phys. **2**, 301–326 (1966). sector decomposition
12. T. Binoth, G. Heinrich, An automatized algorithm to compute infrared divergent multi-loop integrals. Nucl. Phys. B **585**, 741–759 (2000). arXiv:hep-ph/0004013v.2
13. A.V. Smirnov, FIESTA 3: cluster-parallelizable multiloop numerical calculations in physical regions. Comput. Phys. Commun. **185**, 2090–2100 (2014). arXiv:1312.3186. https://doi.org/10.1016/j.cpc.2014.03.015
14. S. Borowka, J. Carter, G. Heinrich, Numerical Evaluation of Multi-Loop Integrals for Arbitrary Kinematics with SecDec 2.0. Comput. Phys. Commun. **184**, 396–408 (2013). arXiv:1204.4152. https://doi.org/10.1016/j.cpc.2012.09.020
15. S. Borowka, G. Heinrich, S. Jahn, S.P. Jones, M. Kerner, J. Schlenk, T. Zirke, pySecDec: a toolbox for the numerical evaluation of multi-scale integrals. Comput. Phys. Commun. **222**, 313–326 (2018). arXiv:1703.09692. https://doi.org/10.1016/j.cpc.2017.09.015
16. I. Dubovyk, A. Freitas, J. Gluza, T. Riemann, J. Usovitsch, The two-loop electroweak bosonic corrections to $sin^2\theta_{\text{eff}}^{b\bar{b}}$. Phys. Lett. B **762**, 184–189 (2016). arXiv:1607.08375. https://doi.org/10.1016/j.physletb.2016.09.012
17. J. Gluza, K. Kajda, T. Riemann, AMBRE - a Mathematica package for the construction of Mellin-Barnes representations for Feynman integrals. Comput. Phys. Commun. **177**, 879–893 (2007). arXiv:0704.2423. https://doi.org/10.1016/j.cpc.2007.07.001
18. J. Gluza, K. Kajda, T. Riemann, V. Yundin, Numerical Evaluation of Tensor Feynman Integrals in Euclidean Kinematics. Eur. Phys. J. C **71**, 1516 (2011). arXiv:1010.1667. https://doi.org/10.1140/epjc/s10052-010-1516-y
19. I. Dubovyk, J. Gluza, T. Riemann, J. Usovitsch, Numerical integration of massive two-loop Mellin-Barnes integrals in Minkowskian regions. PoS LL2016, 034 (2016). arXiv:1607.07538
20. AMBRE. http://jgluza.us.edu.pl/ambre. Backup: https://web.archive.org/web/20220119185211/http://prac.us.edu.pl/~gluza/ambre/
21. K. Bielas, I. Dubovyk, PlanarityTest 1.3, a Mathematica package for testing the planarity of Feynman diagrams. http://jgluza.us.edu.pl/ambre/planarity/
22. K. Bielas, I. Dubovyk, J. Gluza, T. Riemann, Some Remarks on Non-planar Feynman Diagrams. Acta Phys. Polon. B **44**(11), 2249–2255 (2013). arXiv:1312.5603. https://doi.org/10.5506/APhysPolB.44.2249
23. M. Czakon, Automatized analytic continuation of Mellin-Barnes integrals. Comput. Phys. Commun. **175**, 559–571 (2006). arXiv:hep-ph/0511200. https://doi.org/10.1016/j.cpc.2006.07.002
24. A.V. Smirnov, V.A. Smirnov, On the Resolution of Singularities of Multiple Mellin- Barnes Integrals. Eur. Phys. J. C **62**, 445 (2009). arXiv:0901.0386
25. I. Dubovyk, T. Riemann, J. Usovitsch, Numerical calculation of multiple MB-integral representations for Feynman integrals. J. Usovitsch. MBnumerics, a Mathematica/Fortran package at http://jgluza.us.edu.pl/ambre/
26. J. Usovitsch, I. Dubovyk, T. Riemann, MBnumerics: Numerical integration of Mellin-Barnes integrals in physical regions. PoS LL2018, 046 (2018). arXiv:1810.04580. https://doi.org/10.22323/1.303.0046
27. J. Usovitsch, Numerical evaluation of Mellin-Barnes integrals in Minkowskian regions and their application to two-loop bosonic electroweak contributions to the weak mixing angle of the $Zb\bar{b}$-vertex. Ph.D. thesis (Humboldt U., Berlin, 2018). https://doi.org/10.18452/19484
28. MB Tools webpage. http://projects.hepforge.org/mbtools/
29. I. Dubovyk, A. Freitas, J. Gluza, T. Riemann, J. Usovitsch, 30 years, some 700 integrals, and 1 dessert, or: Electroweak two-loop corrections to the $Zb\bar{b}$ vertex. PoS LL2016, 075 (2016). arXiv:1610.07059
30. I. Dubovyk, A. Freitas, J. Gluza, T. Riemann, J. Usovitsch, Complete electroweak two-loop corrections to Z boson production and decay. Phys. Lett. B **783**, 86–94 (2018). arXiv:1804.10236. https://doi.org/10.1016/j.physletb.2018.06.037

31. I. Dubovyk, A. Freitas, J. Gluza, T. Riemann, J. Usovitsch, Electroweak pseudo-observables and Z-boson form factors at two-loop accuracy. JHEP **08**, 113 (2019). arXiv:1906.08815. https://doi.org/10.1007/JHEP08(2019)113

32. J. Fleischer, A. Kotikov, O. Veretin, Analytic two-loop results for selfenergy- and vertex-type diagrams with one non-zero mass. Nucl. Phys. B **547**, 343–374 (1999). arXiv:hep-ph/9808242

33. https://github.com/idubovyk/mbspringer. http://jgluza.us.edu.pl/mbspringer

34. I. Dubovyk, A. Freitas, J. Gluza, K. Grzanka, M. Hidding, J. Usovitsch, Evaluation of multi-loop multi-scale Feynman integrals for precision physics. arXiv:2201.02576

35. X. Liu, Y.-Q. Ma, AMFlow: a Mathematica package for Feynman integrals computation via Auxiliary Mass Flow. arXiv:2201.11669

36. F.F. Cordero, A. von Manteuffel, T. Neumann, Computational challenges for multi-loop collider phenomenology, in 2022 Snowmass Summer Study (2022). arXiv:2204.04200

37. T. Armadillo, R. Bonciani, S. Devoto, N. Rana, A. Vicini, Evaluation of Feynman integrals with arbitrary complex masses via series expansions. arXiv:2205.03345

38. T. Hahn, CUBA: A Library for multidimensional numerical integration. Comput. Phys. Commun. **168**, 78–95 (2005). arXiv:hep-ph/0404043. https://doi.org/10.1016/j.cpc.2005.01.010

39. I. Dubovyk, J. Gluza, T. Riemann, Optimizing the Mellin-Barnes Approach to Numerical Multiloop Calculations. Acta Phys. Polon. B **50**, 1993–2000 (2019). arXiv:1912.11326. https://doi.org/10.5506/APhysPolB.50.1993

40. U. Aglietti, R. Bonciani, Master integrals with 2 and 3 massive propagators for the 2 loop electroweak form-factor - planar case. Nucl. Phys. B **698**, 277–318 (2004). arXiv:hep-ph/0401193. https://doi.org/10.1016/j.nuclphysb.2004.07.018

41. G.P. Lepage, A New Algorithm for Adaptive Multidimensional Integration. J. Comput. Phys. **27**, 192 (1978). https://doi.org/10.1016/0021-9991(78)90004-9

42. G.P. Lepage, VEGAS: An adaptive multidimensional integration program. https://lib-extopc.kek.jp/preprints/PDF/1980/8006/8006210.pdf

43. S. Borowka, G. Heinrich, S. Jahn, S.P. Jones, M. Kerner, J. Schlenk, A GPU compatible quasi-Monte Carlo integrator interfaced to pySecDec. Comput. Phys. Commun. **240**, 120–137 (2019). arXiv:1811.11720. https://doi.org/10.1016/j.cpc.2019.02.015

44. J. Gluza, T. Jelinski, D.A. Kosower, Efficient Evaluation of Massive Mellin-Barnes Integrals. Phys. Rev. D **95**(7), 076016 (2017). arXiv:1609.09111. https://doi.org/10.1103/PhysRevD.95.076016

45. A. Freitas, Y.-C. Huang, On the Numerical Evaluation of Loop Integrals with Mellin-Barnes representations. JHEP **1004**, 074 (2010). arXiv:1001.3243. https://doi.org/10.1007/JHEP04(2010)074

46. U. Aglietti, R. Bonciani, Master integrals with one massive propagator for the two-loop electroweak form factor. Nucl. Phys. B **668**, 3–76 (2003). arXiv:hep-ph/0304028

47. S. Petrova, A. Solov'ev, The Origin of the Method of Steepest Descent. Hist. Math. **24**(4), 361 (1997). https://doi.org/10.1006/hmat.1996.2146

48. A.V. Sidorov, V.I. Lashkevich, O.P. Solovtsova, Asymptotics of the contour of the stationary phase and efficient evaluation of the Mellin-Barnes integral for the F_3 structure function. Phys. Rev. D **97**(7), 076009 (2018). arXiv:1712.05601. https://doi.org/10.1103/PhysRevD.97.076009

49. C. Bender, S. Orszag, Advanced mathematical methods for scientists and engineers i: asymptotic methods and perturbation theory, in *Advanced Mathematical Methods for Scientists and Engineers* (Springer, Berlin, 1999). https://books.google.pl/books?id=-yQXwhE6iWMC

50. R. Wong, Asymptotic approximation of integrals, classics in applied mathematics, in *Society for Industrial and Applied Mathematics* (2001). https://books.google.pl/books?id=KQHPHPZs8k4C

51. F. Pham, Vanishing homologies and the *n* variable saddlepoint method, Singularities, in *Summer Institution, Arcata/California 1981*. Proceedings of the Symposium Pure Mathematical, vol. 40, Part 2, 319–333 (1983)

52. E. Witten, Analytic Continuation Of Chern-Simons Theory. AMS/IP Stud. Adv. Math. **50**, 347–446 (2011). arXiv:1001.2933
53. E. Witten, A New Look At the Path Integral of Quantum Mechanics. arXiv:1009.6032
54. D. Harlow, J. Maltz, E. Witten, Analytic Continuation of Liouville Theory. JHEP **12**, 071 (2011). arXiv:1108.4417. https://doi.org/10.1007/JHEP12(2011)071
55. T. Kanazawa, Y. Tanizaki, Structure of Lefschetz thimbles in simple fermionic systems. JHEP **03**, 044 (2015). arXiv:1412.2802. https://doi.org/10.1007/JHEP03(2015)044
56. Y. Tanizaki, T. Koike, Real-time Feynman path integral with Picard-Lefschetz theory and its applications to quantum tunneling. Annals Phys. **351**, 250–274 (2014). arXiv:1406.2386. https://doi.org/10.1016/j.aop.2014.09.003
57. L. Nicolaescu, *An Invitation to Morse Theory*, Universitext (Springer, New York, 2011). https://books.google.pl/books?id=nCgvt2MY4QAC
58. V. Arnold, A. Varchenko, S. Gusein-Zade, *Singularities of Differentiable Maps: Volume II Monodromy and Asymptotic Integrals, Monographs in Mathematics* (Birkhäuser, Boston, 2012). https://books.google.pl/books?id=1BAGCAAAQBAJ
59. I. Dubovyk, J. Gluza, T. Riemann, J. Usovitsch, Numerical integration of massive two-loopMellin-Barnes integrals in Minkowskian regions. hep-ph....
60. G. Somogyi, Z. Trocsanyi, A Subtraction scheme for computing QCD jet cross sections at NNLO: Integrating the subtraction terms. I. JHEP **08**, 042 (2008). arXiv:0807.0509. https://doi.org/10.1088/1126-6708/2008/08/042
61. P. Bolzoni, S.-O. Moch, G. Somogyi, Z. Trocsanyi, Analytic integration of real-virtual counterterms in NNLO jet cross sections. II. JHEP **08**, 079 (2009). arXiv:0905.4390. https://doi.org/10.1088/1126-6708/2009/08/079
62. P. Bolzoni, G. Somogyi, Z. Trocsanyi, A subtraction scheme for computing QCD jet cross sections at NNLO: integrating the iterated singly-unresolved subtraction terms. JHEP **01**, 059 (2011). arXiv:1011.1909. https://doi.org/10.1007/JHEP01(2011)059
63. G. Heinrich, Sector Decomposition. Int. J. Mod. Phys. A **23**, 1457–1486 (2008). arXiv:0803.4177. https://doi.org/10.1142/S0217751X08040263
64. G. Somogyi, Angular integrals in d dimensions. J. Math. Phys. **52**, 083501 (2011). arXiv:1101.3557. https://doi.org/10.1063/1.3615515
65. Z. Peng. TopicsinN=4Yang-Millstheory, Ph.D. thesis (IPhT, Saclay, 2012). http://tel.archives-ouvertes.fr/tel-00834200

Public Software and Codes for MB Studies

A

A.1 Analytic Software

For a construction of MB representations, analytic continuation in ϵ, expansions, and decreasing dimensionality of the MB integrals, see Chaps. 3–5.

A.1.1 AMBRE.m The Mathematica toolkit AMBRE derives Mellin-Barnes representations for Feynman integrals in $d = n - 2\epsilon$ dimensions. The package is described in [1–3]. Connected useful MB literature is [1–32].
AMBRE versions overview:

- Iteratively to each subloop—loop-by-loop approach (LA): mostly for planar
  (AMBREv1.3.1 & AMBREv2.1.1)

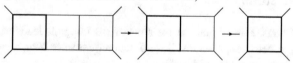

- In one step to the complete U and F polynomials—global approach (GA): general

  (AMBREv3.2)
- Combination of the above methods—hybrid approach (HA)

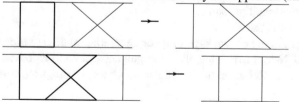

Examples, description, updates, links to basic tools, and literature can be found on the webpage [33].

A.1.2 `PlanarityTest.m`

The `Mathematica` package [9, 33] determines planarity of a Feynman diagram given its propagators (momentum flows).

A.1.3 `MB.m`

The `Mathematica` package [16, 34] allows to analytically continue any MB integral in a given parameter and to resolve the singularity structure in this parameter.

A.1.4 `MBresolve.m`

The `Mathematica` package [24, 29] resolves singularities of multifold Mellin-Barnes integrals in the dimensional regularization parameter ϵ. An alternative to `MB.m`.

A.1.5 `barnesroutines.m`

A tool for an automatic application of the first and second Barnes lemmas on lists of multiple Mellin-Barnes integrals [28].

A.1.6 `MBasymptotics.m`

The `Mathematica` routine expands Mellin-Barnes integrals in a small parameter [34].

A.1.7 `asy2.m`

The `Mathematica` code [35] performs asymptotic expansions of Feynman integrals using the strategy of expansion by regions and recognizes potential regions in threshold expansions or Glauber regions.

A.1.8 `ASPIRE`

The `Mathematica` code [36] which is an alternative to `asy2.m`.

A.2 Multiple Sums

As discussed in Chaps. 2 and 5, taking residues of MB integrals leads to multiple sums over hypergeometric expressions. Several techniques have been developed to manipulate these sums (see, e.g., [37–43]).

Below we list some of the codes which deal with multiple sums:

A2.1 `MBsums`

The `Mathematica` package [12, 33] which transforms Mellin-Barnes integrals into multiple sums.

A2.2 `MBConicHulls`

The `Mathematica` package [44] for generating multiple series representations of N-fold MB integrals using geometrical analysis based on conic hulls, $N > 2$. Relies on the `MultivariateResidues` `Mathematica` package.

A2.3 MultivariateResidues

The Mathematica package [45] for efficient evaluation of multi-dimensional residues based on methods from computational algebraic geometry.

A2.4 SUMMER

The package [46] written in FORM [47] for evaluation of nested symbolic sums over combinations of harmonic series, binomial coefficients, and denominators. In addition it treats Mellin transforms and the inverse Mellin transformation for functions that are encountered in Feynman diagram calculations.

A2.5 nestedsum

The C++ program [48] with GiNaC library for expansion of transcendental functions [37] with the algorithms based on the Hopf algebra of nested sums.

A2.6 XSummer

The package [49] written in FORM [47] for evaluation of nested sums, where the harmonic sums and their generalizations appear as building blocks, originating, for example, from the expansion of generalized hypergeometric functions around integer values of the parameters.

A2.7 HarmonicSums

The Mathematica package [50] for evaluation of S-sums and generalized harmonic polylogarithms.

A2.8 SumProduction, EvaluateMultiSums, Sigma

The packages EvaluateMultiSums, SumProduction, and Sigma [51, 52] deal with multi-sums over hypergeometric expressions. The results are given, if possible, in terms of harmonic sums, generalized harmonic sums, cyclotomic harmonic sums, or binomial sums.

A2.9 HypExp, HypExp 2

The Mathematica package HypExp which allows to expand hypergeometric functions ${}_J F_{J-1}$ around integer and half-integer parameters to arbitrary order [53, 54].

A2.10 Olsson

The Mathematica package [55] which aims to find linear transformations for some classes of multivariable hypergeometric functions. It is based on [56].

A.3   Polylogarithms and Generalizations

In particle physics HPLs $H(\{a\}, x)$ have been introduced in [57] in order to describe systematically QED self-energies and vertices, $\{a\} \in \{1, 0, -1\}$. HPLs with some generalized arguments are introduced in [58]. A generalization of HPLs for three-scale problems (e.g., QED boxes) are the two-dimensional harmonic polylogarithms $G(\{b\}, x)$ (GPLs); the index vector $\{b\}$ has elements $1, 0, -1$, but now also those depending on a second kinematic variable y. This was observed in [59] and worked out in [60] for a planar massless problem. The minimal basis has been systematically worked out in [61]. See also for MPLs studies in [50, 62, 63].

As far as numerical solutions are concerned, the problem is solved for any weight of HPLs and GPLs. Classical and Nielsen generalized polylogarithms are available within both the Mathematica and GiNac [64,65] language as built-in functions (the evaluation for arbitrary complex arguments and without any restriction on the weight). In addition many special packages have been developed.

A3.1 hplog
   The numerical Fortran [66] evaluation of HPLs up to weight 4 with indices $\{0, 1, -1\}$ and the analytic continuation.
A3.2 dhpl
   The systematic numerical treatment of GPLs with indices $\{0, 1, 1 - z, -z\}$, performed in Fortran [67].
A3.3 HPOLY.f
   The Fortran numerical code [68] for HPLs up to weight 8.
A3.4 handyG
   The Fortran library [69] for the evaluation of GPLs, in principle of any weight, suitable for Monte Carlo integration.
A3.5 FastGPL
   The C++ library [70] for efficient and accurate evaluation of GPLs based on work [71], suitable for Monte Carlo integration and event generation.
A3.6 HPL.m
   The Mathematica HPLs package [72] with implementation of the product algebra, the derivative properties, series expansion, and numerical evaluation of HPLs.
A3.7 PolyLogTools
   The Mathematica package [73] for multiple polylogarithms, including the Hopf algebra of the multiple polylogarithms and the symbol map, as well as the construction of single-valued multiple polylogarithms with an algorithm for given symbol combination of MPLs, including the so-called fibration bases.
A3.8 HyperInt
   The package [74] based on the Maple computer algebra system for symbolic integration of hyperlogarithms multiplied by rational functions, which also include MPLs when their arguments are rational functions.
A3.9 CHAPLIN
   The Fortran library [75] to evaluate all HPLs up to weight 4 numerically for any complex argument.
A3.9 HYPERDIRE
   The Mathematica package [76] devoted to the creation of a set of base programs for the differential reduction of hypergeometric functions.

A.4   MB **Numerical Software**

There are at the moment two publicly available packages for MB numerical evaluations:

A4.1 MB.m

MB.m listed in **A1.3** also allows to solve MB representations numerically. It works fine in the Euclidean kinematic region.

MB.m requires the CUBA Monte Carlo and quasi Monte Carlo library [77] for numerical integration and CERNLIB for evaluation of gamma and polygamma functions.

In the original Mathematica package MB [16], the gamma and polygamma functions are obtained from links to the libraries libmathlib.a and libkernlib.a of CERNLIB. Unfortunately, the development and support of the original CERNLIB are no longer available [78] (last update 10 Oct. 2014 and 12 Feb. 2018). This can cause problems when running numeric commands within MB.m and programs based on MB.m.

In the file README.md in [79] we put a short manual how to solve the problems if numerical MB.m output gives NaNs. MB.m is available at [34].

A4.2 MBnumerics.m

The Mathematica package for evaluation of MB integrals in any kinematic point (Minkowskian or Euclidean). MBnumerics.m is available at [33].

A.5   **Other Methods for** FI **Numerical Integrations**

The well-established method of calculations is the sector decomposition method. The programs pySecDec and FIESTA allow now for direct numerical integration of FI in physical (Minkowskian) space. In recent years numerical evaluation of FI based on differential equations has been also developed very much, and there are several private programs or set of connected packages [80–82] with exception of AMFlow and SeaSyde [83, 84]. Some of those rely on DiffExp [85], a Mathematica package for integrating families of FI order by order in the dimensional regulator from their systems of differential equations, in terms of one-dimensional series expansions along lines in phase space, the idea based on the work [86].

A5.1 pySecDec

The project SecDec performs the factorization of dimensionally regulated poles in parametric integrals and the subsequent numerical evaluation of the finite coefficients. The latest version with the algebraic part of the program is written with python modules [87, 88].

A5.2 FIESTA

The program FIESTA is an alternative package to pySecDec; the newest version adds support for graphical processor units (GPUs) for the numerical integration, aiming at optimal performance at large scales when one is increasing the number of sampling points in order to reduce the uncertainty estimates [89].

A5.3 AMFlow

The Mathematica package [83] for numerical computation of dimensionally regularized FI via so-called auxiliary mass flow method. In this framework, integrals are treated as functions of an auxiliary mass parameter, and their results can be obtained by constructing and solving differential systems with respect to this parameter, in an automatic way.

A5.4 SeaSyde

The Mathematica package [84] for numerical computation of dimensionally regularized FI, with internal masses being in general complex-valued. The implementation solves by series expansions the system of differential equations satisfied by the master integrals.

Additional Working Files

B

At the web pages [79] we gathered files with materials and solutions discussed in the book. For viewing the .nb files one can use Wolfram Player https://www.wolfram.com/player/.

1. MB_miscellaneous_Springer.nb
2. MB_SE2l2m_Springer.nb
3. MB_Simpl_Springer.nb
4. MB_V6l3m1M_Springer.nb
5. MB_V6l1MZ_Springer.nb
6. MB_V6l0m_Springer.nb
7. MB_SE6l0m_Springer.nb
8. MB_AMBREnew_Springer.nb
9. MB_Zmatrix_Springer.nb
10. MB_3L_Springer.nb
11. MB_Gamma0_Springer.nb
12. MB_PSint_Springer.nb
13. MB_B5l2m2_Springer.nb
14. MB_B5nf_Springer.nb
15. MB_basicsums_Springer.nb
16. MB_I11_massless_Springer.nb
17. MB_O11_onemass_Springer.nb
18. MB_Decoupling_Springer.nb
19. MB_Euler_Springer.nb
20. MB_symbolicint_Springer.nb
21. MB_B7l4m1_Springer.nb
22. MB_3dimNum_Springer.nb
23. MB_1dimNum_Springer.nb
24. MB_TH1_Springer.nb
25. MB_TH2_Springer.nb

© The Author(s), under exclusive license to Springer Nature Switzerland AG 2022
I. Dubovyk et al., *Mellin-Barnes Integrals*, Lecture Notes in Physics 1008,
https://doi.org/10.1007/978-3-031-14272-7

26. MB_RealPSNum_Springer.nb
27. MB_V6l1m_Springer.sh
28. SD_V6l1m_generate_Springer.py
29. SD_V6l1m_integrate_Springer.py
30. MBDE_Springer.nb

Some MB Talks

C

1. V.A. Smirnov, *The method of Mellin–Barnes Representation*.
   In: School of Analytic Computing in Theoretical High-Energy Physics, Atrani, Italy 2013, https://indico.cern.ch/event/248025/.
2. T. Riemann, *Integrals, Mellin-Barnes representations and Sums*.
   In: Computer Algebra and Particle Physics 2009, DESY, Zeuthen, Germany, 2009, https://indico.desy.de/event/1573.

References

1. I. Dubovyk, J. Gluza, T. Riemann, J. Usovitsch, Numerical integration of massive two-loop Mellin-Barnes integrals in Minkowskian regions. PoS LL2016, 034 (2016). arXiv:1607.07538
2. I. Dubovyk, A. Freitas, J. Gluza, T. Riemann, J. Usovitsch, 30 years, some 700 integrals, and 1 dessert, or: Electroweak two-loop corrections to the $Z b\bar{b}$ vertex, PoS LL2016, 075 (2016). arXiv:1610.07059
3. I. Dubovyk, A. Freitas, J. Gluza, T. Riemann, J. Usovitsch, The two-loop electroweak bosonic corrections to $sin^2\theta_{\mathrm{eff}}^{b\bar{b}}$. Phys. Lett. B **762**, 184–189 (2016). arXiv:1607.08375. https://doi.org/10.1016/j.physletb.2016.09.012
4. J. Gluza, K. Kajda, T. Riemann, AMBRE - a Mathematica package for the construction of Mellin-Barnes representations for Feynman integrals. Comput. Phys. Commun. **177**, 879–893 (2007). arXiv:0704.2423. https://doi.org/10.1016/j.cpc.2007.07.001
5. J. Gluza, F. Haas, K. Kajda, T. Riemann, Automatizing the application of Mellin-Barnes representations for Feynman integrals. PoS ACAT2007, 081 (2007). arXiv:0707.3567
6. J. Gluza, K. Kajda, T. Riemann, V. Yundin, New results for loop integrals: AMBRE, CSectors, hexagon. PoS ACAT08, 124 (2008). arXiv:0902.4830
7. J. Gluza, K. Kajda, T. Riemann, V. Yundin, News on Ambre and CSectors. Nucl. Phys. Proc. Suppl. **205–206**, 147–151 (2010). arXiv:1006.4728. https://doi.org/10.1016/j.nuclphysbps.2010.08.034
8. J. Gluza, K. Kajda, T. Riemann, V. Yundin, Numerical Evaluation of Tensor Feynman Integrals in Euclidean Kinematics. Eur. Phys. J. C **71**, 1516 (2011). arXiv:1010.667. https://doi.org/10.1140/epjc/s10052-010-1516-y

9. K. Bielas, I. Dubovyk, J. Gluza, T. Riemann, Some Remarks on Non-planar Feynman Diagrams. Acta Phys. Polon. B **44**(11), 2249–2255 (2013). arXiv:1312.5603. https://doi.org/10.5506/APhysPolB.44.2249

10. J. Blumlein, I. Dubovyk, J. Gluza, M. Ochman, C.G. Raab, T. Riemann, C. Schneider, Non-planar Feynman integrals, Mellin-Barnes representations, multiple sums. PoS LL2014, 052 (2014). arXiv:1407.7832

11. I. Dubovyk, J. Gluza, T. Riemann, Non-planar Feynman diagrams and Mellin-Barnes representations with $AMBRE$ 3.0. J. Phys. Conf. Ser. **608**(1), 012070 (2015). https://doi.org/10.1088/1742-6596/608/1/012070

12. M. Ochman, T. Riemann, MBsums - a Mathematica package for the representation of Mellin-Barnes integrals by multiple sums. Acta Phys. Polon. B **46**(11), 2117 (2015). arXiv:1511.01323. https://doi.org/10.5506/APhysPolB.46.2117

13. J. Gluza, T. Jelinski, D. A. Kosower, Efficient Evaluation of Massive Mellin-Barnes Integrals. Phys. Rev. D **95**(7), 076016 (2017). arXiv:1609.09111. https://doi.org/10.1103/PhysRevD.95.076016

14. I. Dubovyk, J. Gluza, T. Jelinski, T. Riemann, J. Usovitsch, New prospects for the numerical calculation of Mellin-Barnes integrals in Minkowskian kinematics. Acta Phys. Polon. B **48**, 995 (2017). arXiv:1704.02288. https://doi.org/10.5506/APhysPolB.48.995

15. M. Prausa, Mellin-Barnes meets Method of Brackets: a novel approach to Mellin-Barnes representations of Feynman integrals. Eur. Phys. J. C **77**(9), 594 (2017). arXiv:1706.09852. https://doi.org/10.1140/epjc/s10052-017-5150-9

16. M. Czakon, Automatized analytic continuation of Mellin-Barnes integrals. Comput. Phys. Commun. **175**, 559–571 (2006). arXiv:hep-ph/0511200. https://doi.org/10.1016/j.cpc.2006.07.002

17. A. Freitas, Y.-C. Huang, On the Numerical Evaluation of Loop Integrals with Mellin-Barnes Representations. JHEP **1004**, 074 (2010). arXiv:1001.3243. https://doi.org/10.1007/JHEP04(2010)074

18. A. Freitas, Y.-C. Huang, Electroweak two-loop corrections to $\sin^2 \theta(eff, bb)$ and R(b) using numerical Mellin-Barnes integrals. JHEP **1208**, 050 (2012). arXiv:1205.0299. https://doi.org/10.1007/JHEP08(2012)050

19. B. Tausk, Non-planar massless two-loop Feynman diagrams with four on- shell legs. Phys. Lett. B **469**, 225–234 (1999). arXiv:hep-ph/9909506

20. I. Gonzalez, I. Schmidt, Optimized negative dimensional integration method (NDIM) and multiloop Feynman diagram calculation. Nucl. Phys. B **769**, 124–173 (2007). arXiv:hep-th/0702218. https://doi.org/10.1016/j.nuclphysb.2007.01.031

21. I. Gonzalez, V.H. Moll, Definite integrals by the method of brackets. Part 1. arXiv:0812.3356

22. I. Gonzalez, Method of Brackets and Feynman diagrams evaluation. Nucl. Phys. Proc. Suppl. **205–206**, 141–146 (2010). arXiv:1008.2148. https://doi.org/10.1016/j.nuclphysbps.2010.08.033

23. P. Cvitanovic, T. Kinoshita, Feynman-Dyson rules in parametric space. Phys. Rev. D **10**, 3978–3991 (1974). https://doi.org/10.1103/PhysRevD.10.3978

24. A.V. Smirnov, V.A. Smirnov, On the Resolution of Singularities of Multiple Mellin- Barnes Integrals. Eur. Phys. J. C **62**, 445 (2009). arXiv:0901.0386

25. C. Anastasiou, A. Daleo, Numerical evaluation of loop integrals. JHEP **10**, 031 (2006). arXiv:hep-ph/0511176. https://doi.org/10.1088/1126-6708/2006/10/031

26. C. Bogner, S. Weinzierl, Feynman graph polynomials. Int. J. Mod. Phys. A **25**, 2585–2618 (2010). arXiv:1002.3458. https://doi.org/10.1142/S0217751X10049438

27. N. Nakanishi, GraphtheoryandFeynmanintegrals, in *Mathematics and its applications* (Gordon and Breach, Philadelphia, 1971). https://books.google.ch/books?id=5f7uAAAAMAAJ

28. D. Kosower, Mathematica program barnesroutines.m version 1.1.1 (2009), available at the MB Tools webpage. http://projects.hepforge.org/mbtools/

29. A. Smirnov, Mathematica program MBresolve.m version 1.0 (2009), available at the MB Tools webpage. http://projects.hepforge.org/mbtools/ [24]

30. K. Bielas, I. Dubovyk, PlanarityTest 1.2.1 (Aug 2017), a Mathematica package for testing the planarity of Feynman diagrams. http://us.edu.pl/~gluza/ambre/planarity/ [9]
31. K. Kajda, I. Dubovyk, AMBRE 2.1.1 & 1.3.1 (Aug 2017), Mathematica packages representing Feynman integrals by Mellin-Barnes integrals. http://jgluza.us.edu.pl/ambre/ [8]
32. I. Dubovyk, AMBRE 3.1.1 (Aug 2017), a Mathematica package representing Feynman integrals by Mellin-Barnes integrals. http://jgluza.us.edu.pl/ambre/ [8, 11]
33. AMBRE. http://jgluza.us.edu.pl/ambre. Backup: https://web.archive.org/web/20220119185211/ http://prac.us.edu.pl/~gluza/ambre/
34. MB Tools webpage. http://projects.hepforge.org/mbtools/
35. B. Jantzen, A.V. Smirnov, V.A. Smirnov, Expansion by regions: revealing potential and Glauber regions automatically. Eur. Phys. J. C **72**, 2139 (2012). arXiv:1206.0546. https://doi.org/10.1140/epjc/s10052-012-2139-2
36. B. Ananthanarayan, A. Pal, S. Ramanan, R. Sarkar, Unveiling Regions in multi-scale Feynman Integrals using Singularities and Power Geometry. Eur. Phys. J. C **79**(1), 57 (2019). arXiv:1810.06270. https://doi.org/10.1140/epjc/s10052-019-6533-x
37. S. Moch, P. Uwer, S. Weinzierl, Nested sums, expansion of transcendental functions and multiscale multiloop integrals. J. Math. Phys. **43**, 3363–3386 (2002). arXiv:hep-ph/0110083. https://doi.org/10.1063/1.1471366
38. A.I. Davydychev, M.Y. Kalmykov, Massive Feynman diagrams and inverse binomial sums. Nucl. Phys. B **699**, 3–64 (2004). arXiv:hep-th/0303162. https://doi.org/10.1016/j.nuclphysb.2004.08.020
39. S. Weinzierl, Expansion around half integer values, binomial sums and inverse binomial sums. J. Math. Phys. **45**, 2656–2673 (2004). arXiv:hep-ph/0402131. https://doi.org/10.1063/1.1758319
40. M.Y. Kalmykov, B.F.L. Ward, S.A. Yost, Multiple (inverse) binomial sums of arbitrary weight and depth and the all-order epsilon-expansion of generalized hypergeometric functions with one half-integer value of parameter. JHEP **10**, 048 (2007). arXiv:0707.3654. https://doi.org/10.1088/1126-6708/2007/10/048
41. J. Blümlein, C. Schneider, Analytic computing methods for precision calculations in quantum field theory. Int. J. Mod. Phys. A **33**(17), 1830015 (2018). arXiv:1809.02889. https://doi.org/10.1142/S0217751X18300156
42. A.J. McLeod, H.J. Munch, G. Papathanasiou, M. von Hippel, A Novel Algorithm for Nested Summation and Hypergeometric Expansions. JHEP **11**, 122 (2020). arXiv:2005.05612. https://doi.org/10.1007/JHEP11(2020)122
43. A.V. Kotikov, About calculation of massless and massive Feynman integrals. Particles **3**(2), 394–443 (2020). arXiv:2004.06625. https://doi.org/10.3390/particles3020030
44. B. Ananthanarayan, S. Banik, S. Friot, S. Ghosh, Multiple Series Representations of N-fold Mellin-Barnes Integrals. Phys. Rev. Lett. **127**(15), 151601 (2021). arXiv:2012.15108. https://doi.org/10.1103/PhysRevLett.127.151601
45. K.J. Larsen, R. Rietkerk, MultivariateResidues: a Mathematica package for computing multivariate residues. Comput. Phys. Commun. **222**, 250–262 (2018). arXiv:1701.01040. https://doi.org/10.1016/j.cpc.2017.08.025
46. J. Vermaseren, Harmonic sums, Mellin transforms and integrals. Int. J. Mod. Phys. A **14**, 2037–2076 (1999). arXiv:hep-ph/9806280
47. J. Vermaseren, New features of FORM. arXiv:math-ph/0010025
48. S. Weinzierl, Symbolic expansion of transcendental functions. Comput. Phys. Commun. **145**, 357–370 (2002). arXiv:math-ph/0201011
49. S. Moch, P. Uwer, XSummer: Transcendental functions and symbolic summation in Form. Comput. Phys. Commun. **174**, 759–770 (2006). arXiv:math-ph/0508008
50. J. Ablinger, J. Blümlein, C. Schneider, Analytic and algorithmic aspects of generalized harmonic sums and polylogarithms. J. Math. Phys. **54**, 082301 (2013). arXiv:1302.0378. https://doi.org/10.1063/1.4811117

51. C. Schneider, Modern summation methods for loop integrals in Quantum field theory: the packages sigma, EvaluateMultiSums and SumProduction. J. Phys. Conf. Ser. **523**, 012037 (2014). arXiv:1310.0160. https://doi.org/10.1088/1742-6596/523/1/012037

52. J. Ablinger, J. Blümlein, C. Schneider, Generalized Harmonic, Cyclotomic, and Binomial Sums, their Polylogarithms and Special Numbers. J. Phys. Conf. Ser. **523**, 012060 (2014). arXiv:1310.5645. https://doi.org/10.1088/1742-6596/523/1/012060

53. T. Huber, D. Maitre, HypExp: A Mathematica package for expanding hypergeometric functions around integer-valued parameters. Comput. Phys. Commun. **175**, 122–144 (2006). arXiv:hep-ph/0507094. https://doi.org/10.1016/j.cpc.2006.01.007

54. T. Huber, D. Maitre, HypExp 2, Expanding Hypergeometric Functions about Half-Integer Parameters. Comput. Phys. Commun. **178**, 755–776 (2008). arXiv:0708.2443. https://doi.org/10.1016/j.cpc.2007.12.008

55. B. Ananthanarayan, S. Bera, S. Friot, T. Pathak, Olsson.wl : a *Mathematica* package for the computation of linear transformations of multivariable hypergeometric functions. arXiv:2201.01189

56. P. Olsson, On the integration of the differential equations of five-parametric doublehypergeometric functions of second order. J. Math. Phys. **18**, 1285 (1977). https://doi.org/10.1063/1.523405

57. E. Remiddi, J. Vermaseren, Harmonic polylogarithms. Int. J. Mod. Phys. A **15**, 725–754 (2000). arXiv:hep-ph/9905237. https://doi.org/10.1142/S0217751X00000367

58. U. Aglietti, R. Bonciani, Master integrals with one massive propagator for the two-loop electroweak form factor. Nucl. Phys. B **668**, 3–76 (2003). arXiv:hep-ph/0304028

59. T. Gehrmann, E. Remiddi, Differential equations for two-loop four-point functions. Nucl. Phys. B **580**, 485–518 (2000). arXiv:hep-ph/9912329

60. T. Gehrmann, E. Remiddi, Two loop master integrals for $\gamma^* \rightarrow 3$ jets: The Planar topologies. Nucl. Phys. B **601**, 248–286 (2001). arXiv:hep-ph/0008287. https://doi.org/10.1016/S0550-3213(01)00057-8

61. J. Blumlein, Structural Relations of Harmonic Sums and Mellin Transforms up to Weight w = 5. Comput. Phys. Commun. **180**, 2218–2249 (2009). arXiv:0901.3106. https://doi.org/10.1016/j.cpc.2009.07.004

62. R. Bonciani, G. Degrassi, A. Vicini, On the Generalized Harmonic Polylogarithms of One Complex Variable. Comput. Phys. Commun. **182**, 1253–1264 (2011). arXiv:1007.1891. https://doi.org/10.1016/j.cpc.2011.02.011

63. M. Czakon, J. Gluza, T. Riemann, Harmonic polylogarithms for massive Bhabha scattering. Nucl. Instrum. Meth. A **559**, 265–268 (2006). arXiv:hep-ph/0508212

64. C. Bauer, A. Frink, R. Kreckel, Introduction to the GiNaC Framework for Symbolic Computation within the C++ Programming Language. J. Symb. Comput. **33**, 1 (2002). arXiv:cs.sc/0004015

65. Ginac. https://www.ginac.de/

66. T. Gehrmann, E. Remiddi, Numerical evaluation of harmonic polylogarithms. Comput. Phys. Commun. **141**, 296–312 (2001). arXiv:hep-ph/0107173. https://doi.org/10.1016/S0010-4655(01)00411-8

67. T. Gehrmann, E. Remiddi, Numerical evaluation of two-dimensional harmonic polylogarithms. Comput. Phys. Commun. **144**, 200–223 (2002). arXiv:hep-ph/0111255

68. J. Ablinger, J. Blümlein, M. Round, C. Schneider, Numerical Implementation of Harmonic Polylogarithms to Weight w = 8. Comput. Phys. Commun. **240**, 189–201 (2019). arXiv:1809.07084. https://doi.org/10.1016/j.cpc.2019.02.005

69. L. Naterop, A. Signer, Y. Ulrich, handyG —Rapid numerical evaluation of generalised polylogarithms in Fortran. Comput. Phys. Commun. **253**, 107165 (2020). arXiv:1909.01656. https://doi.org/10.1016/j.cpc.2020.107165

70. Y. Wang, L.L. Yang, B. Zhou, FastGPL: a C++ library for fast evaluation of generalized polylogarithms. arXiv:2112.04122

71. J. Vollinga, S. Weinzierl, Numerical evaluation of multiple polylogarithms. Comput. Phys. Commun. **167**, 177 (2005). arXiv:hep-ph/0410259. https://doi.org/10.1016/j.cpc.2004.12.009

72. D. Maitre, HPL, a mathematica implementation of the harmonic polylogarithms. Comput. Phys. Commun. **174**, 222–240 (2006). arXiv:hep-ph/0507152. https://doi.org/10.1016/j.cpc. 2005.10.008

73. C. Duhr, F. Dulat, PolyLogTools — polylogs for the masses. JHEP **08**, 135 (2019). arXiv:1904. 07279. https://doi.org/10.1007/JHEP08(2019)135

74. E. Panzer, Algorithms for the symbolic integration of hyperlogarithms with applications to Feynman integrals. Comput. Phys. Commun. **188**, 148–166 (2015). arXiv:1403.3385. https:// doi.org/10.1016/j.cpc.2014.10.019

75. S. Buehler, C. Duhr, CHAPLIN - Complex Harmonic Polylogarithms in Fortran. Comput. Phys. Commun. **185**, 2703–2713 (2014). arXiv:1106.5739. https://doi.org/10.1016/j.cpc.2014. 05.022

76. V.V. Bytev, M.Y. Kalmykov, S.-O. Moch, HYPERgeometric functions DIfferential REduction (HYPERDIRE): MATHEMATICA based packages for differential reduction of generalized hypergeometric functions: F_D and F_S Horn-type hypergeometric functions of three variables. Comput. Phys. Commun. **185**, 3041–3058 (2014). arXiv:1312.5777. https://doi.org/10.1016/j. cpc.2014.07.014

77. T. Hahn, CUBA: A Library for multidimensional numerical integration. Comput. Phys. Commun. **168**, 78–95 (2005). arXiv:hep-ph/0404043. https://doi.org/10.1016/j.cpc.2005.01. 010

78. The CERN Program Library. available at https://cernlib.web.cern.ch/cernlib/

79. https://github.com/idubovyk/mbspringer. http://jgluza.us.edu.pl/mbspringer

80. M.K. Mandal, X. Zhao, Evaluating multi-loop Feynman integrals numerically through differential equations. JHEP **03**, 190 (2019). arXiv:1812.03060. https://doi.org/10.1007/ JHEP03(2019)190

81. M.L. Czakon, M. Niggetiedt, Exact quark-mass dependence of the Higgs-gluon form factor at three loops in QCD. JHEP **05**, 149 (2020). arXiv:2001.03008. https://doi.org/10.1007/ JHEP05(2020)149

82. C. Brønnum-Hansen, C.-Y. Wang, Contribution of third generation quarks to two-loop helicity amplitudes for W boson pair production in gluon fusion. JHEP **01**, 170 (2021). arXiv:2009. 03742. https://doi.org/10.1007/JHEP01(2021)170

83. X. Liu, Y.-Q. Ma, AMFlow: a Mathematica package for Feynman integrals computation via Auxiliary Mass Flow. arXiv:2201.11669

84. T. Armadillo, R. Bonciani, S. Devoto, N. Rana, A. Vicini, Evaluation of Feynman integrals with arbitrary complex masses via series expansions arXiv:2205.03345

85. M. Hidding, DiffExp, a Mathematica package for computing Feynman integrals in terms of one-dimensional series expansions, Comput. Phys. Commun. **269**, 108125 (2021). arXiv:2006. 05510. https://doi.org/10.1016/j.cpc.2021.108125

86. F. Moriello, Generalised power series expansions for the elliptic planar families of Higgs + jet production at two loops. JHEP **01**, 150 (2020). arXiv:1907.13234. https://doi.org/10.1007/ JHEP01(2020)150

87. S. Borowka, G. Heinrich, S. Jahn, S.P. Jones, M. Kerner, J. Schlenk, T. Zirke, pySecDec: a toolbox for the numerical evaluation of multi-scale integrals. Comput. Phys. Commun. **222**, 313–326 (2018). arXiv:1703.09692. https://doi.org/10.1016/j.cpc.2017.09.015

88. S. Borowka, G. Heinrich, S. Jahn, S.P. Jones, M. Kerner, J. Schlenk, A GPU compatible quasi-Monte Carlo integrator interfaced to pySecDec. Comput. Phys. Commun. **240**, 120–137 (2019). arXiv:1811.11720. https://doi.org/10.1016/j.cpc.2019.02.015

89. A.V. Smirnov, FIESTA4: Optimized Feynman integral calculations with GPU support. Comput. Phys. Commun. **204**, 189–199 (2016). arXiv:1511.03614. https://doi.org/10.1016/j.cpc. 2016.03.013